I0464618

Quantum Earth Simulation

by T J Hegland

The above scenario represents the alternate Earth Simulation –
a 3D Earth Construct, as VR Sphere, with internal mechanics
and an energy shield Quarantine (i.e., Charles Fort's *Gegenschein*)
with the universe contained in a larger Matrix-like Creation.

The Simulation vs Construct issue is addressed in Chapters 1 & 13.
The Flat Earth alternative is covered in Chapter 11.

Categories: Metatags: Anunnaki, Nephilim, Djinn, interdimensionals, ET, UFO, Mankind, origins, Genetic engineering, Greys, abductions, hybrids, DNA, Quantum Physics, Maya, Egyptian, extraterrestrial influences, soulless, sociopaths, auras, OPs, NPCs, creation, evolution, Darwin, reptile, Sumeria, Catholic Church, Religion, Fortean phenomena, Jacques Vallee, John Keel, Stuart Wilde, Anatoly Fomenko, Robert Monroe, OBE, Control System, Virtual Reality, Simulation, holograms, Holodeck, Angels, Beings of Light, demons, souls, Scripts, karma, reincarnation, recycling, Déjà vu, Earth Graduate.

Cover design:	**The Flat Earth in Simulation**
	source: Bing Images/
	fopunkt.com/alexandria/donkology/flat-earth.html
	(Excellent website – please see it.)
Back Cover:	source: Bing Images/ winterpatriot.com
Frontspiece underlay:	Bing Images: the-matrix-movies-2229943-1024-768
	Boscosgrindhouse.com

Book text in Garamond 11 font.

Author may be reached at TJ_cspub14@yahoo.com

ISBN – 13: 978- 1514178621

Other Books by the Author

*See last two pages at book's end.

Table of Contents

Quantum Earth Simulation

Introduction

Due to a number of VEG reader requests for information on how Earth could be a Simulation, and how that would work, this book will pull together key elements from Chapters 12-13 of <u>Virtual Earth Graduate</u> (VEG) and Chapter 8 of <u>The Transformation of Man</u> (TOM). This book will not be a rehash of the information in those two books, but is more of a comprehensive connection of the dots – coupled with several related things to consider.

With this revision, **version 9**, a new look at **The Flat Earth Theory** has been added as it oddly enough has interesting evidence to support the initial premise of this book: the **3D Construct as a VR Sphere**. This book examines two main things: (1) Earth as a Simulation and (2) <u>What</u> is being simulated – a Flat Earth or a VR Sphere? The answer is worked up to logically in Chapter 11. Additionally, there is new input from the Physicists and Astronomers who still cannot admit to God and the Soul, thus they promote Simulation and Earth as a real rock circling the Sun to explain it all. Of course, they throw in Darwin because we are all just Apes on that rock circling the Sun – with no purpose because there is no God. If you don't like that view of the world, Chapter 11 will give you cause to believe in God, that Earth is special, and that you do have a purpose.

Simulation

This book will examine what the scientists are saying about Earth being a Simulation, a very sophisticated one, and the reader will discover that their points are well-taken… yet their evidence and the **Anomalies** can also be applied to a 3D Earth Construct, instead of a real 3D planet Earth in a real solar system. There is not much quantum evidence to support a real 3D planet, so the issue is <u>what</u> is being simulated? Quantum behavior supports Simulation.

While this book probably amounts to **Brain Candy** for some people, and is not to be blindly believed, it is something that presents <u>reasonable input</u> from the scientists and philosophers. It thus examines **3 sets of Anomalies**, odd occurrences, inexplicable objects, glitches and high-level Simulation coding, **noise from the edge of the universe** (Chapter 8), and presents a general, coherent, all-in-one-place overview of the Simulation Theory. **At the layman level.** This book also examines VR, Simulations, and Holograms, how they could work, who would run them, and whether they can coexist in a real 3D world.

Whereas the scientists have postulated that a Simulation would be run by <u>us</u> somewhere in the future, for whatever reason, that is not the point of view of this book. Nor does this book consider advanced ETs to be running a Simulation on a virtual hominid lifeform different than theirs. Again, Simulation has to be your most viable option if you can't accept that some form of a Higher Being has created Earth and Man. And even that will eventually challenge your take on Reality, as explored in Chapter 11.

The great number of **Anomalies** also suggests that we are in a Simulation… and an imperfect, finite set of computer coding that has limitations and displays rounding errors (glitches).

In the event that a reader discovers this book and has not read either VEG or TOM, this book will still suffice to reveal a complete, coherent picture of what Earth is today alleged to be. VEG and TOM reveal many other things that are of value and the reader is directed to them to better understand Earth History, Errors in Science, Health, the InterLife and why Religion was devised to control Man.

Are you really living on the planet you think you are?

In short, there are a lot of physical anomalies, odd scientific discoveries, and metaphysical issues to strongly suggest that Earth is not what we have assumed it to be. Remember the Flat Earth belief 500 years ago? Remember that the Church was sure that the Sun revolved around the Earth? We somehow survived the Dark Ages and yet those issues never completely disappeared. The Flat Earth is still popular with some people, and it will be seen that it might have been true at a time long ago. In any event, the Flat Earth issue is gaining popularity again.

Flat Earth

New to this revision, and the main reason for it, is a review of what the original Flat Earth (FE) concept meant to early Man, and why cultures all over the world believed the Earth was flat. What is fascinating is that many people today have not stopped believing in a Flat Earth – and believe it or not, there is some very substantial evidence to wonder if our ancestors of long ago saw a physically different Earth which led them to believe it was flat. And if that belief was plausible then, that is one antecedent for some people believing so today. There are several solid, serious pieces of FE evidence that cannot be easily explained away, and some of them substantiate the VR Sphere and they are presented in Chapter 11.

So it was my intention to prove the Flat Earth paradigm wrong – to sustain the VR Sphere concept.

It has to be emphasized right at the beginning that Chapter 11 was added to examine FE briefly **and prove it wrong**. This book promotes the VR Sphere as the Earth reality, as that was what I thought I was given back in 2008. Then a scientific reader sent me an email 'proving' the FE and I could not let that go – I diligently researched what he sent me and I could not prove it wrong, nor was the logic at fault. That caused me to dig further and today it remains such a mystery that instead of disproving FE, I am giving it serious weight with the VR Sphere... until the chapter conclusion.

Back in 2008 I was given the information that led to the VR Sphere as the Earth structure. It has to be said that I was given only **3 things**:
      (1) The 3D Construct [Chapter 1], (2) the Gegenschein,
          and (3) Earth as a Simulation.
I assumed They meant Earth was a sphere as I was not told anything about an FE concept (which I had always discounted). While the VR Sphere concept already says Someone is watching over us, if the FE scenario is true then we have that proof in spades.

While it cannot be 100% proven that the Earth is flat, neither can it be 100% proven that it is **a 3d Construct in Simulation as a VR Sphere**. (But 98% may be good enough.) Modern Man has accepted today's scientists at their word – and VEG Ch. 8 demonstrated that they are <u>not</u> 100% correct. You will see this in the case of **Sir Isaac Newton** herein who had to <u>guess</u> about Gravity, and <u>assume</u> he was right.

Yet today's scientists have taken Newton at his word, and built **CERN** still looking for *gravitons*, and <u>they may not exist</u> – there is reason to think that something else may be at work – even something as simple as "the apple fell because it is heavier than air!" Newton admitted that a Gravity <u>Force</u> was his <u>assumption</u>. So today's scientists are still trying to prove Newton's <u>guesstimate</u>! How scientific was he? Well, we know that he was an Illuminati, Alchemist and a follower of Quabbalah, [1] so... was his unprovable <u>assumption</u> based on an esoteric, metaphysical supposition?

> Man has not been able to discover what **Gravity** is, and that may be because Newton just said it had to exist (the apple again) so that he could promote his other ideas, and yet you'll see that he didn't have a clue – and today's physicists have not made any real progress in identifying what Gravity is or how it operates, either... Curiously, according to a clue in the Flat Earth Theory (which says Gravity doesn't exist because it isn't necessary), we may be sitting on the answer without knowing it.

Newton could not have known about Quantum Entanglement or Tunneling, and it is possible that Gravity is related to the sub-quantum 'attraction' of objects – including an esoteric aspect of the **Casimir Effect**. What if Gravity is the result of Dark Matter/Energy quantization at work, called **Loop Quantum Gravity**, effecting the attraction? Gravity would be a **property** of Space-Time itself – not a force. Newton's Classical Physics does not explain Gravity, and Quantum Mechanics seems to be where the answer lies. (Chapter 11 explores some of this at the layman level.)

Planet Earth is unique and at the end of Chapter 11 you'll see just how special it is. It does demonstrate some definite Simulation aspects... and that is why there is a final Chapter 13 – to suggest a solution. Why bother?

Earth School

Instead of quarreling with other people, fighting for our way, scrambling to be King of the Heap, on top of the pile, have the right clothes, car and spouse, live in the right neighborhood, collecting more and more toys, fighting with other people over whose concept of God is right or wrong, and fearing Death… when you finish this book, you may have a new insight to the **Earth School**, sit back and decide what is really important to you and yours, and stop the unimportant games. Or keep playing them if that is your choice. But you will be a little more enlightened – even if you don't believe the Simulation Theory. Remember the classic saying:

The Truth will set you free….
but first it may p*ss you off!

You will be free to think outside the box that others created for you. When your eyes are open and you know the Truth, you cannot be manipulated. You will know what is really important in your life, and no doubt you will make wiser choices… unless you choose to stick your head back in the sand, and act ignorant.

The goal of this book is to reveal more of what could be our world and why we are here. No matter how it was created, and what it really is, **the Earth School is real**.

So for those readers who did read VEG, and found some of it to be unbelievable, hang on to your hat -- there is more. Again, this should all be seen as **catalyst** – not something to be blindly believed, but something to be thought about, perhaps researched (via footnotes and links), and the last chapters will evaluate ways that the new realization can be a blessing to you.

> Note: book titles are abbreviated as set forth at the bottom of the Copyright page… VEG is Virtual Earth Graduate.

Chapter 1: In the Beginning

The English Bible tells us that

> In the beginning was the Word …
> and the world was without form…
> -- John 1:1 (paraphrased)

In reality, what existed to create this planet and set it up as a School and a Biosphere was indeed a bit weird… to normal, everyday humans who don't have a clue what this place is. With all due respect to King James and his Bible translation, the 'gods' <u>did</u> converse and their Word was executed to create Earth and the School and everything in it.

Overview

> Note: the following is written as if the Simulation is a fact.

Originally Earth was a 4D planet and was designed as a unique Biosphere with many types of flora and fauna (biota). It was in 4D for millennia, and as is explained later, other 4D beings came to Earth and experimented with the genetics here and over time the original **Experiment** began to run off-purpose. So after AD 900, 4D Earth was finally replicated as a **3D Construct** (still in 4D) and while the original Earth still exists in 4D, the Earth and Moon that we see are Simulated with a very sophisticated HVR [Holographic Virtual Reality] Bio-plasmic 4D computer (according to my Source, see VEG).

3D Earth in 4D Quarantine
(source: www.gafnews.com/sites/)

Earth in AD 900 was terraformed and maintained as a Biosphere, a **3D Construct** with many different types of flora and fauna. It was a copy of the 4D Earth. It was built in 4D and contained in a sort of energy **Quarantine** – to protect it from other 4D sentient beings who would explore it out of curiosity and could disrupt the ecosphere attempting to get resources and biota for themselves…
…but that wasn't enough.
(See Chapter 5.)

Those who designed and now manage the 3D Earth VR Sphere are referred to **as 'the gods'** (see Glossary) whereas in reality they are Higher Beings in the 4th and 5th level Realms. It is they who advise and guide Souls on their choice of experiences related to soul-development – a myriad of experiences designed to develop one's strength and knowledge. (See TOM.) The Soul is a precious, **eternal being** and will live in many realms throughout eternity, collecting knowledge and ability as it progresses. (The Soul is neither male nor female, but can choose to experience a male body or female body on Earth for the different strengths and weaknesses found in each one.)

VR Sphere Mechanics

The 3D Construct is also called the **VR Sphere** and consists of the Earth Biosphere and the Moon. To isolate it and protect it, the Sphere was **phase-shifted** 90° (a form of **superposition**) so that no external (3D or 4D ET) lifeforms are present.

> This explains why when Robert Monroe (Chapter 4) did his
> Out of Body journeys around Earth and into the surrounding
> space, he found no 3D life forms – All life in the Multiverse is
> 4D and above. (VEG, Ch. 12.)

What we see in the night sky is the <u>replicated</u> 4D universe that surrounds our **3D Construct.** It is reminiscent of a VR Game where you can see the background but you can't go there. The Stars, Solar System, our Galaxy and other Nebulae are a replicated part of the real 4D Realm… simulated <u>around</u> the 3D Construct – **which means it is contained in a larger Konstruct.** (That was the only way the gods could do it and Chapter 5 goes into the why and what of the setup more in detail.)

The VR Sphere concept is this: at the core is the 3D Earth Construct itself with about 7200 miles of space to the **first 'Shell'** – which often reflects the *Gegenschein*. The Sun and Moon operate close within the **Cspace** which extends several million miles to the **outer 'Shell'** also called the **Konstruct** – in which are the simulated stars and constellations.

Perhaps the significance of the following enigmatic structure will now make more sense. A replica stands outside the Vatican (and other places around the world) and has always puzzled visitors… A world within a world – and both are damaged…? It makes a perfect symbol for a damaged Earth (inner sphere) within an outer sphere, or Konstruct, that is itself suffering damage – from the negative vibrations (energy) generated on Earth. If a Simulation starts to fail, the outer shell will also fail… (See Chapter 4 – Fort and Monroe.)

The Earth or 3D Construct is inside the first Shell (aka Firmament).

The outer Shell is the Konstruct.

The Cspace is the distance (not to scale) between the two Shells.

Sphere Within a Sphere at the Vatican: Two Shells
(Credit: Bing Images: Panoramio.com)

The <u>superposition</u> of the **3D Earth Construct within a Konstruct** can be thought of as similar to shifting the Earth into another Timeline. The effect of phase-shifting the whole Konstruct has the same effect as shifting the Konstruct into an adjacent Timeline – It is not available in normal 4D where it originated.

Rewind: The 3D Earth Construct is contained within the inner Shell (aka Firmament) which is contained within the larger Konstruct Shell and between the two Shells is the **Cspace**. The simulated solar system and stars of the universe are pictured inside the Konstruct (outer Shell). **The Sphere within a Sphere in the picture above is not to scale** – there is quite a bit of space (3-10 million miles) between the inner and outer Shells. And that space will be referred to as **Cspace** (for ease of reference).

When we send **space probes** off into the "solar system," the (benevolent) gods that run this Simulation permit the probes to travel into and transmit from the Cspace area where time and space are **fractally** manipulated. Quite a large Simulation. (More on Probes in Chapter 3.)

> Whereas this boggles the mind even further, the Simulation does have the ability to expand, even fractally, as needed. And it can replicate itself – creating a Timeline split.

Scenes outside the Earth Sphere in the Cspace can be simulated much as in a Video Game where the player moves thru many scenes, but they are constructed and deconstructed (as needed) as the main Video character moves thru the settings. Sun and inner Solar System, for example, may be 4D *simulacra* surrounding us (in the Cspace) and some objects may be real, but we on Earth are contained by an (inner) energy 'Shell'. Thus Sun and stars appear to be part of the energy 'Shell' containing the 3D Earth and on occasion, the <u>inner energy Shell</u> reflects the Sun's light and is called the ***Gegenschein.*** Charles Fort figured that out and his discoveries and conclusions will be explored more in Chapters 4 and 11.

The following is another NASA picture of the *Gegenschein*:

Another View of the *Gegenschein*
Credit: NASA: http://apod.nasa.gov/apod/archivepix.html

Other Players in the Earth Drama

In <u>Virtual Earth Graduate</u>, (VEG, Ch. 13), Earth as a self-contained Simulation resembling a Video Game showed that the main character will encounter many **NPC**s (Non-Playable Characters) that are not controllable, but which are essential to the operation of the Video Game. In the similar Earth Drama these same humans who come and go in one's life are often called **OP**s (Organic Portals) and they are manipulate-able by the gods to effect whatever experience or lesson the ensouled human needs. (See Glossary.)

As a Soul incarnates, he often brings members of his Soul Group with him to work out a specific scenario, but that doesn't mean that everyone in his life Drama has a soul… The OPs have no soul because even the gods will not manipulate an ensouled human – but the OPs (or NPCs) are fair game. (Appendix D, TOM.)

In a different time, several cycles removed before this current **Era** we find ourselves in, souls were exposed to incredible behemoths called dinosaurs. Not many humans survived the encounters and the gods removed the dinosaurs as no one had the ability to fight them and win. Most animals since then were down-sized, and that is why large bears, Mastodons, Pterodactyls, Sabre-toothed tigers and large Apes (*Gigantopithecus.blacki*) became smaller and more manageable for smaller, weaker Man. **Gigantopithecus** is on the left, and to the right of Man (below) is an Orangutan:

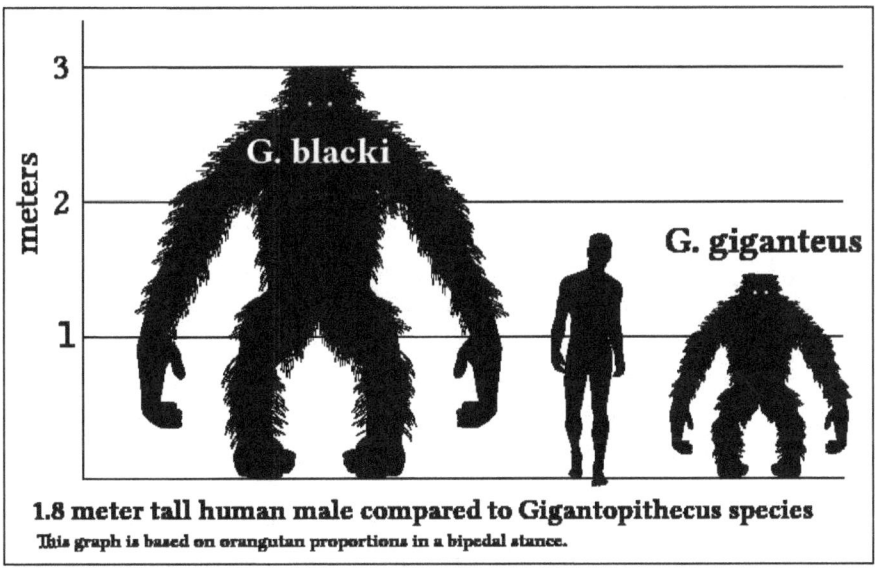

Courtesy: Wikimedia Commons: "Gigantopithecus v human v 1" by Discott.

Earth Eras

At this point, it would be relevant to clarify what has been meant by 'Era' inasmuch as the periods that Man has been living in are generally consecutive, time-wise, but the Eras are separated by the oft-suggested **"Wipe and Reboot,"** or Reset, thus inserting gaps in the chronology.

> **Note:** that the Sphere operates in Eras of indeterminate length.
> Also note that Eras A – D were in normal 4D, and Era E has been
> in the Sphere (below):

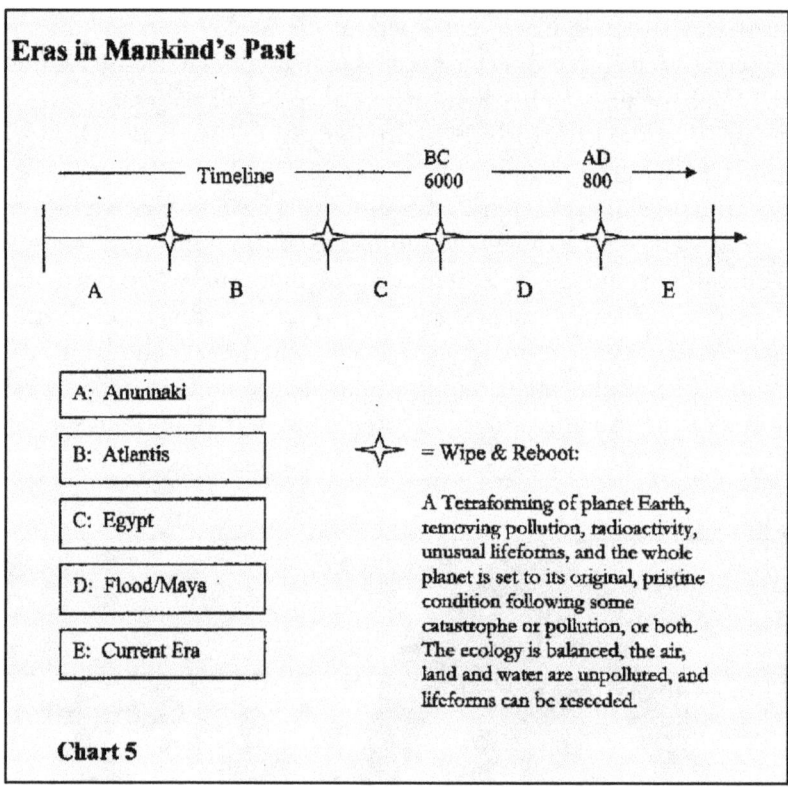

Eras in Mankind's Past

Timeline — BC 6000 — AD 800 →

A   B   C   D   E

A: Anunnaki

B: Atlantis   ⟡ = Wipe & Reboot:

C: Egypt   A Terraforming of planet Earth,
                     removing pollution, radioactivity,
                     unusual lifeforms, and the whole
D: Flood/Maya   planet is set to its original, pristine
                     condition following some
                     catastrophe or pollution, or both.
E: Current Era   The ecology is balanced, the air,
                     land and water are unpolluted, and
                     lifeforms can be reseeded.

Chart 5

When humans are too rough on the environment and each other (i.e., Pollution or War), the gods may stop the Drama, terraform the Earth (or lately the Sphere) – or better yet, RESET the scene -- and nowadays reload the Sphere with new humans for new adventures.

The current Era was begun around **AD 800-900** (by our time reckoning) and it was necessary to back-date many things so that it looked like we had a more complete, linear history – back to BC 2000, for example. Along the way, the gods **inserted** objects, like the **Antikythera** (Chapter 9, Anomalies III) found in the ocean off Greece, and hammers deep in coal strata, and footprints in shale beds – all to get Man to wonder more about himself and his world. (VEG, Ch. 10.)

The gods often **insert** lights in the sky (Orbs) that zip around at 90° angles and 2000 mph. They also **insert** huge hairy hominids which we call Bigfoot, or unique creatures in the ocean called Mermaids.

Things like the **Stonehenge** monument and the Great Pyramids of Egypt were not inserted but built by humans in former Eras (sometimes with inserted hybrid help – yes, even the Greys who are **bio-cybernetic androids** are inserted), and some objects are left in place each time the Earth is Reset and Restarted.

As an added attraction, the **Moon** was positioned to exactly eclipse the Sun. And it is concurrent with the mathematical principles running this Sphere that the Moon affects the oceans, creating tides. It can also be an observation platform.

And that is just a quick overview of where this book is going. Subsequent chapters will explain the Sphere, how it works, who works it, why, and what we are doing here. Chapter 11 will evaluate whether that holds water, or not.

Eras and Scripts

Each Era is under the control of not only the Control System (examined in Chapter 4), but a **Greater Script** for that Era – i.e., what is Man to basically experience and learn for the current Era?

Man is incarnated into the Sphere, on Earth, with his own Script that dovetails with and cooperates with the Greater Script , much as MS Word must 'fit' within the Windows XP operating system, for example. Man's Script is like an application that runs within the Era's operating system, or Greater Script. For example, a man's Script cannot have him fighting dinosaurs if dinosaurs are not present in the current Era… he would have to incarnate in the Era where dinosaurs existed.

Just think, if someone could see what is in the Greater Script, they would know what is coming… and that is the secret to Nostradamus and most **prophets** – they are let in on what is about to happen. That can be given to them in a dream or by an inserted human. This is also evidence for us being in a Simulation – since **simulations are scripted**, anyone who is let in on what the Greater Script says, will be able to accurately **prophesy**.

> This is not an argument for Fate – no Era Script is ever scripted so tightly that humans do not have **freewill** – What would be the point of having a School in which to learn and make mistakes if there were no freewill?

As the book progresses, there will be brief looks at some significant Dramas from past Eras that we tend to think are actual parts of our linear history. They were in fact subsets of the planet's overall history, in different Eras, and while some of them were in fact real 4D events (remember Earth was originally in 4D), they really are nothing more than the Drama for that Era, scripted and orchestrated for the benefit of those players.

Historical Aspect Examples

Drama 1: The Anunnaki Arrive

According to popular press nowadays, into this magnificent Garden of Lifeforms and ample physical resources, came explorers from the Orion and Sirian systems.... Draconians, Pleiadians, Lyrans, Zetans, Oannes, and Denebians to name a few. Some looked human, some didn't, but they were all represented in the Solar Council and were admonished to respect the special rules to protect and wisely use the Earth's resources. Earth was **an Experiment**.

... or so goes the story promoted by **Zechariah Sitchin**, Erik von Daniken, Michael Tellinger and others... including <u>Virtual Earth Graduate</u> which attempted to clarify who the Anunnaki were and what they did. Later research has shown that they were part of a former real, 4D Era, and were not inserts into the Sphere. They actually came to Earth back when Earth was physically in 4D and not in the Sphere.

> Remember that until about AD 900 the Earth was not in Simulation. Earth was in 4D where most sentient life in the universe is – according to Robert Monroe (Chapter 4).

In this scenario, the Anunnaki <u>were</u> responsible for modifying the DNA and human lifeforms on 4D Earth ... and they were from another 4D planet. Things took their course until the Anunnaki in the ships in Earth orbit came down and had relations with the newly created Earth women... creating a mess. The Solar Council terminated things and a **Flood** myth was left to account for why Man started again.

The Anunnaki and Man were (about 8,000 BC) watched to see how things played out. This time, Enlil was to teach and develop the humans, and that led to the development of Religion to try and control rowdy, ignorant humans... Again, things had to be adjusted and Man wound up under the tutelage of the Greek gods.... which gave way to being the Roman gods... And that didn't work.

Hence, the Roman Empire was dissolved, and when we now look back on what we think was Ancient History, we don't understand why the Anasazi disappeared, the Maya disappeared, as well as some Asian cultures, <u>all at about the same time</u>. The Maya were at the height of their civilization around 800 AD when they just vanished. The Reset, according to a Dr. Fomenko (profiled in VEG, Ch. 10) occurred about **AD 900** and the Church, via an insert called Scaliger, was tasked with the responsibility to backdate history so that suspicion was allayed that something had happened and Man had a big gap in his linear history on the planet…. This is when Earth was put into the Sphere (more in Chapter 5).

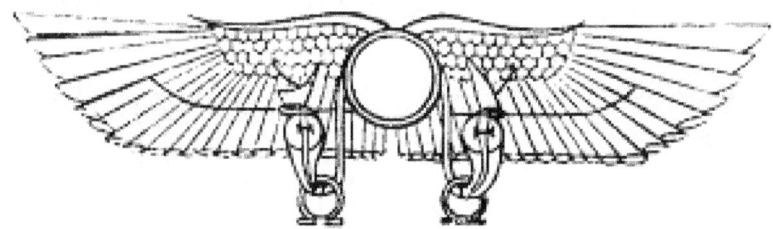

The winged Sun was a symbol used by the Anunnaki, the Sumerians and the Egyptians. It represented not only the Sun but the Anunnaki craft because they were shiny and they flew (see <u>Anunnaki Legacy</u>, at the end of Ch. 2).

Drama 2: The Aldebarans Come to Earth

About 20,000 BC a small group of colonists allegedly left their 4D planet Aldebaran. They were definitely human beings, but advanced, and they launched a ship to explore and develop another habitable planet. Due to some malfunction or accident in space, they lost their navigation and communication equipment, and drifted – eventually finding 4D Earth. (This is the essence of TEW.)

Up to this point, the Anunnaki had only created the Black race in Africa, and the Brown and Red races in the Americas. The Yellow race had been briefly started in the area now known as China by the Denebians, and the Anunnaki had gotten along well with them, but the two had left each other to their own pursuits. Not so with the Aldebarans.

The Aldebarans were an advanced White race, with a similar technology that matched that of the Anunnaki – rockets, lasers, computers, genetic skills and nuclear capability. They landed on a small land mass to the west of Africa and founded **Atlantis**. There was constant friction between the Anunnaki and the Aldebarans for resources and push came to shove and war broke out – the Aldebarans destroying outposts in Peru and Bolivia (i.e. Puma Punku), and the Anunnaki destroying Atlantis. The war also destroyed several cities in Western India, and the spaceport in the Saudi Peninsula was destroyed. All were nuked.

As if that weren't enough, the Aldebarans attacked the Watcher component in orbit around Earth, and they declared war on the Anunnaki on Mars – resulting in the large scar on the planet's surface (***Valles Marineris***)…

Valles Marineris Scar

A huge scar stretches over 4000 miles across Mars and there is no natural geologic reason for it. It is 120 miles wide and over 7 miles deep.

This is what a large plasma disruptor weapon does.
(credit: Bing Images)

And as a result of the space war, the destruction of the planet between Mars and Jupiter really got the Solar Council's attention and it was decided to put a stop to wayward activities on Earth.

It is said that the Solar Council got involved and contracted with a powerful group of beings, allegedly across the Galaxy, who just wanted to be left alone, but who were finding the fabric of space upset by the war on and around Earth… so they agreed to help out. The superior Eridani were tasked with confronting and forcing more Anunnaki to go home, and their numbers were cut down to less than 200 (underground). The Aldebarans lost their technology with Atlantis and fled to **Northern Europe** – but first erected a multi-stone henge as a signal to any rescue ships that might one day come from Adebaran looking for survivors… The structure, later called Stonehenge, is allegedly a copy of a similar temple on Aldebaran.

The Aldebarans thus inserted the White race onto the Earth, and populated Scandinavia and Northern Europe. Of course, Anunnaki science officer Enki noted all this and abducted some Aldebarans for genetic material, studied it, and added it to the mix when he recreated Man after the Flood. This was a superior version of *Adapa*, which became **Cro-Magnon**.

The Anunnaki and Enki, their genetics scientist, actually first created humans on 4D Earth and even though Earth was later brought into the Sphere, Man came along with it all, plus the pyramids(all over the planet) and Stonehenge. The existing human-like hominids on Earth provided 'containers' in the Sphere into which Souls could incarnate.

Anunnaki Manipulation

Thus, the first manipulation of Man was done by the Anunnaki – creating **slave workers** to work the mines, till the fields, and build their buildings. And because they were designed to be all brawn and few brains (just enough to be able to follow directions), there were problems controlling them. The workers were a mix of **Homo *erectus*** (below), a wild and rowdy hominid, and the Anunnaki genetics were said to be lusty and war-like – thus early humans were noisy and often uncontrollable. This is what Enlil considered unworthy of continued development as a race and that is why he sought to wipe them out when their usefulness was at an end (Hint: The Flood).

Homo *erectus* looked like the following:

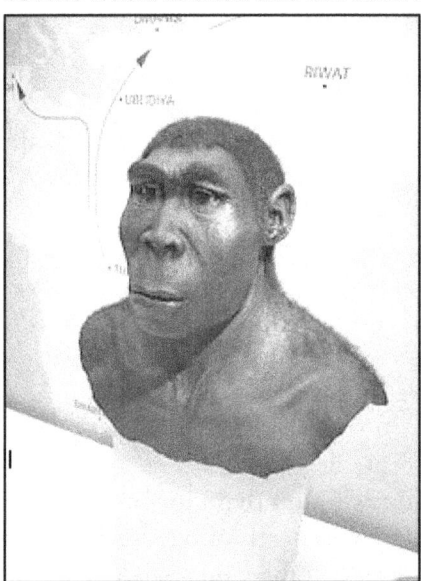

The first skeleton was found in found in Dmansi, Georgia ... not far from Mesopotamia where the Anunnaki were creating all sorts of humanoids.

It is suggested that this was in fact the source of *Adamu* that the Anunnaki created.

Credit: Wikipedia: Homo erectus

Mankind's Heritage

One thing is for sure: the successive forms of Man on the planet are mute witness to the fact that Man <u>was</u> changed – from **Homo *erectus*** (above) to **Neanderthal** to **Cro-Magnon** to **Homo *sapiens*** (below), and lately to **Homo *noeticus*** (Indigo children).

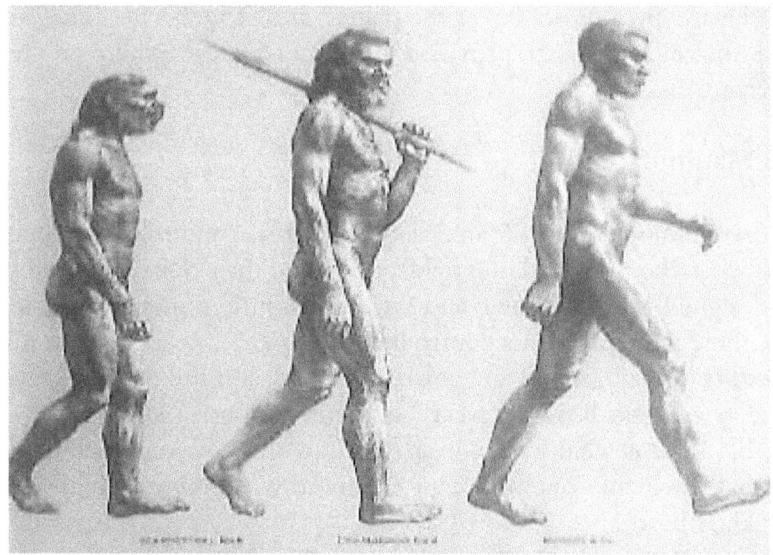

Man: Neanderthal, Cro-Magnon, & Homo Sapiens
(source: http://www.wilderdom.com/evolution/HumanEvolutionSequencePictures.htm)

The Anunnaki could have modified Homo *erectus* when they created *Lulu*, or their sterile worker human (the 1st creation). The improvement, by Enki, into *Adamu* could have been closer to Neanderthal man (the 2nd creation). Enki would later 'personally upgrade' *Adamu* to *Adapa* -- much like Cro-Magnon or Homo *sapiens*.

> And **the Greys** are behind the latest genetic change to Homo *noeticus* (the Indigos). Neanderthal was one of many experiments that didn't work out and was replaced by Cro-Magnon. In a similar way, there is an upcoming replacement or insertion -- a **replacement of Homo *sapiens* with Homo *noeticus* (i.e., Hybrids)**.

As will be mentioned in a later chapter, after Enki developed a better, more intelligent version of Man, around 2500 BC it was found that humans could be scared into behaving better by threatening them with death or disease. And that worked for a while, but was counter-productive when the Council ordered the Anunnaki to be responsible for the development of their sentient beings… killing them was no longer an option. So the outpost in Africa was discontinued (in Zimbabwe) and the few humans there were let go to fend for themselves, and they dispersed into the wild.

In Sumeria, it was a different story – Enki had developed some humans there into semi-learned beings (much to his brother Enlil's displeasure), and Enki hit on the idea to teach Man Religion, and ethics (with the **Hammurabi Law**) and take his best *Adapa* humans and make them **priests**… to shepherd the rest of the humans – again so that the Anunnaki didn't have to do it. The Anunnaki would train and even 'program' the best ones, and set them up as priests and later as kings, to rule over the rest of the humans.

Thus the Anunnaki had to try and educate the humans, in agriculture, medicine, accounting, math, and astronomy. Unfortunately, they also taught the humans warfare – and got humans to fight some of the Anunnaki wars for them!

Thus it was that law, ethics, and religion were promoted among the humans. And since the humans already saw the Anunnaki as gods, it was a simple step to emphasize that heaven was above, where the Anunnaki came from, and if humans didn't behave, they would be sent below into a world of hellfire and eternal damnation – an attempt to scare them into better behavior!

As a matter of fact, the humans were subjected to disease by the Anunnaki, something called ***Suruppu* Disease** and after decimating their numbers, and after Noah [Utnapishtim] interceded for their relief in the *Atra Hasis*, Enlil recanted and set the cure on them. It is thought that the Plague, or Black Death of the mid-1300s was another such 'discipline' sent to cull mankind's numbers. (It would be interesting to know whether HIV/AIDS is also their creation….)

The Anunnaki were always concerned with the proliferation of Man on the planet, and as such the **Georgia Guidestones** (erected in 1980) reflect the same kind of thinking… cutting mankind's numbers to a more manageable number, as well as instituting what appear to be common-sense planetary controls – to better manage Man.

(Credit: Bing Images: folkmaktnu.wordpress.com)

The whole Guidestones English page looks like the following:

What is interesting is that the original structure was built in March 1980 with a **notch** left undone… and then in 2014 the notch was filled... (see below)

Notch

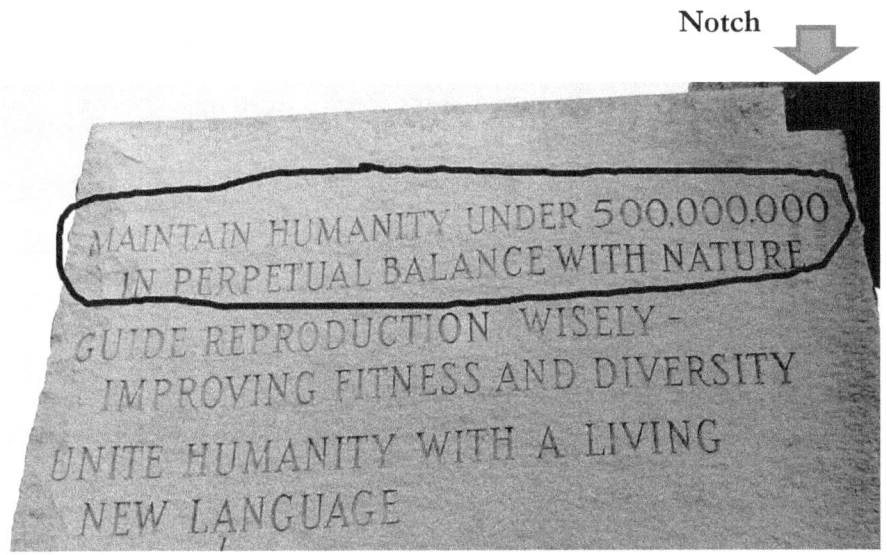

Section of English Text on Georgia Guidestones
(source: http://en.wikipedia.org/wiki/File:Georgia_Guidestones)

In 2014 a new feature was added to the cutout seen in the upper right corner of the picture above: a smaller stone with the 'date' 2014. --- see URL:
https://www.youtube.com/watch?feature=player_embedded&v=j_jz5c3GVVg

The Right-angled stone with "20 14" is the addition.

YouTube caption:

"The mysterious Georgia Guidestones, which some see as an elite manifesto for neo-eugenics and population reduction — have received a strange update."
Credit: Facebook: @ https://www.facebook.com/Paul.J.Watson.71

One of the Guidestone's 'suggestions' reads (4th line above):

Improving **fitness** and diversity

Note that nothing is said about sustaining health or being healthy. And yet one of the major diseases on Earth today is cancer. Just being fit will not stop or eradicate that bane on society. But there may be a way in better understanding our link with the hologram around us… If we live, move and have our being in a dynamic holographic world (as scientists throughout the book will suggest), it would behoove us to find out (1) that it does exist, and (2) how to synch up with it – for better, sustained health.

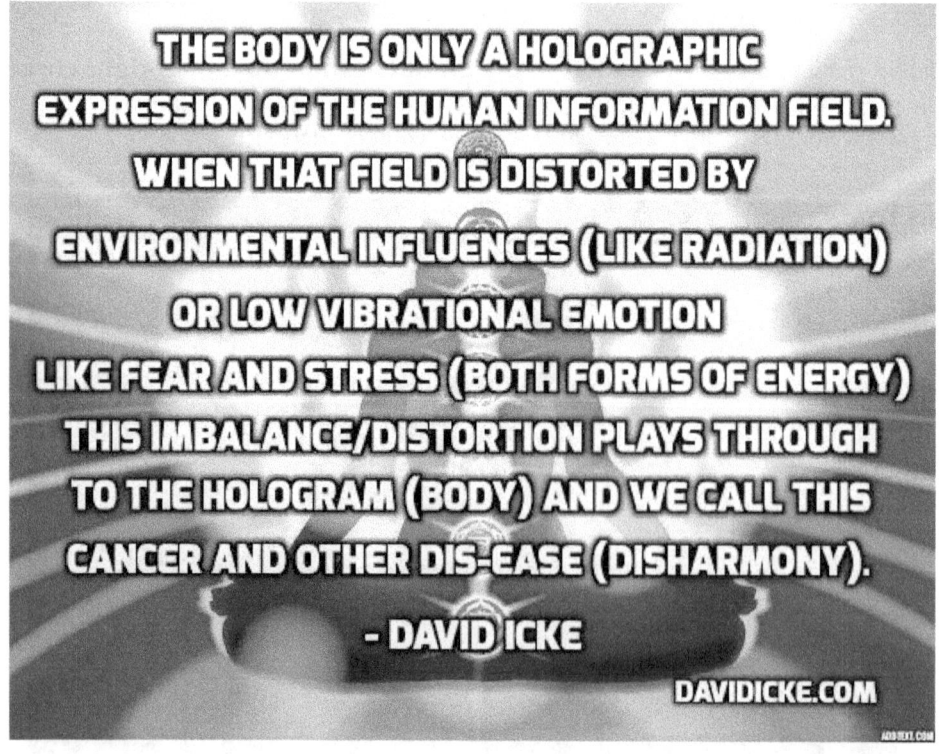

THE BODY IS ONLY A HOLOGRAPHIC EXPRESSION OF THE HUMAN INFORMATION FIELD. WHEN THAT FIELD IS DISTORTED BY ENVIRONMENTAL INFLUENCES (LIKE RADIATION) OR LOW VIBRATIONAL EMOTION LIKE FEAR AND STRESS (BOTH FORMS OF ENERGY) THIS IMBALANCE/DISTORTION PLAYS THROUGH TO THE HOLOGRAM (BODY) AND WE CALL THIS CANCER AND OTHER DIS-EASE (DISHARMONY).
- DAVID ICKE
DAVIDICKE.COM

Again, the sages of the world, as well as the Amerindians, have suggested throughout the centuries that we learn to live in harmony with Nature and be at peace with all

around us. Simply ask: Do heavy metal music, violent video games, or blood-and-guts shoot-em-outs on TV contribute to a harmonious peace in our bodies/minds? And how does it affect the planet's energy… and that of the Sphere?
(See TOM [Ch 4], VEG [Ch 14], and TSiM [Ch.9] for a more complete view of health issues.)

Recap

As can be seen, the VR Sphere is a 3D Construct created by Higher Beings in the 4D-5D Realm, operating under limited 3D rules but controlled by 4D processes. There is more power in 4D to create and manipulate matter at a lower level, and thus a controlled subset of 4D processes, called a self-contained 3D Construct, is what the VR Sphere is alleged to be.

The plan for Earth was not always a School. It was the decision of the Solar Council and the gods to isolate Earth and make it into a School to educate and train the growing number of souls incarnating in human bodies – without 4D ET interference (thus the quarantine, or energy barrier, off which the *Gegenschein* reflects).

The reason for doing this about AD 900 was to provide a place to train souls and force them to think and act responsibly – not with their normal 4D abilities. In 3D they would not be able to use their advanced 4D abilities (clairvoyance, telekinesis, telepathy…) they would have to operate in a more "bare bones" level and that would provide the discipline that the Higher Beings sought to impart to the souls who would enter the Earth School.

The VR Sphere structure would be run by a Master preset Script and Beings of Light (Angels) would oversee the guidance of incarnate souls. Each soul would have a script and a rôle in the Earth Drama – making Shakespeare's comment that the **Earth is a Stage** a reality. Individual scripts can be executed by the equivalent of an Operating System, fractally for one soul, if need be, or a whole group, thus furnishing appropriate experiences which translate into lessons for each soul or group.

> The individual scripts leave a lot of room for freewill – the soul's script is just a plan for what should be learned and it identifies tests and exit points. (More in Transformation of Man [TOM].)

Now we'll take a look at various aspects of the Sphere and examine the what, who and how of the Simulation. This will be more in-depth than what was covered in Virtual Earth Graduate (VEG), although some key material has been repeated for coherence, and because some readers may not have read VEG.

Chapter 2: Is It Real or...?

Before examining all the data and reasons for Earth being a Simulation, it is very helpful to take a look at a few relevant examples.

Scenario 1

John sits at his PC playing a new video game, Castles and Damsels. The game, as always, is seen from the viewpoint of John who holds a spear.

The scene starts outside a castle which faces a forest 100' away. There is a cry for help from the forest, and John moves forward, careful to avoid the rocks. As he enters the forest, the scene smoothly transitions to all trees and bushes around him. Slowly, cautiously making his way forward, he pushes tall weeds out of the way with his spear, and some of the Saw Grass cuts his left arm. He comes to a clearing and there is a stone altar with a pole in the middle, and a blonde maiden tied to it.

There are two serfs with spears guarding the altar and John throws his spear and nails one of the men who cries out, and slumps to the ground... The other serf sees John and charges him, and John whips out his dagger, throws it and dispatches the 2nd man. He rushes forward, climbs the altar and finds he cannot untie the maiden, so he jumps back down and retrieves his dagger and goes back up and cuts the ropes binding the maiden to the sacrificial pole.

He has rescued the maiden and they must now exit the woods and try to gain the safety of the castle, but John has gotten turned around and has lost the way back. He and the girl head off to the left and are faced with a large pit with a big water moccasin in it, hissing at them. The maiden yanks John back from the edge, John swings around, and they re-enter the altar area, the dead serfs are still lying there, John grabs his spear, and exits with the girl off to the right – back into the words again.

They move thru new territory until at last they come to a small peasant hut made from sticks and straw... They hear the sound of barking dogs and men shouting – someone is chasing them, so they dash into the hut.... John sees food on the table but no one is there. The girls says it smells wonderful and moves toward the table... John knows they can't stay there, they will be discovered, so he kicks a hole in the back wall of the hut, grabs the girl by the arm and yanks her away from the rough-hewn table, but in so doing, she gets a splinter in her hand.

They exit the hut and run thru the woods, just a few yards and come upon a lake. Off to the right is a small boat with oars... They run to the boat, look back and the scene shows the woods, the hut and one of the pursuing dogs... a Mastiff. They hop into the boat and push off. They hear arrows fly overhead. Some hit the boat with a thunk, but John and the girl escape across the lake.....

Possible video game, right?

Note the scene changes, the sounds, the 'smells', and even the players (run by the Game's Operating System) act like normal people – i.e., they are not superhuman, they don't fly, the girl gets a splinter, John cuts his arm on the SawGrass..... everything follows a 'programmed' action. John has the choice of where to go, which way to run, and he could have left the girl alone, but he is making choices which the System recognizes and it challenges him to survive and make further choices.

Wherever he turns and runs to, the System changes the scenery to match the pre-programmed area and inserts whatever challenge is needed... such as the altar. Every time John returns to a previous scene, it is re-created for him as it was left. The program is coherent that way, and the girl he takes with him always looks the same – except for the addition of the splinter in her finger.

Now let's look at this in real life...

Scenario 2

Bob is a soldier in Vietnam and has gotten separated from his platoon. He has a knife and an M16. He knows he must get into and thru the jungle in front of him without the Viet Cong seeing him. He cautiously but quickly enters the jungle, avoiding some rocks in his way. He advances thru heavy grass, using his rifle to push the grass away, and some snaps back and cuts his arm.

Bob comes upon a couple of Viet Cong who have tied a local village girl to a tree, and he shoots both and runs to the girl to rescue her. Knowing the sound of his M16 might attract nearby Cong, he uses his knife and cuts the girl free and runs off with her to the left, thru more jungle.

They run around a clump of bushes and almost fall into a large pit with sharp stakes and a viper in it... a nasty trap that cannot be seen until you come upon it. The girl yanks Bob back and saves him. The Viper hisses at them, and they quickly move to the right again, but their way is blocked by a large water pit. They have to go back the way they came, and they run back thru the clearing, the dead Cong are still lying there, but the girl picks up one of the Cong knives, and they exit the other side of the clearing....

They move thru new territory until at last they come to a small peasant hut made from sticks and straw... They hear the sound of barking dogs and men shouting – someone is chasing them, so they dash into the hut.... Bob sees food on the table but no one is there. The girls says it smells wonderful and moves toward the table... Bob knows they can't stay there, they will be discovered, so he kicks a hole in the back wall of the hut, grabs the girl by the arm and yanks her away from the rough-hewn table, but in so doing, she gets a splinter in her hand.

They exit the hut and run thru the jungle, just a few yards, and come upon a large river. Off to the right is a dugout [boat] with oars... They run to the boat, look back and the scene shows the jungle, the hut and one of the pursuing dogs... a Pitbull. They hop into the boat and push off. They hear bullets fly overhead. Some hit the boat with a thunk, but John and the girl escape down the river.....

Not too much difference from Scenario 1, huh?

Virtual or Reality?

Could Scenario 2 actually be a very sophisticated Simulation, run by a super computer in the Astral?

Holographic Space Scene
(Credit: Bing Images: briankoberlein.com)

All that was missing from Scenario 1 was a sense of touch and smell. And that is now available thru a new form of video game wherein the player wears sophisticated goggles which provide 100% of his vision and hearing, the player holds a joystick to control action and direction, and the player can be wired (in laboratory trials) to receive sensory input – touch and smell. How is this any different from Scenario 2?

US Army Paratroopers now use HMD (Head Mounted Display) to simulate and practice landings. US Army Apache helicopters now use head-mounted gear to track the pilot's eye movements so that wherever he looks, the 50-mm cannon points where he is looking!

Oculus Rift by Oculus VR is another headset with inclusive sensory feedback. All the developers have to do is 'wire' the player's brain with 2 sensory taps to the areas stimulated by smell and touch, and the experience will be complete.

The following picture shows VR gear for the eyes, ears and something that traps hand/finger motions, replacing the need for a joystick? She also can speak commands into the VR scenario… (kind of like the movie *Johnny Mnemonic.*)

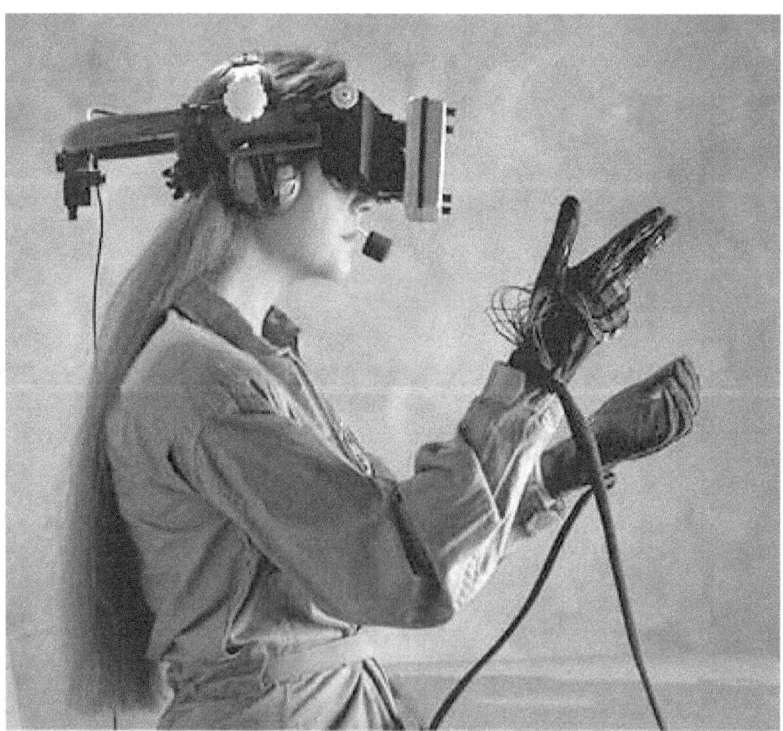

Courtesy: Bing Images: Virtualreality.net.au

Man is not that far from developing something like the Holodeck found in *Star Trek*. A complete sensory experience – with no goggles or headsets required.

Earth as a Holodeck

Those who watched *Star Trek* years ago, during the Captain Kirk era, should remember that aboard the Starship U.S.S. Enterprise NCC 1701, there was a big room called the Holodeck. It was empty, and had all black walls with yellow or orange calibration (grid) lines running up and down and across it. It was about 30' wide by 60' long and maybe 20' high. The crew used it to run computerized **simulations** of some event or process that a crew member wanted to try out.

What was interesting, and this is very relevant, was that the crew could activate a computer-supplied simulation of a jungle, or forest, and the trees were real – one's hand did not go through them. And coming to a precipice, the drop was more than to the floor of the Holodeck, and if one came to the top of a hill, the view across the landscape extended way beyond the 30' wide walls of the Holodeck. Last but not least, one <u>could</u> get seriously wounded, killed, or trapped in the computer simulations if the security/safety protocols failed.

Star Trek Holodeck
(**credit:** www.crystalinks.com)

The overall scenery (on the walls) was said to be **holographic**, and the parts you could touch, like the trees and rocks, were provided by the *replicator* technology. And somehow these two technologies were coordinated by the ship's super computer to provide a 3D realistic experience.

Evidently, the Holodeck created a fractal, dimensional subset of reality within its walls... so that one could fall off a cliff, or run 5 miles in a straight line. That is in fact what this book is suggesting is our reality, and will examine the plausible aspects of that.

Virtual Reality and the Hologram

Unless we see some revealed aspect of the hologram, like a glitch or a bubble, we can't tell if it is real or whether our reality is a holographic virtual reality – so good, in fact, that it appears to be real..... and that is just what this book and the many scientists and philosophers suggest.

(credit: Bing Images: Pravda-tv.com)

We are beginning to create our own simulations, so far without simulated humans, but that is next.

Simulation Games

Well, in today's world, we have stalk-it-and-kill-it video games, as well as a game that began for sociological planning purposes – it let city managers and urban planners simulate cities and additions to them – wherein the models were useful as they highlighted problems with roads, sewers, etc...

The latest versions of these modeling tools have gotten very complicated. In fact, they are sometimes played by people who are not urban planners... Such packages are called **Sim City, Sim City 3000 and Sim City 4.** Of course these simulations are void of simulated humans in a 2D version, and the 3D version, **SimCity Societies**, still lacks people. But it is just a matter of time before one of these versions can simulate people.

And these are quite sophisticated packages, generally applicable to modeling new communities with a full range of civic services, economic and ecological concerns, and the latest Sim version offers societal values to consider: productivity, creativity, prosperity, spirituality, authority and knowledge. [2]

Create Paris the way you want it…

(**Credit both: Bing Images**)

Sim City 5

Sim City 2

… or invite a monster to romp through your city.

This borders on creating your own video game!

And in 2013, **Sim City 5** was released with new technology to simulate traffic jams and we are beginning to see people in the scenario now – on a limited basis. Whereas prior versions would accumulate statistics regarding water, waste and population, the new GlassBox Engine

>…replaces those statistics with **agents,** simulation units that represent objects like water, power, and workers; each graphic animation is directly linked to an agent's activity. For example, rather than simply

displaying a traffic jam animation to represent a simulated traffic flow problem, traffic jams are instead produced **dynamically** by masses of Sim agents that simulate travel to and from work...

The citizens in the game are also agents and do not lead realistic lives, they go to work at the first job they can find and they go home to the first empty home they find. [3] [emphasis added]

GlassBox is the engine that drives the entire game – the buildings, the economics, trading, and also the overall simulation that can track data **for up to 100,000 individual Sims** inside each city. There is **a massive amount of computing** that goes into all of this, and GlassBox works by attributing portions of the computing to EA servers (the cloud) and some on the player's local computer.

But this is a good start. In future versions, it will be possible to set parameters and characteristics of chosen agents to behave in specific ways...

Sim City Expansion

SimCity: Cities of Tomorrow cover art.

An expansion called *Cities of Tomorrow* was released on November 12, 2013 and is set 50 years in the future. It features new regions, technology, city specializations, and transport methods.

The new features in *Cities of Tomorrow* are divided into three categories – "MegaTowers", "Academy" and the "OmegaCo".

The **MegaTowers** are massive buildings built floor by floor with each floor having a specific purpose, being residential, commercial or to provide services like schools, security, power and entertainment. Each floor can provide jobs, services or housing for hundreds of citizens at the same time.

The Academy is a futuristic research center that provides a signal called "ControlNet" to power up structures and improvements developed there and the **OmegaCo** is composed of factories used to produce an elusive commodity only known as "Omega" to increase the profits from residential, commercial and industrial buildings alike and manufacture drones to further improve the coverage of healthcare, police, fire services or just be used by citizens to perform shopping in their places, thus reducing traffic.

> The expansion also supports "futurization", in which futuristic buildings tend to **"futurize"** the buildings, roads, and services around them by significantly blending the roads and buildings to simply make them look more futuristic, such as differences in traffic lights (they have a different sprite), turning service cars more futuristic (futurizing a police station will significantly change the cars and architecture), and so on. Buildings that will futurize the vicinity are distinguished with a hexagon pattern at the lower part of the building when viewed in the Construction screen. [4]

All of that to say – Look where it is going. Is it popular?

> *SimCity* sold over 1.1 million copies in its first two weeks, 54 percent of those sales were of the download version of the game. As of July 2013 the game has sold over two million copies….
>
> In Q4 2012 …. a completely new version of the treasured classic, includes deep online features. More than 100,000 people played the *SimCity* beta last weekend…. and the critical reception is shaping up well. [5]

They send a 'footprint' to load on the user's PC (or Mac) which is just enough software to initiate and run the application – from the vendor's servers. One can play with the building and managing concepts oneself, but more often the Game is to have rival cities, and run economics and trade between cities and see if you can become rich and bust the other team's economy/city.

Resource Intensive

Lastly, on a more serious note, the online version of the game suffered initially from too many people logging on and overloading the servers – So more servers were added, and yet there was another problem when it came to designing the Sim City package to let the user build a huge city – While the developers have done that, it is a monstrous download and **very few PCs receiving the download could run it**… It is a real resource hog.

Significance?

Guess how big the Bio-Plasmic HVR Computer that runs the Earth Simulation would have to be!!? And that doesn't count what will have to be the storage and server requirements when agents, and people, are coded to move and act out a scenario.

The fact that we are beginning to simulate the real world, for whatever purpose, is significant according to Dr. Nick Bostrom (Chapter 7). It means: if we can do it, so can They.

Job: A Comedy of Justice

Ever wonder just who was socking it to poor Job in the Bible? Job was a Man of God, trusted Him in all his ways, and was 100% true blue. Supposedly Satan asked God for permission to test, tempt and afflict Job – to see if Job would renounce his faith and turn and curse God. Well, he didn't and God then increased Job's holdings… and Job was blessed with finances, sons and daughters born to him…

> It is a very powerful allegory, as opposed to something that actually happened.
> Metaphysicians would say that this is a subtle justification of Karma and we get what we deserve, so don't bitch about it, just handle it. It amounts to a test anyway you look at it.

This whole scenario set a super Science Fiction writer, **Robert Heinlein** to wondering… Could the Game-players be brought to justice and made to answer for what they did? Suppose it wasn't God and Satan who were playing with Job, but two brothers, Yahweh and Lucifer ..? And what if Job was stuck in a simulation while the gods socked it to him?

The main hero (Alec) is jerked around from scene to scene on Earth and just as he settles into a scenario and begins to make something of himself or his circumstances, the rug is pulled out from under him and he, as the main character, is translated to another drama…. All thru the drama, he writes and maintains a **manuscript**… like a diary.

The relevance of Heinlein's book is that the gods are playing with Man. And while the main character in **Job: A Comedy of Justice** (Alec) is being driven nuts with the constant jerking around, we see a similar potential in any Simulation. The gods can change the scenario on us, insert/remove items and people, and create synchronicities and bless us or curse us – and Heinlein's Job goes thru it all.

In essence, the two brothers, Yahweh and Lucifer, had a bet and the human on Earth, Alec, was the victim. And then their Father finds out what they were doing, and steps in to stop it and rectify the drama. Everyone is dragged into His Office and He chews them out for violating a cardinal rule in the Earth Drama: **Consistency**. He knows what is in Alec's manuscript, and He calls his sons to account… and an interesting ending ensues.

Heinlein appears to be aware of the issue with Earth being a simulation, or at least a Stage for a larger Drama, as he uses the words simulacra, scripts, and deception. Says one of the key characters in the book's end, explaining things to Alec:

> I did *not* say that the world was created 23 billon years ago; I said that was its age. **It was created old**. Created with fossils in the ground and craters on the Moon, all speaking of great age. Created that way by Yahweh, because it amused Him to do so. One of those scientists said, 'God does not roll dice with the universe.'
>
> Unfortunately not true. Yahweh rolls loaded dice with His universe…. to deceive his creatures.' [6]   [emphasis added]

Wow, somebody also knows or suspects what is going on.
Heinlein's book is also humorous in places and very much worth a read!

The main character experiences many **anomalies** besides his being shifted from scene to scene on Earth. And of course these get his attention because his world is not a consistent one. What if that happened in our daily lives?

Anomalies, Part I

In point of fact, it did happen and the following are actual incidents that happened to the author. To repeat, these are not included for anyone's amusement. They actually happened, and there are other people out there who have had similar things happen, but don't talk about them for fear of being ridiculed. Maybe it is time to 'fess up.

January 1965: Returning to Syracuse, NY to finish college, from Alabama, the author was struck with a savage Flu that the local doctor called the Grippe. Registration and dorm were cancelled and I hit the road with my parents who came up to help me move back down to Huntsville, Alabama. We were headed West on the NY State Thruway. The Winter blizzard really socked in so bad off Lake Oneida, that our cars were starting to get stuck in the snow drifts across the road, so we had to pull off the Thruway and spend the night in a HoJo motel. Mom called Unity Church's Silent Unity for prayer support (I was really bad and didn't know I was close to dying).

The next morning, the Sun was out and the plows were clearing a 1-lane path on the Thruway, so we skipped breakfast and hit the road in our two cars. I had the radio on in my car, and the weatherman said the storm was to intensify and last for two days… I looked in the rearview mirror and the storm was closing in on us as we drove – about 1 mile behind us.

We made it to Huntsville, Alabama and I was immediately put in the hospital. I had a single room, and the nurse at 9pm locked my door. I fell asleep, not knowing one lung was full and the other ½ full – the mucus had really accelerated and the doctor prepared my folks for the eventuality: I was going to die. (Hence the single-occupant room and the locked door.)

Sometime during the night, I was awakened by a bright light at the foot of my bed, but I didn't open my eyes. I assumed it was the night nurse. I was asked how I felt. I said Ok. Then I was asked, **"Do you want to continue?"** I said Yes, and was immediately asleep again.

Credit: Bing Images: Light Being

The next morning I awoke hungry as a horse. I looked at the room clock, and it was about 9 am. Where the heck was my breakfast? I slipped out of bed, and went to the door, and opened it. I stood in the doorway and got the attention of the nurse at the station across from my room. She shrieked! I said "Good morning to you, too! Where is my breakfast?" I went back and sat on the bed. No one came to see me.

I later learned that Mom had called for prayer support again that night.

About 1 hour later, my doctor showed up, checked me out, mumbled something and shook his head, and ordered me down to x-ray. The x-ray showed nothing in my chest. No Flu, no Grippe, no mucous… nothing! And while I had had 7 bouts of serious pneumonia before I was 15, I was never subject to pulmonary problems of any sort, after that.

And by the way, night nurses don't ask a patient if they want to 'continue.' I realized that months later when I was thinking about the event… What had me really thinking hard was that **I had a choice to continue in this life**, or had I said No, I would not be here. I suspect it was a Being of Light that showed up and asked me – I had a choice?! I have wished I had opened my eyes to see who/what that was asking me the questions to this day! And I don't think I'd have seen much if it is just an amorphous white light sphere. Are Angels inserts?

Anyway, the doctor kept me in the hospital another 4 days because somehow it was a case of "spontaneous remission" and might come back. It didn't.

By the way, the doctor took to smoking and drinking after that.

May 1982: (This one is a corker and to this day I still wonder about it.)
I was living and working in Orange County, CA and needed a new car. I knew the people at the Buick dealership and they gave me a great deal on a sporty new 1982 Skyhawk – Black with tan interior, gold pinstriping and trim. It was Buick's answer to Chevy's Cosworth Vega. For a 4-cylinder it really ran. It was to be my "city car," and the Blazer was for the Sierras, backpacking and romping thru the backroads north of San Bernardino.

After two weeks it stopped running well and started missing and became hard to start. So I called the dealer, arranged a loaner car and took it into the shop. Because the Service Manager, Sam, knew me, he took it home Friday night just to see what it was doing, and it ran fine… So next day he prepared to take the family and dog to the park… still road testing it. The wife and kid got in it, but not the dog. Sam gets out and grabs the dog and heads for the car, and his dog tries to bite him! He could not get the dog into the car. Odd because Susie really like riding and going to the park. So he locks her up in the backyard and goes to start the Skyhawk. It won't start.

Long story short, he gives up and the wife and kid go on in their car. Sam gets it going and drives it to the repair bay at the dealership and the mechanic pulls it partially into a lift bay. They both leave it there. Sam stays at work and goes back to his inside office, to finish some paperwork.

Two hours later, the mechanic runs into Sam's office and tells Sam he'd better come see this! They both go to the service bay where the Skyhawk was parked... what was left of it. Somehow the lift had come down, full force, onto the hood of the Skyhawk and smashed it. Sam knows this will count as a 'totalled' car. No one knew what had happened... before anyone could get to work on it, they had heard a loud crunch and the lift had apparently lost hydraulics and crashed onto the car. (Later inspection showed nothing wrong with the lift and there was nothing to fix.)

Sam gave me a call on Tuesday and said I had better come down and checkout my car. I drove down there, and could not believe the mess. The whole front end plus windshield was destroyed. Sam and I walked back to his office... on the way, he explained his dog's unusual reaction to the car, and now this and he turns and looks at me, "Jay, this car is cursed." I looked at him like he had three heads... "Seriously, the dog loves riding in a car and she tried to bite me – she would not get into the car, snarled and fought me like something scared her!" I still wasn't convinced. He then explained that they examined the lift and it now works fine... and as we entered his office, and he shut the door, he went to the file cabinet and pulled out the car's manufacturing specs in the dealer folder... "Here read this," and handed me the file folder.

The engine had been manufactured in Brazil outside a small town, and someone had penned a note on the waybill – "Car is jinxed." Chills ran down my spine. Sam said he did some checking with the plant in Brazil and told them about this unusual car, describing it to the Engine Assy Mgr who told Sam to get rid of the car. Sam was not superstitious, and neither was I, but the local-born Brazilian manager said that sometimes they have to fire local workers for repeatedly attaching feathers, amulets and trinkets to the engines, and he said that Santeria was a big thing down there.

Sam said he'd recommend to GM that the engines be built somewhere else and he walked me over to a line of new cars, and said, "Here, I'm sorry. Pick out whatever you want from this line!" Unfortunately, it was the end of the model year, and no Turbo Sport Regals were left... so I settled for a Sport Skylark. The paperwork was updated (I had financed thru the dealership) and I was off into the sunset. Sam later told me that the Latino District Rep came in to see the mess and had it trucked to the Orange County Iron Salvage Yard and turned into a 'box' of metal and Sam had heard that it was buried in a mountain of compressed vehicles in a corner of the Yard. (You gotta wonder if mine was the only one...?)

Significance? If this is a Simulation, why would the gods 'curse' the car and play that game? They didn't. They cannot stop advanced humans who have discovered how to manipulate 'unseen energy' which is provided along with the Simulation...its ambient energy cannot be suppressed and some people have discovered how to mentally channel it – for healing or mischief!

So, what could a Santeria priest invoke that would imitate or generate a curse attached to the car? In my opinion, this is evidence for Santeria priests somehow getting nasty entities in the Astral (Poltergeists or Discarnates) to attach to the car and wreak whatever havoc they can... That means that the nasty entities as Discarnates are not prohibited by the Simulation... (as Monroe discovered in his Our Of Body journeys! – Chapter 4) So the priest does not have 4D powers, but can invite entities, as he has the right to [mis-]use his freewill, to get Discarnates to do what witches have done for centuries – Get unseen entities to do the naïve humans' bidding, and then when the human witch dies, s/he "owes" them, and they don't want your money!

> So, yes, this can happen in a Simulation. It means we just don't know about all aspects of the Simulation. I'd be willing to bet that if we abuse the "use of unseen energies" (See TOM, Ch 10-12) we will have to answer for the misuse of our freewill... and the energy.

Feb. 1987: My California company had sent me to Huntsville, Alabama on an audit for their big aerospace division still supporting the Redstone Arsenal in support of the Saturn V missile development that von Braun and his boys had started years ago. While there, I came down with a hellacious sinus infection that responded to nothing I did and so I looked up my old doctor, and he was still in business... and when I walked into his office (still where it used to be), I thought he was going to have a heart attack! Long story short, he also could do nothing for me and now I had missed two days of work because the tickle and sinus drain was so bad I could not stop blowing my nose, and I could not wear my glasses... the pressure from them kept my nose draining. My boss in California threatened to fire me, even though my teammates vouched that I looked bad and could not even get back on a plane to fly home! Thus it was that I went to bed on Friday night, said a brief prayer for healing, and crawled under the covers.

I awoke Saturday morning, and I was lying across my large bed, sideways, on top of the covers, and my pajama bottoms were on backwards. (I didn't go to bed that way.) It wasn't until I went into the shower that I noticed that my sinuses were clear and that whatever it was [fungus] that had attacked me was gone. I finished the audit and had no more problems with anything and flew home a week later.

> Being healed **twice** in Huntsville, Alabama led me to quip in future years that anytime I was sick with anything, I would fly to Huntsville as my own Lourdes, for healing.

By the way, in 1987 that doctor was still drinking and smoking, according to his nurse – and his waiting room reeked of cigarette smoke. He let patients smoke in the waiting room, because he did, too! Some doctor! (I'm glad Someone else was there to heal me!)

July, 1998: I was in Sedona, AZ enjoying a bit of hiking among the red cliffs and sitting in the vortices. In fact, my favorite spot was Boynton Canyon, and I had hiked back in all the way, among the cliffs at the end of the Canyon, and up to the right onto a ledge that was my favorite spot... the energies felt really good there.

While I was meditating, I heard a crashing thru the brush behind me and wondered what idiot was walking down the slope, thru the brush, and over the boulders!... And how was he going to get thru the very thick bush that covered most of the ledge!? As I turned to look, a very large, grey lizard emerged from under the bush... at least 2' long but not like the geckos (which were all around!) It stopped and looked at me, and I wondered if it was going to attack... Was I in its territory...? Did I have any fruit bars left in my sack to offer it...a peace offering?

It walked over directly in front of me, stopped sideways and looked at me. I then thought it looked like a small Komodo lizard. I said "Hi!" and then wondered why I said that...

Credit: Bing Images

It opened its mouth, closed it and walked slowly off to the left where there was just one large boulder and a clear slope up the side of the hill. It shuffled over to the rock, looked at me, and went behind the rock and I expected it to climb up the slope. It didn't. Nothing happened. I wondered where it could have gone, so I crawled over to the boulder and looked behind it – expecting to see a hole into which it must have crawled. There was nothing, no lizard, no hole… and nowhere it could have gone. Creepy, I thought.

So I decided to leave, as shadows were starting to fall across the high-walled Canyon and I had been warned by the locals to not be in the Canyon when darkness fell. I had been told tales about things moving thru the bushes, large grey owls watching hikers from the nearby trees, and there was alleged to be a portal off to the side in

the Canyon. Hikers had seen other beings walking near the cliffs at sunset.

The north end of the Boynton Canyon, which is almost inaccessible now since the Park Service put up a heavy chain-link fence at the entrance.

Credit: Bing Images: sedonahikingtrails.com

I returned the next day, about 10:30 am, and from the side of the Canyon, a pretty brown-skinned gal met me on the trail on the way back to the ledge. I couldn't tell if she was Amerindian, Hindu, Latina, or what…but she was real pretty.

She had some snacks and we sat on the vortex meditation ledge together. I don't remember exactly what we talked about, but it was all metaphysics and ETs, and I do remember that several hours went by, other hikers went by us on the path below and up onto the heavily bouldered slopes below the cliffs… And all of a sudden, the shadows began to fall, and my watch said 3:30 pm…! We descended the slope and began making our way out of the Canyon, and I paused to retie my boot, and she slowly walked on, and when I got back on the trail, about 20 seconds later, she was nowhere to be seen… I ran down the trail hoping to catch her… She had not even waited for me. I never saw her again. I realized I had not even gotten her name…

June 2001: Driving to a seminar on the freeway in Dallas, I was in the slow lane, and we were all doing about 35-45 mph, congested traffic, and all of a sudden a small white car with two Pakistanis in the center lane, changes lanes right into me! Crunch! My truck was dark green and I could imagine the white paint and dent on the side! I motioned to them to pull over, they pulled in front, and we stopped on the shoulder. I was angry – Dallas is a crazy place drive anytime, but I was up to here with such stupidity, and I decided to punch the other driver out! I had really had it with bad Dallas drivers! (North Tx has the third highest auto insurance rates in the USA, according to my insurance man!)

I got out and started to walk up to the other driver who wasn't sure he wanted to get out… His passenger friend had already jumped out and ran down behind the other side of my truck and up behind me – I was ready for a rear attack! I stopped about 10' from the driver, who was ignoring me and listening to his friend who was babbling something about Allah … I turned to look for the first time at my truck.

There wasn't a mark on my truck. Not even a scratch. No dent. Nothing. Just as if it never happened…

How the hell can I be mad and do damage to another driver who hasn't even scratched my truck!!? I was stunned. And grateful. The two Pakistanis continued to babble, neither spoke English, and the driver ran back to look at the side of my truck, and they both said something about Allah again. I had to give it up. The driver bowed quickly to me, pointed to the sky, and they both got back in their car and drove off.

Unreal. If anomalies are there to stop us and make us wonder… it worked. I wondered if I was seeing things… a lizard, and an impact with no damage? Had it really been real… or what? Easy to produce if we are in a Simulation…

Then I started to wonder about how we see… What is vision?

Chapter 3: How DO We Really See?

One of the first things we need to examine is Vision. How do we really see in the first place? What is sight?

It has been suggested for years that we open our eyes, look at a tree, and the tree's images is recorded upside down on the back of the eyeball (**retina**), and magically the tree's image is converted to digital pulses that travel to the back of the brain where they are 'displayed' on a 'screen' – rightside up.

> If you believe the standard scientific explanation above, I have some ocean-front property in Montana I want to sell you.

There is no 'area' of the brain that uses a group of cells to 'see.' It is all neurons transmitting, and somehow the brain 'assembles' the visual signals (from the **optic nerves** which actually cross in the brain) into a coherent 'picture.' And that is about as much as standard science knows.

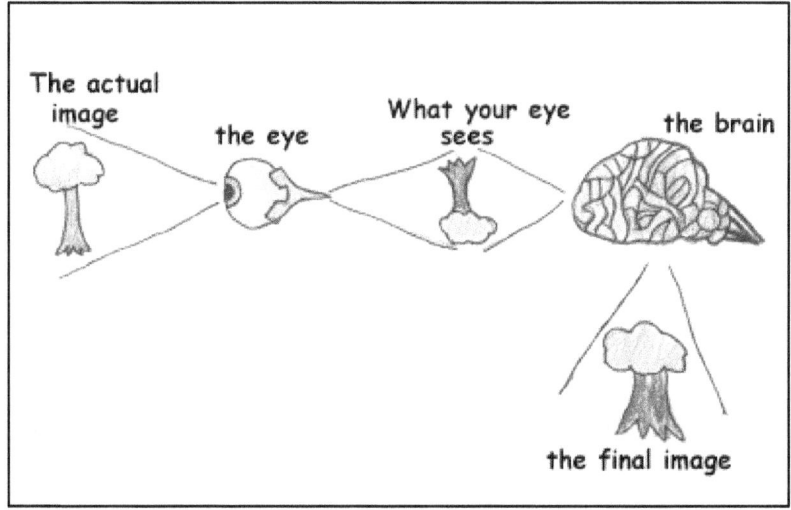

Courtesy: cogsci.stackexchange.com

That isn't much of an explanation. So let's look at perception in a new light...

Perception

Being more aware of <u>how</u> we perceive is critical to an understanding and accepting of the revelation of what this Earth is. We take what we see for granted, not really considering how we see, what perception is, how perception is related to the holographic essence of the world, and how one's perception can be manipulated (especially by hypnosis).

Besides the obvious DNA-based effect of color blindness, there is also the aspect that our DNA coding has 'permitted' us to see only a limited range of the light spectrum. We humans cannot see what dogs and cats see that cause them to stare across the room, snarl, hiss or cower at what looks like empty space to us. There are parts of the world around us that we cannot see <u>and</u> we may not really see what we think we do see...

> *Remember that shapeshifting is not moving one's molecules into a different configuration; it is all about mentally controlling what the observer 'sees.'*

Let us take a look at holograms, hypnosis and perception in this light. It will help to understand what this place really is.

> *For additional examination of Vision and Holograms, see Appendix B in Virtual Earth Graduate.*

Holograms

Holographic images are a fascinating aspect of the real world around us. Holography signifies "the whole in every part." While it is not necessary here to go into what holograms are and how they operate, it is enlightening to hear that our **vision and memory operate in a holographic way.**

Vision is reportedly holographic as there is no way for the actual images 'out there' to be displayed on a 'screen' inside the back of our heads (in the brain's so-called "vision center"). In fact, there is no "vision center" nor is there a one-to-one correspondence between the object we see and the image's representation in the brain.

> [Dr. Karl] Pribram discovered that not only did **no** such one-to-one corres-
> pondance exist, there wasn't even a discernable pattern to the sequence in
> which the electrodes [sensing brain activity in volunteers] fired. He wrote
> of his findings, "These experimental results are **incompatible** with a view
> that a photographic-like image becomes projected onto the cortical surface." [7]
> [emphasis added]

Pribram also discovered that memory, like vision, was holographic – or more precisely, distributed. This meant that **the brain was using some kind of internal holographic processing** and there would be no more correlation between brain electrical activity and what was being seen, than there would be any meaning in the interference patterns seen on a piece of holographic film. He found that the neural activity in the brain operates as a wavelike phenomenon "… creating an almost endless and kaleidoscopic array of **interference patterns**, and these in turn … give the brain its holographic properties." [8]

So the image of Tank sitting before all those screens in *The Matrix* with their green, vertically flowing symbols was not so far-fetched. The concept is correct; our brains interpret whatever the holographic symbols are 'out there' in the world around us and, get this, **we spatially create the image of the object 'out there' as if we are projecting it in front of us** – how else could we navigate to it to touch it? We're part of the hologram. [9]

> *The brain decodes the hologram and **the mind** enmeshes and actualizes us into what we 'see.'*

Allegedly, the patterns of holographic interference work as the following brilliant drawing illustrates…

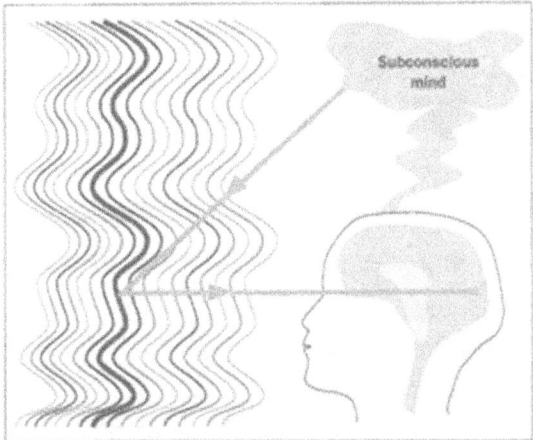 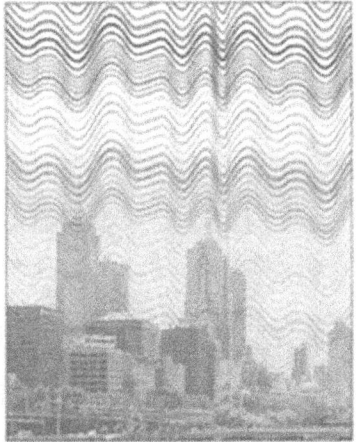

Figure 48: Subconscious mind creates the wave or thought patterns and the conscious mind 'observes' them into the holographic illusions that we take to be the 'real' world. It is only an illusion, a figment of our implanted belief and imagination

Figure 49: If we could see the 'world' before it enters our eyes it would be a mass of wave patterns – thought fields. Through the collective mind we transform these fields into an agreed 'reality' – the landscape we think we see all around us. In fact, it is within us, within our own minds

Holographic Vision
(Source: David Icke, <u>Tales from the Time Loop.</u> p. 351)

What if there is nothing out there but clouds of holographic **"interference patterns"** that our brains decode and 'construct' as reality in front of us? Maybe it is our mind that decodes the brain's wave patterns? David Bohm, a foremost physicist, thinks that the entire universe is a hologram. If that is true, then how does a camera take a picture of an object: Is it the actual object on the film or a set of the holographic interference patterns? The **camera has no brain to filter and decode the patterns.** Thus, according to the holographic theory, the camera also photographs the interference patterns and like the real object, our brains decode the swirls on the photo and we 'see' the object... [10]

Tricky Vision

As is already known, the eyes consist of multiple parts, some not touching each other, and if any key component is missing, vision with the eyes doesn't happen. As Virtual Earth Graduate (VEG) explains (in Appendix B), some people can 'see' with other parts of their bodies. But the real 'smoking gun' as far as the true nature of vision goes, is that the eyes have a **blind spot** – both eyes have a small section of the retina where there are no rods/cones and thus no image is relayed to the brain from that area – that should produce a 'hole' in the image we see. It doesn't because **the brain 'fills in' the picture** and we see a complete image!

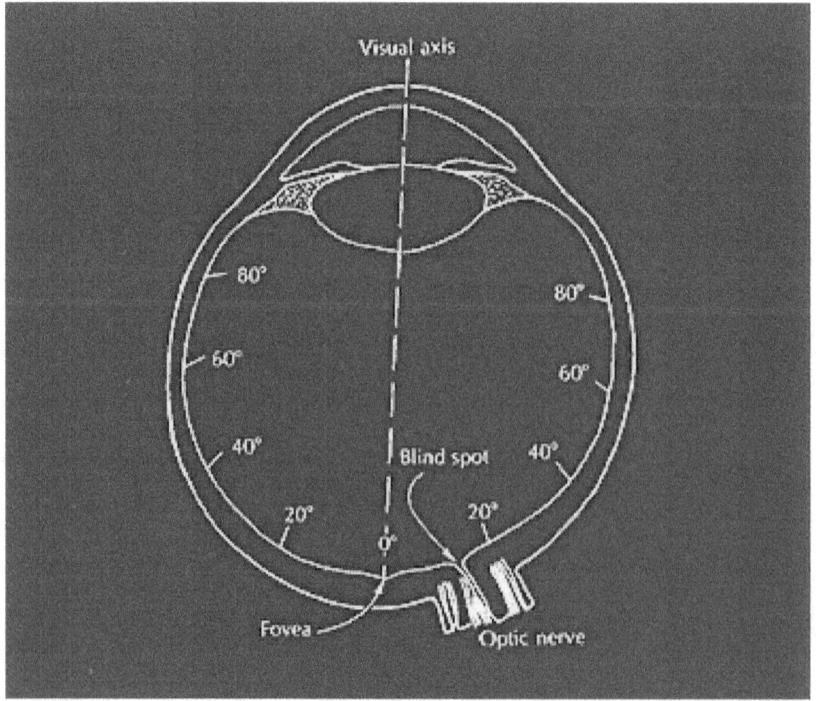

Blind Spot in Eyeball
(Credit: Bing Images.)

Vision is much more than electrical pulses transferred to the back of the brain, into the visual cortex, where there is NO movie screen upon which the world is displayed – and by the way, the scientists tell us that the image in the brain is replicated upside down… and **the brain rights it** for us! What is the upside-down image displayed on? How did the brain 'know' to right the image? And since the optic nerve does not transmit photons, how are electrical pulses converted to images?

> So far, the brain is doing <u>at least two</u> manipulations of what we see… Is there a third? Is the mind using the brain to translate the holographic "interference patterns" as well? Do we really know <u>how</u> we see?

Cat's Eye Vision

The accepted view of vision is that the eye 'sees' by having a photographic image of the scene or object reproduced onto the cortical surface of the brain… something like an internal movie projector. And the belief is that the optic nerve, running from the eyes to somewhere in the back of the brain, carries analog impulses to be decoded in a "vision center."   Dr. Karl Lashley would disagree after running a very interesting experiment:

> ….in a number of experiments, Lashley had discovered that you could sever virtually all of a cat's optic nerve without apparently interfering whatsoever with its ability see what it was doing. To his astonishment, the cat apparently continued to see every detail as it was able to carry out complicated visual tasks. [11]

Chinese EHF Vision

Just a basic word will be said here about the psychics in China, mostly young children, who can see with their nose, their fingers, and other body parts – again proving that vision is not solely the province of the physical eyes. [12]

> The major review of Chinese Psychic Vision was in Appendix B of VEG.

In China, instead of saying ESP, they say Exceptional Human Function (EHF), and in the case of those who 'read' with their nose or fingers, they claim they see the words in their mind. It is called **'non ocular vision'** and has been prevalent in several thousand children, and is rarely found in adults. [13]

If consciousness is distributed in the body, and we are in the **Holographic Universe**, as Dr. Pribram claims (Chapter 8 and Dr. Penrose), then vision would be a function of consciousness. The object being read would just transfer the text to the brain (to the same place the eyes do) and the person would be reading – without the use of their eyes.

> The holographic axiom is: "The whole in every part" and if vision and memory are holographic, and most parts of the body have meridians channeling *chi* in the Bionet, then the fingers would have meridians to channel images/sensory input just as the eyes do.

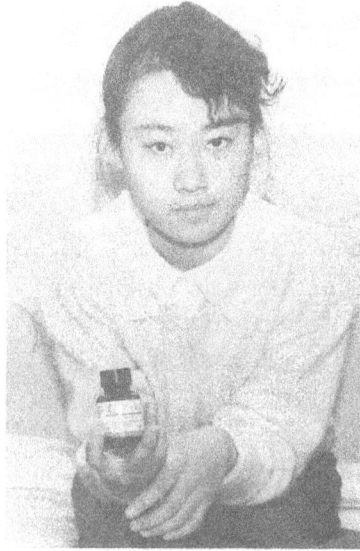

Fig. 11–1
Yao Zheng is using mind power to make vitamin pills come out of a bottle.

In addition to mentally changing the molecular structure of water, and taking pills out of a sealed medicine bottle, there are those who can manifest an apple out of thin air. In addition, there was one young girl who could wave her hand over flower buds that had yet to open, and they would spring into bloom. [14]

(credit: China's Super Psychics, Ch. 9-11)

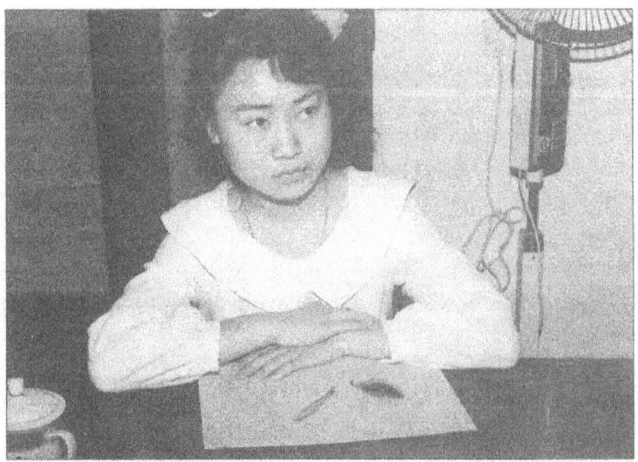

Fig. 11–2
Yao Zheng breaks a soup spoon by EHF. It made a noise when it broke.

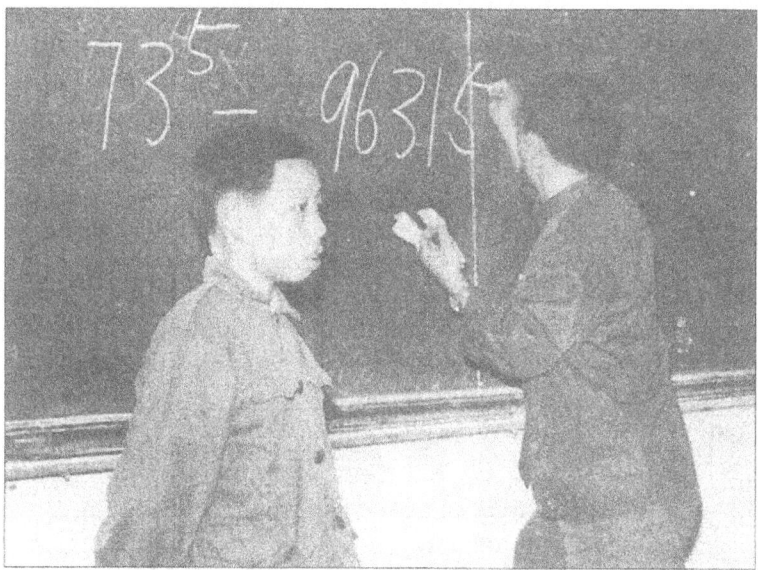

Fig. 9–2
China's "supercomputer," Shen Kegong, a thirteen-year-old prodigy who takes twenty seconds to give an answer to $625^9 = 14,551,915,228,366,851,806,640,625$.

(credit: China's Super Psychics, Ch. 9-11)

(Four pictures' quality reflects graininess of Chinese photos.)

Fig. 11–6
High-ranking executives of a U.S. oil company are watching Yao Zheng open a flower bud with her mind power. The young lady next to her is removing pills from a bottle by EHF. The photo was taken in Beijing, 1992.

Paul Dong's book on Chinese EHF is highly recommended.

Amazing, Yes, and if we live in a Simulation and are tune with The Field, as Lynne McTaggart calls it (Chapter 8), then we humans are more than we think we are and not only can we manifest apples, but we can manifest our own healing… it all comes from The Field, or what is called the Simulation.

Do we need special training or genetics to do these things? Well, barring either one of those, which probably DOES help, humans are capable of amazing things while under a state of hypnosis. It has been documented that a person under hypnosis can be touched with an eraser-end of a pencil and told that it is a hot iron, and their skin will immediately blister! Or, just as strange, the hypnotized person can be touched with a red-hot poker and be told that it is a pencil, and their skin will not react…. No blister or burn!

What is Man that a person with **MPD** (Multiple Personality Disorder) can manifest different things while each 'self' is the dominant person? For example, in the same body, Sue is shy and timid, can't eat chicken and her eyesight is very bad. When the persona changes to Vicki, she doesn't need glasses, sees perfectly, loves chocolate, but has asthma! When Judy comes to the fore, a third body change is noted and Judy has hives, limps, hates chocolate and is deaf in one ear. [15]

The same body in 3 different states… What is that?

> In point of fact, it is evidence that the soul affects the body (MPD above is 3 souls sharing 1 body). It is a form of Psycho-Cybernetics (qv. Dr. Maxwell Maltz).

Hypnotism and Perception

An interesting question to ask oneself is: Are there things 'out there' in our reality that we cannot see (1) because our DNA does not permit seeing anything but the standard visible light spectrum, or (2) because we have been subconsciously conditioned to not see them? Or both?

Case in point is the story of the hypnotist who demonstrates our ability to see 'through' solid objects, or see things that are not there. In this case, Tom was hypnotized and told there was a giraffe in the room. He gazed in wonder at it (obviously creating it in his reality). Later, and more interesting, was when still under hypnosis, the hypnotist had Tom sit on a chair, and he put Laura right in front of him, standing up. But Tom had been told she was not in the room, and when asked if he could see her, despite Laura's giggles, Tom said no.

Then the hypnotist went behind Laura so he was hidden from Tom's view and pulled an object out of his pocket. He kept the object carefully concealed so that no one could see it, and pressed it against the small of Laura's back. He asked Tom to identify the object. Tom leaned forward as if staring directly thru Laura's stomach and said that it was a watch. The hypnotist nodded and asked if Tom could read the watch's inscription. Tom squinted as if struggling to make out the writing and recited both the name of the watch's owner (which happened to be a person unknown to any of us in the room) and the message…. Tom had read its inscription correctly. [16]

Consensus Reality

Another interesting aspect of hypnosis is that it tends to connect people – the hypnotist with the subject. Or two hypnotized people who share the same reality.

Case A
Sir William Barret hypnotized a young girl and suggested that she would be able to taste whatever he put in his mouth. He blindfolded her so that she could not see what he was tasting. When he put salt, sugar, pepper, mustard and ginger (separately) in his mouth and she correctly identified each of the substances. [17]

Case B
Anne and Bill were accomplished hypnotists and they agreed to hypnotize each other. They initially reported everything was grey, but soon they saw vibrant, brilliant colors and found themselves on a beach of 'unearthly beauty.' The psychologist who was present to observe could not see what Anne and Bill saw, but from what they described, it was obvious that they shared the same hallucinated reality – or were they hallucinating? [18]

What if a shared reality is just plugging into the **Reality Fields** that Dr.s Pribram, Bohm and Tiller have spoken about? And what if the connection with same is a feature of the Simulation, enabling humans to 'synch up' with a common reality – and which could be a reason for us not being able to see what is really around us – Is it possible to 'suppress' certain aspects of the Simulation so that they are all around us but we can't see them, like Tom could not see Laura?

Case C

If there are **Reality Fields** and we can plug into them, and share a certain consensus reality, one upon which we have all come to agree (almost like a *meme*), is it then possible for physicists around the world to plug into this Field and observe the same subatomic phenomena? And if a physicist in America doesn't like what others in Russia have discovered regarding neutrinos, would it result in the American physicist not being able to see what his Russian counterparts have already seen? [19]

Anomalons

This raises the issue of **anomalons** – subatomic particles that the physicists around the world seem to 'create' because they are looking for them, and the subatomic particles appear with the parameters that the scientists are expecting to see. Are subatomic physicists <u>creating</u> the subatomic world instead of discovering it? [20]

There does seem to be a certain aspect to the **"observer affecting what is observed"** as Physics maintains, and yet, as the final section (in this chapter) relates, 10,000 people could not blink out a light on top of a building. Perhaps we only affect the subatomic world because our thoughts carry more energy than do quarks, electrons, etc.

> Indeed if the universe is a holodeck [as Dr. Tiller suggests], all things that appear stable and eternal, from the laws of physics to the substance of galaxies, would have to be viewed as **reality fields**, will-o-the-wisps no more or less real than the props in **a giant, mutually shared dream**. All permanence would have to be looked at as illusory, and only consciousness would be eternal, the **consciousness of the universe** [or may we suggest, the consciousness of the Simulated universe?]. [21]

What do we really know about our ability to see? Do we really see the world around us as it actually is? You will soon find out we don't. And because of that, it is plausible that this whole reality we live in is an <u>Orchestrated</u> Simulation.

How would anyone prove that it isn't?

And as was said in VEG, memory appears to be holographic and DNA processes and communicates via waves of light or biophotons – interacting with the cells of the body, including those in the brain. So **vision is holographic because the real world 'out there' is holographic**, <u>and</u> we store and retrieve memories holographically. So, we are part of the hologram.

Thus the Simulation is holographic.

The Holographic Universe

Then it can be said that the physical universe might appear to be a giant hologram to us, as physicist **David Bohm** said. [22] He would suggest, in keeping with Occam's Razor, that our world is 3D physical, but our perception of it (i.e., vision in the brain) is akin to holographic processing. **That makes us part of the 'hologram,'** unconscious of the actual perceptive mechanism, but able to function physically within it.

Credit Bing Images: fractalenlightenment.com

Physicist **William Tiller** thinks our reality is similar to that of the Holodeck – he thinks the universe is a kind of **Holodeck** "…created by the 'integration' of all living things." [23] He suggests that the universe is comprised of **Reality Fields** sustained by a flow of Consciousness – and consciousness again seems to be the key to particles existing… but, <u>whose</u> consciousness?

> **Bohm** rejects the idea that particles don't exist until physicists look at them…. He believes that consciousness is a more subtle form of matter…[and the connection between the two lies]deep in the implicate order. Consciousness is present in various degrees

> .... in all matter.... and to divide the universe up into living and nonliving things also has no meaning. Even a rock is in some way alive...[because] life and intelligence are present in ... all matter... as well as the "fabric of the entire universe".... Just as every portion of a hologram contains the image of the whole, every portion of the universe enfolds the whole. [24]

Consciousness [capital C] is another way of scientifically allowing for the reality of the God Force.

Again there is the concept that if the universe is a hologram, and we are part of it, and consciousness is found in the universe as well as Man, then separateness ceases to exist, and Man's consciousness becomes part of the trees, flowers, rocks and the ocean. And then "...reality itself becomes little more than **a shared dream .**" [25]

> **And would a 'shared dream' resemble a Simulation in which we all share and interact with each other? Entanglement?**

Reality Fields

Having mentioned the concept several times, we come to a point where it can be said that the Reality Fields include the Zero Point Energy, holographic interference waves, and there must be something like a **Gridwork** on which the physical elements in our world are constructed – i.e., a framework to which trees, rocks, buildings, mountains etc. are anchored so that they stay put and don't drift.

Even video games program all scenery and action within a grid, and the character moving on the grid is what tells the Game (Master Kernel) what the character is doing and where he is headed so that he doesn't walk off the screen and disappear – it tells the Kernel that it will have to load the adjacent scenery [just beyond and adjacent to Grid location 22C] so the character can appear to move. In our case in the Earth Simulation, the whole Earth is replicated and a Grid much like Latitude and Longitude is used as a reference point, to plant scenery, move characters...

In addition, the Earth Sphere, sitting in 4D, is also part of an energy grid that fills the universe, often called **Dark Matter/Dark Energy**. It is suggested that the Dark Matter filling the universe has a similar matrix (if not exact component replicated inside the 3D Construct) in which the Earth sits.

The Dark Matter 'dance' of the virtual particles could be pictured this way:

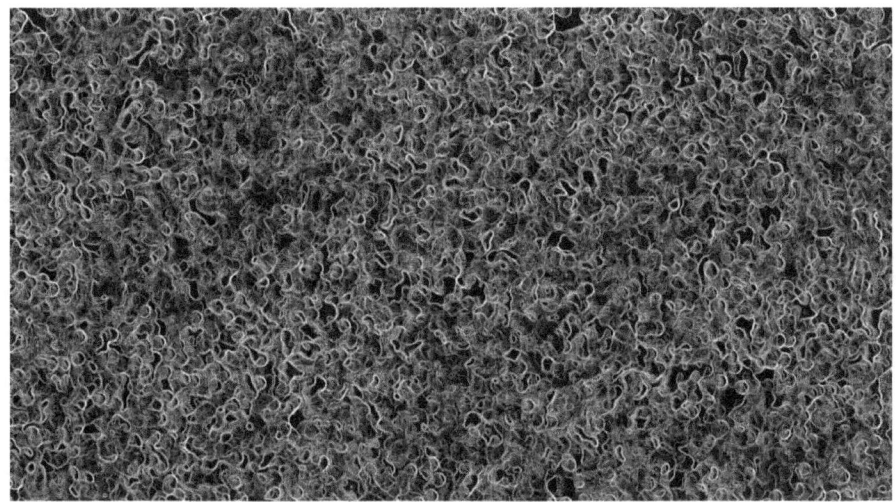

Dark Matter/Energy of the ZPE Field
(source: Bing Images.com)

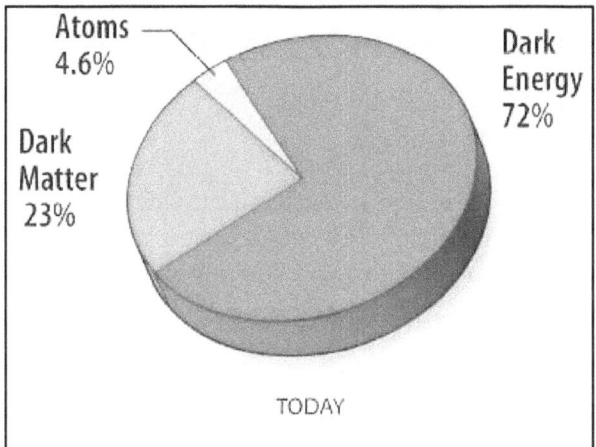

TODAY

Scientists are saying that they know something is out there filling space and consists of the following breakdown:

72% of space (universe) is composed of Dark Energy. Atoms at 4.6% is the physical world that we see and touch!

Dark Matter/Energy Proportion (est.)

Physicists reason that the Dark Matter must exist as something seems to be causing the galaxies to spread out and at the same time, something is preventing them from centrifugally flying apart. They theorize that the **Dark Energy** is causing the expansion of galaxies across the Universe, and **Dark Matter** is inhibiting the galaxy from unwinding. (Not everyone agrees the universe is expanding. [26])

And they also think the Earth 'floats' in a **Field of Dark Matter/Energy**: (Inside and outside the Sphere.)

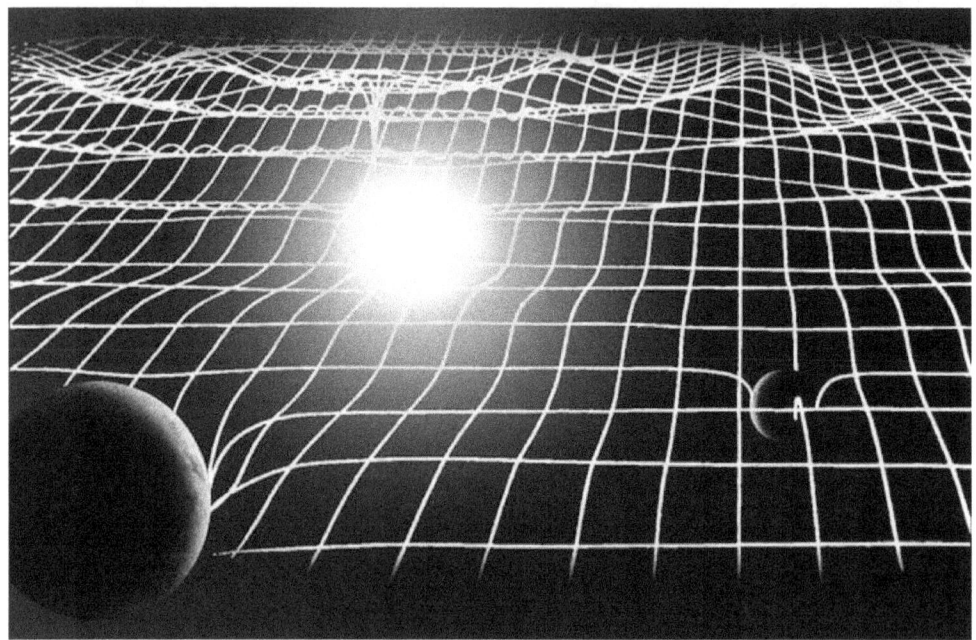

The Sun, Earth and Moon in the Field
(credit: Bing.images.com)

The above satisfies the 'warping of space' according to Einsteinian physics and the bending grid (above) suggests the toroidal fields around objects in space. Space is not empty but the old concept of **Æther** was replaced with **Dark Matter** which contains an energy more often called Zero Point Energy (ZPE).

This whole concept can be referred to as Reality Fields... since reality sits in the Field, or Æther, or Dark Matter... outside the 3D Construct.

So what is it the scientists think they are observing or measuring? If we are in the Sphere, they cannot measure whatever is outside the Sphere. Keep in mind that the **Cspace** is the area between the Earth Sphere and the outer shell, or Konstruct. And in the Cspace is where the observable galaxy and universe are simulated. It is conceivable then that the scientists are observing and trying to measure the substance that supports and contains the stars in the Cspace... and perhaps that is the same things as the Dark Matter we hear them speak about – just the consistency of the stuff that constitutes the make-up of the Cspace.

> Don't stop reading yet. I will be the first to say 'nonsense' if that is the case, but the Physicists and Philosophers have much to say that is plausible in Chapters 7 & 8. And if we are in a Simulation, then what was just said about the Cspace has to be true.

A Simulation has to be contained within something, it has to have a Grid, energy and look real, and it has to have a purpose. Man has also got to be able to see the stars which would be on the other side of the Sphere's containment barrier. And if there is no second barrier, then the stars we see are the stars of 4D and there is no Cspace…. **But** – what did Robert Monroe hit when he went out of body (next chapter), and why does the Sphere Within a Sphere artwork in front of the Vatican agree with what Robert Monroe discovered? And how would we be able to run measurements of space on the other side of the barrier, which must be an energetic barrier that neither lets our SETI signals out, nor lets any ET signals in? (See Chapter 12 for fields around Earth.)

It is very possible, unfortunately, that we are contained and **the Cspace is expertly manipulated** to appear that all other galaxies and solar systems are incredibly far away, as if to say "Don't even try to get there!" Has anyone wondered why we are so isolated? The isolation would be a required feature of our Simulation as the beings who simulate our universe would not want us trying to get to the nearest solar system with crude rocket-power and discovering that, like Truman in *The Truman Show*, we hit a wall with our ship!

Space Probes

I can hear the biggest objection already: "Oh, but we sent space probes to Mars and beyond and they sent back pictures, so we know they really went there!"

Really? Remember: the Simulation is expert, very advanced, and very good.

Here is how They could have done it:

> When the space probe leaves Earth, it passes thru the 1st barrier –
> the one the *Gegenschein* reflects off of.
> The probe continues on into the Cspace, past the Moon, and let's say
> the Cspace is 5 million miles between the inner and outer barriers.
> If the probe goes to Mars, it actually heads out into the Cspace where
> our radar and tracking systems cannot physically see it.
> It halts. Maybe 3 million miles out.
> We know it is Ok because it still transmits back to Earth and we assume
> it is still moving (we can't physically track it) BUT by our mathematics,
> we know each day just about where it should be. It had a heading and a
> velocity so that **vector** will take it to Mars, or Saturn…
> The Simulation gods also know where it is headed and They compute the
> day/hour for it to begin transmitting that it's nearing the target (or the
> on-board radar [if it has such a thing] signals that Mars or Saturn is near)…
> And the Simulation engineers pump digital pictures and data into the

probe – if They are sophisticated enough to build and run a super Simulation, They certainly know how to feed data to the probe – even though it hasn't moved from its position in the Cspace!
And everyone back on Earth thinks that the probe is circling Mars, or doing a fly-by of Jupiter or Saturn…
Even allegedly landing a Rover on Mars would entail the same scenario – the Probe/Rover sits stationary in the Cspace and is fed all its input; the landing, digging in the soil is simulated, pix of the horizon are simulated, etc…

> Here's a really far-out scenario. Suppose that NASA knows about the Simulation barrier and knows we can't physically go to Mars, and so they fly a Mars Rover down to the Chilean Altacama Desert (telling those that see it and set it up that this is just a practice run), change the color settings so the sky and ground look more reddish, and have the Rover move around – simulating Mars. They'd have to be careful that trash doesn't blow across the scenario – such as a white 'rabbit' and 'blueberries' we have seen.

What do we really know for sure?

These are things to consider as we journey through the remaining chapters.

Concluding Ideas

There is the possibility that anomalous events, those that defy easy, scientific explanation, are the **Reality Fields** and that the universe is sustained as a stable construct that is **not** subject to modification by our individual consciousnesses. That is, the reality fields occur within a limited conscious agreement with other sentient beings – if we all see a purple elephant in the room, we can all agree that that is part of our reality. Unless of course, some Higher Beings are manipulating what we see, hear and think, and entraining us into a group (mind) **Simulation** – for agendas that we are not aware of.

What the consensus of quantum physicists dealing with this issue seeks to promote, knowing that it can't be empirically proven (yet), is that **there is no reality beyond that which is constructed by the "integration of all consciousnesses"** and that the holographic universe could be a **construct** reflecting our coherent sculpting by our subconscious minds. [27]   But here again, that assumes that we in 3D reality have some power to create and manifest with our minds, but even with 100,000+ minds 'efforting' to create something collectively does not mean that it is possible for humans in this 3D reality… it just means that we mutually agree on our observed, coherent Reality.

Mass Visualization

> Back in 1959, a radio station in Massachussetts asked all of its listeners over the period of 2 weeks to participate in a mind experiment. On a given day, they were all to focus on the light atop the John Hancock building in Boston and when the DJ gave the countdown, 3 – 2 – 1 and at 0 everyone was to picture the light going out. Visualize it black.
> People on the street at 9 pm that night were also in on the test. It was estimated by random sampling that about 10,000 people participated.
>
> **Nothing happened**. The light didn't even flicker.

All that to say that the intention of higher beings (in 4D), who probably run this reality, are very powerful, and we minor beings do not have the power we wish we had.

Vision

So the best that forward-thinking scientists have come up with so far is that vision is a projection 'out there'. What occurred to Pribram was that when we see something, it isn't being seen in the back of our head somewhere. We see it in three dimensions and out in the world.

> It must be that we are creating and projecting a virtual image of the object out in space, in the same place as the actual object, so that the object and our perception of the object coincide. This would mean that the art of seeing is one of **transforming.** In a sense …. We are transforming the timeless, spaceless world of **interference patterns** into the concrete and discrete world of space and time…. As with a **hologram**, the lens of the eye picks up certain interference patterns and then converts them into three dimensional images… it requires a [form of] **virtual projection** …. If we are projecting images all the time out in space, our image of **the world is actually a virtual creation**. [28] [emphasis added]

And now for the mechanics… when you see an apple…

According to Pribram's theory, when you first notice something, certain frequencies resonate in the neurons in your brain. These neurons send information about these frequencies [interference patterns] to another set of neurons. The second set of neurons makes a **Fourier translation** of these resonances and sends the resulting information to a third set of neurons, which then begins to construct a pattern that will eventually make up the **virtual image** you create of the apple out in [personal] space... [29]   [emphasis added]

Now aren't you glad you asked...?

The **Fourier transform** decomposes a function into the frequencies that make it up, similarly to how a musical chord can be expressed as the amplitude (or loudness) of its constituent notes.

The brain talks the language of wave interference – phase, amplitude and frequency. We perceive an object then by 'resonating' with it, or getting 'in synch' with its vibrations. Chapter 8 goes into this a bit more... and also ASOM book.

Weather Anomaly

This doozy has been saved for last as it is heavy evidence that the Simulation DOES exist and **interferes with our satellites, and our weather**. The pictures below were taken over Australia in 2010 by a weather satellite, and in many cases a strange disk (or 'wheel') appeared over the regions being photographed... and each was followed by extreme weather.

All Images: credit: http://www.news.com.au/lifestyle/real-life/bueau-of-mereology-cant-explain-mysterious-patters-on-radar-system/story-e6frflri-1225848774377

And in every case, the appearance of the 'wheels' was followed by unusually severe weather – Each wheel seemed to correspond to a particular type of weather... but if it was a warning, the weathermen didn't get the message.

Over S/E Australia

Black spokes with clockwise motion.

Over N/W Australia

Black spokes with no motion. Appeared to shimmy or vibrate.

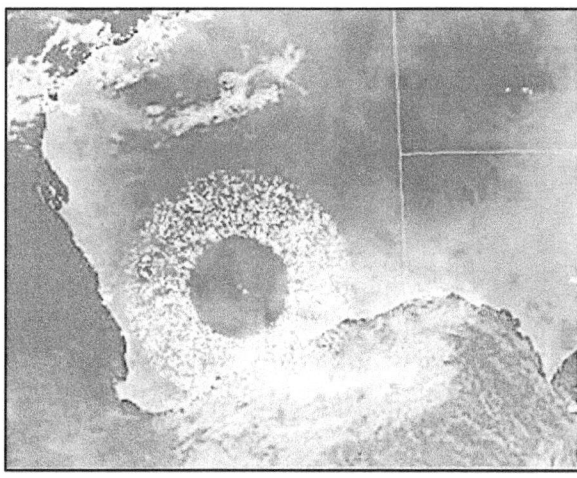

This 'wheel' was over S/W Australia.

White granular ring. Disssipated to the right across Australian mainland.

The weathermen and the scientists examined their computer software and hardware receiving the satellite images, as well as the satellite photographing system, and neither was found to be the culprit. It was not a prank and it was not the result of a computer virus, but the weathermen could not bring themselves to accept that something external to their system was trying to warn them of impending severe weather.

At last report, The Australian Weather Bureau is currently investigating ways to reduce these interferences seen as 'anomalous propagation of errors.'

Of course conspiracy buffs assigned the blame to HAARP for creating both the 'wheels' and the bad weather... yet these wheels have not been seen in other locations... due mainly to the type of satellite imaging and radar that is used. The Australians were using a unique technical process which revealed the wheels, and when they changed their hardware, the extreme weather continued but they could no longer catch the wheels on film. [30]

> The point in showing several examples is that (1) they are all different, and accompanied different and severe weather, and (2) this was not just a 1-time fluke. It only 'stopped' when the hardware was changed.

If the Simulation engineers had found a way to communicate with our weathermen and warn them of impending extreme weather, it is a classic human response to ignore something we don't like and shut the door on what we don't understand.

There is the possibility that the 'wheels' were inserted by the Simulation to create the weather, and each type of weather corresponded to a specific wheel design.

In Sum

Lastly, regarding Simulation, it is reasonable that we are in containment... the issue remains to be seen just <u>how</u> we are contained. That will give a clue as to <u>why</u>.

Chapter 4: Those Who Knew

Three major writers have presented us with what they found when they first examined the anomalies around Earth – Charles Fort, Robert Monroe and Stuart Wilde. Each has an input that should be considered before looking at the opinions of today's Quantum Physicists.

Fortean Sphere

The famous **Charles Fort** often questioned the scientific dogma of his day, 1912. He wondered where all the strange objects falling from the sky came from, and he wrote 4 books documenting the oddities and speculating on their source. During his writings, he also kept coming back again and again to the idea of **a 'shell' surrounding this Earth**.

> …whether there be a shell-like, evolving composition,
> holding the stars in position, and in which **the stars are openings**,
> admitting light from an existence external to the shell, or not, all
> stars are at about the same distance from this Earth as they would
> be if this Earth were stationary and central to such a shell, revolving
> around it. [31]   [emphasis added]

This does give a new meaning to Copernicus' idea that the Sun did not revolve around the Earth – What if he were wrong? Intriguing, but nonsense you say. And Charles Fort would agree with you – with a wink! It was an idea which really fascinated him, and he was close to the truth, as we now close in on what even the Quantum Physicists of today are considering: **A 3D Construct run in a very Sophisticated Simulation.**

Fort was no uneducated man, totally mystified by science and the world. It was his unique ability to go toe-to-toe with the astronomers and scientists of the day – demanding answers to the oddities he had recorded. He was considered an *enfant terrible* of science – questioning everything. In math and astronomy, he could hold his own. And yet, he repeatedly questions the presence of <u>something</u> surrounding the Earth:

> The **Gegenschein** -- Now we have indication that there is such a
> shell around our existence. The Gegenschein is a round patch of
> light in the sky. It seems to be reflected sunlight, at night, because
> it keeps position about opposite the Sun's position. The crux:
> Reflected sunlight – but reflecting from what?
>
> That the sky is a **matrix** in which the stars are openings, and that,
> upon the inner, concave surface of this celestial [transparent energy]

shell, the sun casts its light, **even if the earth is between**… [32] [emphasis added]

Note his use of the word, 'Matrix' – 100 years ago, and the Wachowski Brothers had not even conceived of the 1999 movie, *The Matrix*. (In fact, no one had conceived of them, yet!)

The Gegenschein

The Gegenschein
(credit: NASA: http://apod.nasa.gov/apod/archivepix.html and below)

It is recommended that the reader check out the above NASA link to three samples:
2008 May 07: The Gegenschein over Chile. (sunlight)
2006 December 26: The Gegenschein. (sunlight)
June 25 1999: The Gegenschein. (sunlight + Sun)

Interesting that in his book <u>New Lands</u>, written in the 1920's, he used the word **matrix**. But he is making a very interesting point that science still today cannot answer:

Suppose the Gegenschein could be a reflection of sunlight from anything at a distance less than the distance of the stars. It would have **parallax** against its background of stars.

Observatory, 17-47: **"The Gegenschein has no parallax."** [33]

> Parallax is the apparent change in shape or size of an object if the observer changes his position.

So Fort is saying that since its perceived shape and size does **not** change with any change in our position from which it is viewed (i.e., a parallax), that it must be reflected off the surface of something consistent in shape. If it were reflected off dust in the atmosphere, it would have size and shape distortion depending on the viewing angle – but it doesn't.

Fort was no neophyte to astronomy and certainly would have known the difference between sunlight reflecting off dust in the upper atmosphere, and swamp gas, Moon dust, or whatever the standard argument of the day is for what he personally observed.

Note that **NASA** has pictures of the Gegenschein (above) – Fort wasn't imagining things. As will be seen shortly, Fort was on to something, and while he knew the Shell wasn't a metal thing, and that there were no holes in it to simulate stars, it is real, and is also what bounces back our radio and TV transmissions:

> …not enormously far away, there is **a shell** around this earth…
> According to data collected by the Naval Research Laboratory
> [1925], there is something, somewhere in the sky, that is
> deflecting electro-magnetic waves of wireless communications,
> in a way that is similar to the way in which sound waves are sent
> back by the dome of the Capitol, at Washington. The published
> explanation is that there is an "ionized zone" around this earth…
> the [term] "ionized zone" is not satisfactory… From Norway
> [there were] short-wave transmissions… reflected back to earth…
> as if from **a shell-like formation**, around this earth, not
> unthinkably far away. [34] [emphasis added]

And yet, we keep up the ruse that we are looking for extraterrestrial life with our SETI radio telescopes… [shut down in 2011 – Ed.]   Wouldn't any ET radio transmissions to us be reflected back, too, from their side of the same 'barrier?'

> *As Robert Monroe shortly discloses from his OBE ventures, there are no*
> ***3D*** *lifeforms out there… most sentient life in the universe is in the 4D*
> *and above realms. And that makes sense if you understand that the soul*
> *has almost unlimited potential 'out there', yet is constrained here on Earth.*

There is a very advanced (energy) Sphere around this Earth as we will see, functioning in 4D+ but reflects light in 3D. It is interesting that an open-minded skeptic of the 1912-1932 Era first voiced the concern that Science was ignoring the *Gegenschein*, and not giving us all the answers…

In fact, knowing Fort from his writings, he was 'baiting' the scientific establishment with the idea of a Shell whose holes were the stars, but the bait was not taken. And as will be seen, there **is** something around the Earth, and another researcher, Robert Monroe, ran into the Sphere's walls – literally (in the next section).

Fortean Clues

Here are some of the things that Charles Fort discovered and documented.

1. **Stone throwing**.
Since 1841 people in the English countryside have reported stones being thrown at their houses and windows by unseen beings. Sometimes the stones just fall from the sky onto the roof, and at other times they enter the house, breaking a window.

Despite being vigilant and watching each others' houses for a while in a small, rural l town where the stones were commonly 'thrown', no one ever saw anybody doing it.

To argue that it must have been the "little people," or fairies and elves, raises issues beyond the scope of this book.

2. **Falling Fish and Toads**
Sometimes fish, alive and sometimes dead, were found falling on rural houses in parts of England... too many to have been thrown there by vandals. In addition, some hailstones were found to contain frogs, some alive, some dead, when they were picked up and thawed out. Sometimes just frogs fell and hopped away.

The usual explanation is that some sort of cyclone scooped up the fish or frogs and blew them inland. Fort did not accept that since some locales were hundreds of miles from the nearest water – <u>and</u> some of the fish were fresh-water types and there had been no tornadoes or such reported in connection with these anomalies.

The two above are recorded in Mr. Fort's <u>Book of the Damned</u>.

3. **Disappearing People**
This one is a real oddity.
A man is sitting on his front porch, watching the Sun set. His wife goes in to get her sweater, and when she gets back 20 seconds later, her husband is gone. The rocking chair is still rocking, the pickup truck was not taken, and nothing is missing except her husband. He was never seen again.

On another account, a farmer out in his field calls to his wife who is on the back porch of the farm house, she waves, he waves, and he starts across the field, as it is supper time. As she watches, right before her eyes, he disappears. She runs out to where she last saw him, and there is no hole, no evidence that he was ever there.

In a third account, a waitress leaves a restaurant, steps out to the street to put another quarter in the parking meter for a customer, and she is never seen again.

And as was reported in Ch 14 of VEG, sometimes lakes and whole rural villages have disappeared – a lake in Peru and a village in Siberia.

These above accounts are principally from Mr. Fort's book called <u>Lo!</u>

The point is that these are not normal occurrences, but they could be glitches in a global Simulation, of the sort that Dr. Brian Green suggests in a subsequent chapter.

While Robert Monroe (founder of the Monroe Institute and Hemi-Synch brain training fame) never reported anomalies such as were reported by Charles Fort, Mr. Monroe discovered some very interesting things about the realm we live in – when he went Out of Body (OOBE). His discoveries corroborate our being contained by something.

Out of Body Experiences

A trip into the unexplored aspects of the real world around us would not be complete without looking at what **Robert Monroe**, of The Monroe Institute fame, had to share with his experiences of our reality. His major discovery, that the Earth realm consists of layers of reality and is all enclosed in a kind of Quarantine, particularly concerns us.

It should be understood that Monroe became a true seeker and an accomplished OBEr and kept a journal over the years that he traveled, noting where he went, what he saw and did, and some of the interesting things he learned about our reality. His ability to leave his body was not intentional at first; it was something he discovered was happening <u>to</u> him, and in an effort to learn more about it, and control it, he began to investigate it.

It turned out that what he thought in the beginning were just lucid dreams were much more than that. We are <u>not</u> in a dream-like reality, much as some aspects appear to be that way, but as another scientist-researcher, **Dr. Jacques Vallée**, said, we are in a kind of **Control System** run by Higher Intelligences. And you cannot have a Simulation that is not controlled.

Different Realms

The first thing Monroe noted (especially in his first book, <u>Journeys Out of the Body</u>) was that there are levels of reality (Locales) once one has left the body, and those nearest this Earth experience are less than wonderful. According to his classification system (see Chart 4 following):

Locale I/Earth is the one we live in – the here and now, the world of the physical senses.

Locale II is the first level one sees when initially out of the body – the immediate Astral world, and it has

> ....depth and dimension incomprehensible to the finite, conscious mind. In this vastness lie all of the aspects we attribute to Heaven and Hell.... It is inhabited, if that is the word, by entities with various degrees of intelligence with whom communication is possible.... You think movement and it is fact...Communication is instantaneous. [35]

> This is also where ghosts (discarnates) reside.

Locale III is the upper end of Locale II and is where the 'next timers' reside... getting ready to reincarnate/recycle but not yet leave the Earth School.

Locale IV is the uppermost area contained within the HVR Sphere, assumed to be home to the Interdimensionals, as well as 'last timers' or graduating souls.

HVR = Holographic Virtual Reality (described later).
This is also called the 3D Construct or Sphere.

Locales and Barriers

Chart 4 has been borrowed from VEG as it clearly shows a structure to our unseen world and it reflects the locales and barriers between levels (or locales) that Mr. Monroe encountered.

Locales I – IV (as given above)

B1 and **B2** are symbolic of a change in energy state, a functional barrier, between L2 and L3 and L4, effectively containing entities to a region commensurate with their level of development.

B3 is a significant physical 4D barrier that disallows casual passage to/from 3D Earth within 3D Construct, and keeps most entities in the 4D+ realms from interfering with humans in 3D Sphere. This is the barrier that quarantines **HVR**

Earth from the rest of the 4D+ Cosmos where most life is. It is the **Konstruct** outer shell.

The **Gegenschein** that Charles Fort spoke of would have been shining off the **barrier B1** as the innermost barrier, also related by Monroe that he hit several times in disoriented OBE attempts to return to his body. The Moon would be the other side of the *Gegenschein* (reflective barrier) since the reflection appears <u>between</u> Earth and the Moon. B3 would be the Konstruct from Chapter 1.

This suggests that the *Gegenschein* may be reflected off the Van Allen Belts.

> **Note that the distance between B1 and B3 (not to scale) is what has been called Cspace. The Moon sits between B1 and B2.**

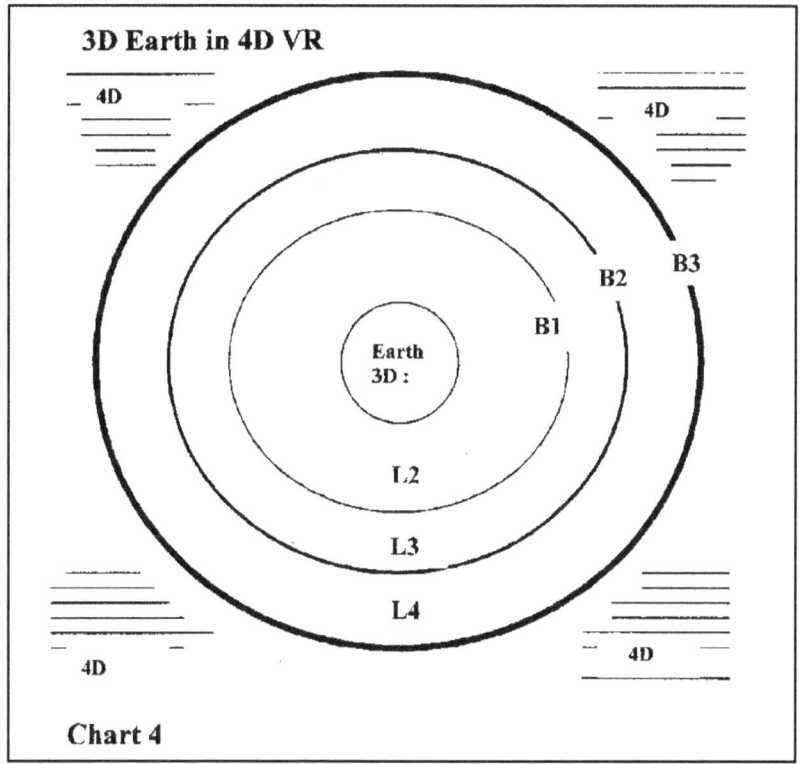

The above chart is a conceptual view of the HVR Sphere including the elements described in this book, and attempts to conceptualize the levels that Robert Monroe encountered.

Note that the chart is greatly simplified but it adapts Monroe's Locales/Barriers concept, and the **"fields upon fields"** concept that Morpheus described in the movie *The Matrix*, to better depict the HVR Sphere.

A Personal Simulation

Monroe mentions another experience he had while out of body, and that was with what he called an *Inspec* (short for Intelligent Species) – they closely resemble the Beings of Light, or angels of the Light. They have a way of teaching that is something like a **virtual reality:**

> A favorite quick and learn-forever method of theirs was simulation. It was based on their ability to create and place into a human consciousness – mine – **an earth-type situation so real and so overwhelming that I could not tell reality from illusion**. I don't know the limits of such simulation talent or technology. Nor do I know the extent to which they employ the technique…. **once into the simulation, it became absolutely and totally real – and I lived it.** [36]   [emphasis added]

At any rate, going through the simulation, if he did it wrong, the simulation would stop and automatically reset itself to the beginning again, and run once more – until he got it right and chose the correct action. If he got it right, the simulation automatically ended. [37]   Reminiscent of the movie ***Groundhog Day***… and in agreement with the principle that if you don't make progress in this lifetime, you will repeat Earth School**.**

One important aspect of Monroe's simulation, like *Groundhog Day*, was that he knew he was repeating, even though it was very real… his seeing the same scene over and over again amounted to… **Déjà Vu**. And that is a clue to being in a simulation… which means that reliving a simulation means you have been **recycled** back into the same simulation (maybe fractally – i.e., just for you).

Thus, Déjà vu means that you have been thru the same Simulation before – i.e., you were recycled back into it because you didn't do it right. You didn't complete something. You are repeating it as the same person, not as a reincarnated, different person who lived some time ago, and you are allegedly tapping into their memories. You are actually reliving the exact same lifetime… If you didn't pass 3rd Grade in school, do they promote you to 4th? Or do you redo the same Grade, even with the same teacher? (Examined further in VEG and TOM.)

If the Inspecs can manipulate a simulation for Monroe, so can the Higher Beings who run this HVR Sphere and oversee the Inspecs.

Key Concept

What is very important about Simulation is that it takes place in a virtual reality, which can be holographically created and manipulated.

And the *Inspecs'* putting Monroe through a simulation that was so real that he could not tell he wasn't in the physical, real, solid, 3D world is very revealing. Just like us: we think we're in a solid 3D experience on Earth… but are we? How would we know?… unless we occasionally see pixels or gridlines in the landscape…

Scene from The Thirteenth Floor
(credit: The Thirteenth Floor, Columbia Pictures, Roland Emmerich. 1999)

… and objects that have been part of our world are suddenly not there anymore, or in a different location…? Missing time is a big clue. People you know whom no one else remembers…

The mystics have said for centuries that all is *maya*, or illusion…probably holographically created. Could a Virtual Reality on the scale of the Earth be created holographically? Why not? What is scale to ascended beings?

Solid to the Touch

Of course, you will say that when you touch a tree, it is solid… but what if your sense of touch is programmed to interpret the subcodes within the holographic 'tree' vibration as 'solid?' Is this a partial key to how the avatars and yogis of the Far East walk through walls – because they know that they aren't really solid?

How do we know we're not in a simulation on Earth? Robert Monroe, and the late Stuart Wilde, gave us some more answers in following sections… but first it is necessary to complete the overview of Monroe's discoveries as they help to describe where we are.

The Barrier

On several OBE trips, Monroe encountered a **barrier** (which had to be B3 in Chart 4), that extended in all directions – he was trying to get back to his physical body and ran into an impenetrable wall or force field of some sort. He tried to go around, under, over and got nowhere. So he stopped and even asked for help but none came. No angelic help. No *Inspec*. No one came to his aid. He finally got the idea to go in the opposite direction and that worked, but why it did he never resolved, and he was never told what the barrier was. [38]   It seems he somehow got disoriented and hit the barrier going in the wrong direction. (He also occasionally hit the B2 barrier.)

> **That means that even as a soul, we cannot get out of this HVR Sphere – unless we die (or "graduate out") or Shift out to another Timeline.**

Obviously, B3 was a barrier to something, keeping him from leaving the HVR realm in which he found himself… which is why no one came to help him; he was expected to think, turn around and go back. Since **he didn't find any 3D beings on any of his journeys,** and he hits a barrier, it appears that <u>the barrier is keeping him inside something</u>, and that Sphere will be very shortly discussed.

It is suggested that Monroe's barrier B1 is the 'Shell' of Charles Fort, and the Sphere shortly mentioned by Stuart Wilde (next section). Note that Monroe hit a barrier (B1) while doing an OBE in 3D (still within the HVR Sphere) – it is invisible in 4D but reflects light in 3D. In the last section on "Limits" Monroe told us that **all life in the Multiverse occupies 4D and above, and he found no sentient life in 3D**. Thus the barrier is designed to work for 4D+ entities, but might not stop 3D physical beings from going to/from planet Earth and the Moon within the HVR Sphere. There is a potential problem with this as we'll soon see...

Nonhuman Intelligences

Monroe met numerous intelligences that were also in their Second Body (Light-energy body) and they could **shapeshift** into something more comfortable for him to look at, but **the great majority wanted nothing to do with a human**. The only beings that sought to assist and protect him were Beings of Light, which he called "intelligent species" or *Inspecs*. And the (once human) *Inspecs* usually steered him away from the nonhumans as they explain: [39]

> Inspec: *And there are also the nonhuman intelligences. We have tried*
> *to steer you away from them, as much as we can.*
> *These are the* **Interdimensionals** *[Djinn].*

Monroe: Why?

> *Some early encounters with some of us did not work out well. They*
> *do not regard humans in the way we thought they might. They have*
> *a sense of superiority because they have evolved in a different way.*
> *(VEG discusses the Djinn.)*

So there are no big brothers in the sky?

> *Not in the way we humans dream that there are. The difficulty is that*
> ***these intelligences have abilities in the manipulation***
> ***of energy that we cannot yet conceive of.*** *And they use*
> *them without the restraints we put on ourselves....*

I see .... Are there a lot of these intelligences?

> *Too many in the physical universe. Trillions perhaps.*
> *And there's the other one.*

The other one? The other nonhuman intelligence?

> *Would you believe that in all of our history, ours and yours, we have*
> *encountered* ***only one nonhuman intelligence*** *with an origin not*
> *in time-space...* [40] [emphasis added]

> Monroe was not clear whether this was a reference to God, Djinn, insert or not... "Non-human" ... as in Bio-computer intelligence which would be running this Simulation?

What is interesting about this exchange, to repeat and emphasize, is that **most of the life in the Multiverse exists not in our 3D realm, but in the 4D+ dimensions**.

> (The Inspecs continue…)
> *And some of them are not too friendly to Man. How many different species there are, no one really knows. Some apparently are from the same galaxy as we are. Others seem to be **from other energy systems and times**… All have certain elements in common … they have scarcely any interest in who and what we [humans] are, and finally, communication with them is an impossibility because we don't understand their methods of doing so.* [41]
> [emphasis added]

So it is apparent that Man is not alone and there are others here in the 3D Earth realm, even on the Earth, and probably underground as the 'stagehands' who help run this Drama. The significance of the non-human intelligences' treatment of Man is that (1) that is how Man treats lesser lifeforms on Earth, and (2) since Man doesn't respect himself, why should Others respect Man? They treat Man like we treat ants.

So, he did run afoul of entities that inhabit the space close around the Earth (in the lower Astral realm). And he did encounter the Barrier.

But the question remains, <u>What</u> is this place that we and they inhabit?

Holographic Virtual Reality -- HVR Sphere:

The Earth is contained, quarantined in an energy Sphere off which Fort's *Gegenschein* reflects, conceptually shown below:

The Earth in Inner 3D Construct (Sphere)
(source: www.gafnews.com/sites/)

In addition to the HVR Sphere, which has already been revealed, there is another part without which the Sphere cannot operate – the **Control System**. Every School must have a Curriculum, and the Control System is the Curriculum, delivering lessons (catalyst) to the souls-in-training here in accordance with their Scripts.

These two aspects, plus the Quarantine, and the Simulation are further examined in <u>Virtual Earth Graduate</u>, and Chapter 5 – next.

Wilde Corroboration

The last major introduction to our reality comes from a brilliant English writer, the late **Stuart Wilde**, who first examined the idea that we are contained in a sort of Sphere. While he was not the first to theorize its existence (**Rudolph Steiner** holds that honor), Wilde developed the idea and examined its implications. Similar information to that from Robert Monroe, Charles Fort and Stuart Wilde, has been found in the writings of David Icke and Michael Talbot, and lately from Nick Bostrom [42] and Jim Elvidge [43] (in Chapter 7).

The realm to which Man normally belongs is <u>not</u> in this 3D Construct; his normal home was <u>not</u> visited by Robert Monroe in his OBE journeys. This replicated Earth could be a version of the original 4D world allegedly visited by the Anunnaki, and Sirians, etc. centuries ago. Thus it is **a very sophisticated Simulation**, and

holographic replication, set in a 3D realm still within 4D, and so the Anunnaki 'history' and cuneiform tablets are still part of the Earth scene.

Key Concept

> In addition, **the Earth is basically quarantined from the normal cosmos by being a 3D holographic-type Construct within a 4D Virtual Reality**.

Or, a Holographic Sphere for short, if you will. If one connects the dots and resolves the minor differences in what the other seven men have been saying, this is the result. And it is tending to be amazingly accurate as will be seen.

By the way, Robert Monroe, in running the personal simulations presented to him, was <u>alone</u> even when he thought the other people in his drama were real. [44] They were part of his simulation, and can be considered **Non-Playable Characters** (NPCs) as in a video game. The counterpart to our human world is the presence of OPs (Organic Portals) and this was examined in much detail in VEG.

According to a well-known physicist, Brian Greene (featured in Chapter 5):

> One future day, a cosmic census that takes account of all sentient beings might find that the number of flesh-and-blood humans pales in comparison with those made of chips and bytes [OPs], or their future equivalents. And, [Nick] Bostrom reasons, **if the ratio of simulated humans to real humans were colossal, [say, 60%?] then brute statistics suggests that we are not in a real universe**. The odds would overwhelmingly favor the conclusion that you and I and everyone else are living within a **simulation**… [emphasis added] [45]

And as was said in Ch. 5 of VEG, the number of OPs on Earth DOES appear to be about 60% of the population. The next chapter further addresses Dr. Greene's ideas supporting the proposition that we are living in a <u>very sophisticated</u> Simulation.

In fact, Charles Fort knew we were 'contained' by something, John Keel suspected it, and the late Michael Talbot delivered the message that many aspects of our world/universe have holographic explanations to them… David Icke correctly identified the low-level STS consciousness aspect of the "Vibratory Prison" (as he calls it) or 3D Matrix surrounding us, Nick Bostrom suggests that our lives are lived in a type of computer simulation (**not** a version of the *The Matrix*), and Engineer Jim Elvidge (Chapter 7) suggests that

> … I would have to conclude that we are not biological willing
> participants but rather spiritual entities [souls] occupying an
> **advanced machine** somewhere in the universe. [46] [emphasis added]

And next Stuart Wilde will introduce us to what he called The Sphere and this chapter calls the 3D Construct … but first we need to remember that our Sphere is **contained**, by Monroe's Barrier which reflects Fort's *Gegenschein*.

The Morphing Sphere

Wilde had been a student of higher consciousness, or enlightenment, and while traveling it had a surprising effect on him as one day he noticed his hotel walls morphing from solid to fluid, changing color, and beings occasionally slipping in and out of perceived reality. [47]

If you're not on drugs, and he wasn't, that will get your attention. Says Wilde,

> …I have had many very strange experiences: beings
> morphing in and out of my perception, some human
> looking, some not; walls that bend and wobble;
> the strange unexplained scents of flowers; flashes of light;
> and doorways, endless doorways. **All around us are
> worlds more intricate than you can ever imagine**.
>
> There is dark and light. The tales we have been told of
> the spirit worlds and the afterlife are vaguely true in part,
> yet wildly inaccurate in other aspects. [48] [emphasis added]

There have also been people who reported seeing **gridlines** stretching over the countryside and trees and buildings missing where they used to be… as if the fabric of our world morphs and is not solid but somehow plastic. This may sound like something out of *The Twilight Zone*, or the *Star Trek* Holodeck, but such is a part of the underpinnings of our reality that some gifted people occasionally see.

Wilde describes the Sphere thus:

> ….a prison created by our minds and for our minds. In the
> olden days, it was called the **Reflective Sphere** [Rudolph Steiner]….
> It's everywhere.
>
> It is deep inside all our religious teachings and our New Age
> philosophies – it is in every spiritual practice that was ever
> invented.

(continued…)

> The Sphere tricks you mercilessly. It's very callous.
> The world is lying to you.
> The trick is unbelievably clever. It is so total, so complete.
> It's all around you as a diamond-shaped **net in the etheric**….
>
> The **transdimensionals** [Djinn?] in nearby etheric dimensions
> also trap you…. [Monroe's Interdimensionals]
>
> **Almost everything you have been taught is 'round backwards.**[49] [emphasis added]

Sound familiar? **David Icke**, another Britisher, said something similar in his books and called it the "vibratory prison."[50] According to Icke, this prison is run by a form of higher consciousness from the 4th dimension; entities who seek to 'manipulate' this 3rd dimension by controlling the human mind. The attempted control is allegedly performed by 4D STS [negative] entities, who are manipulating humans into wars where they can experience a veritable 'feeding frenzy' off human energy [aka *Loosh*]. But in reality, **benevolent HVR Sphere control is maintained by the Higher Beings through their designees, the Beings of Light and the Neggs, via the Script for each soul – set up before every soul incarnates**.

This was developed in much detail in <u>Virtual Earth Graduate</u>.

Dr. Jacques Vallée

Lastly, we come to **Dr. Jacques Vallée**. And so as to not repeat much of VEG (Appendix C), it is his significant contribution that there is a **Control System** that oversees what we are doing, and guides us (as a sophisticated form of catalyst). While he could not quite get to the idea of us being in a Super Sophisticated Simulation, his concept is valid and fits the control aspect of the Simulation.

He was studying UFOs and determined to resolve the mystery of what they really were. After several years of collecting and analyzing data, he determined that:

1. There are too many landings. And they happen mostly between 6 pm and 6 am… mostly at night.
2. Most UFO landings are in unpopulated areas.
3. Most sightings simply vanish, as if the craft just 'wink out.' (This is the behavior of an image or a **holographic projection**.)
4. Real, physical objects do not make 90° turns at high speed.
5. Some of the appearances appear to be **staged**.[51]

During the abduction account of Betty and Barney Hill [1961], the UFO occupants spoke English with a German accent (!) and Betty's recall <u>under hypnosis</u> of a **star map** on the wall (inside the craft) is a *non-sequitur*! Would real aliens from across the Galaxy have a star map <u>in paper</u> mounted to the wall?! This suggests a terrestrial source and thinking behind that UFO account.

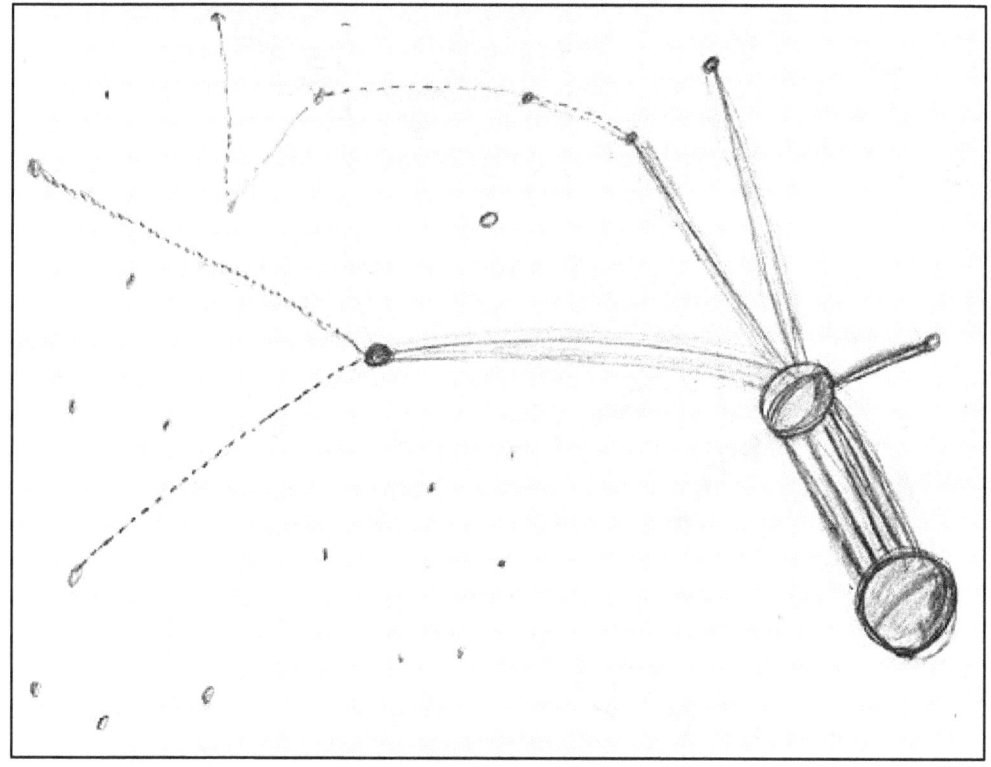

Betty Hill's drawing of the Star Map
(Credit: Bing Images: Ufoevidence.org)

As they said in *Star Trek*,
>"Lay in a course for Alpha Centauri, Mr. Sulu… and
>Don't forget to verify it with the star map on the wall!"

And on another occasion, he offered a unique opinion.

> I believe that when we speak of UFO sightings as instances of space visitations, we are looking at the phenomenon on the wrong level. We are not dealing with successive waves of visitations from space. We are dealing with a **control system**.[52] [emphasis added]

And again…

> I propose that there is a **spiritual control system** for human
> consciousness and that paranormal phenomena like UFOs are
> one of its manifestations…. [perhaps] under the power of some
> superhuman will. It may be entirely determined by laws that we
> have not yet discovered. [53]   [emphasis added]

And any Simulation running in the 4D Realm, producing a 3D Construct, would
certainly operate under different and more powerful laws. Dr. Brian Green will
address this issue of laws and formulas more in Chapter 5.

Dr. Vallée's point is that Man is being guided in his growth (sounds like the
Anunnaki again, huh?). And the only way you could control, manipulate or manage
that growth is if the School is controlled, the curricula is designed and administered
in a way that the ensouled humans cannot avoid it (but they can choose to ignore it –
at their own peril).

Dr. Vallée then says, "I suggest that it is human belief that is being **controlled and
conditioned."** And then he catches himself, and makes a very unusual statement:

> When I speak of a control system for planet Earth… I do not want
> my words to be misunderstood. I do not mean that some higher
> order of beings has locked us inside the **constraints** of a space-bound
> jail, closely **monitored** by psychic entities we might call angels or demons.
>   [emphasis added] [54]

He has almost said it! He obviously figured it out (as did John Keel), and then disses
the whole concept… Or is he just saying he eschews the religious aspect of the
Simulation? Beings we call Angels are responsible for guiding us… and that is not
meant in a religious way. The Angels are really **Beings of Light** (they don't have
wings and are a lot like Monroe's *Inspecs*), who help administer the School… and this
was examined in detail in VEG and TOM.

At any rate, his observations are correct, and a logical conclusion available to anyone
who assiduously examines the Earth situation. We are not alone and there is really
only one possible answer to what Earth is, and why we are here. Such is the
revelation of this book.

Chapter 5: HVR Sphere & Simulation

It is time to take a look at the definition and workings of the Simulation, and then in a subsequent chapter, the scientists will add their input about the Simulation concept. It will be seen that the idea has plenty of input, pro and con, and amazingly, it answers some deep questions about the Earth:

1. **Why are some Eras in history totally unknown?**
   Because Eras are set and reset sometimes with gaps.

2. **What happened to the Maya and the Anasazi?**
   Sometimes cultures do not work out and are removed. Or they are just here for brief observation.

3. **Why does the Moon exactly eclipse the Sun?**
   Because it was designed and positioned that way.

4. **Where did the 5 races come from?**
   They were designed here by ETs who were visiting the Earth before it went into the Sphere.

5. **Why are there no intermediate transition fossils if Evolution is true?**
   Because the dinosaurs did not morph into another form; e.g., Birds did not evolve from Raptors.

6. **Why has no Bigfoot been caught, and not even the bones of any of them have ever been found?**
   Because they are projections into this 3D Construct, and do not live here.

7. **Where did Religion come from?**
   It was given to Man by those who were responsible for his development.

8. **Did Man originate with the apes?**
   The DNA of Homo *erectus* and Anunnaki hominids were mixed to create the first humans. Evolution was assisted genetically.

Many of these questions were answered by Virtual Earth Graduate (VEG). Some are answered in this Chapter 13. The questions that particularly arise in this book are:

9. Why is Earth so far from other solar systems?
10. Was Earth created looking old?
11. Why are there different versions of Man and yet Evolution is not still going on today? Did something else happen?
12. Why is Man here? What is he supposed to do?

Those questions and more will be obvious after reading this book. It is not intended to replace the VEG tome, but there were aspects of the Simulation concept in the areas of Religion, History and Science that could not be dealt with in VEG, due to the fact that it was already approaching 700 pages in length, nor could the issue of Timelines be adequately dealt with. Thus this book is a sequel to Chapters 12-13 in VEG, and Chapter 2 in Transformation of Man (TOM). In short, a few more dots can be connected.

Elements of a Simulation

So what is the definition and nature of the Earth Simulation? That is best answered by revealing (1) what the definition of our world is, and then (2) how/why the Simulation was created.

HVR 3D Earth Simulation Definition

The 3D Holographic, replicated Earth is real, contained within the quarantine "shell" (*Gegenschein*) as a 3D construct, actually sitting inside a larger Konstruct. The "computer" running the Show is a semi-sentient, Bio-Plasmic organic computer-like organism with a feedback Control System as part of its main Operating System. Objects on Earth are quantized and materialized within the holographic framework by a subatomic process (Replication) similar to what was described in the *Star Trek* Holodeck. To protect Man, the Earth and its immediate surroundings (the Sphere within a Sphere) are enclosed in a high energy barrier allowing very controlled exit and entrance, like a Quarantine.

Man is inserted as a soul into this environment for lessons and comes in with a Script that says generally what he cannot do, what his tests are, and what exit points may be taken. Anything else is up to him, and is called freewill. He interacts with other simulated humans who have been called OPs or who play roles (NPCs) in the Earth Drama as they would in a video game. Beings of Light and Neggs help administer the individual soul's Script within the framework of the larger Greater Script – which reflects the purpose for the HVR Sphere in each Era.

Note that a soul's Script is controlled within the Greater Script (Operating System Pre-programmed Drama), just as individual application programs on a PC are controlled by the Operating System. For example, MS Word must operate within the confines and structure of the MS XP operating system and use its resources. Anything not prohibited by the Script to the soul is considered "freewill."

Each Era has a purpose or Plan that is the overriding Goal of the Era – What is Man supposed to experience and handle now? This is not Fate – it is just that each Era has a Theme that sets the Stage for the Drama that Man will undergo … with a hoped-for outcome. He has freewill to meet and handle (or not) the tests and challenges and either succeed or fail. Failure guarantees a repeat of the experience, even fractally for just that human who either avoided the 'lesson' or handled it incorrectly.

The HVR is subject to manipulation by Higher Beings ("They" aka the gods) who monitor phases of the Greater Script, and Man's individual Scripts, coordinating all with the Beings of Light and the Neggs. They can insert objects, people and events as deemed necessary… sometimes called (respectively) anomalies, geniuses, and synchronicities. They can also disappear things and remove people as needed, and They can perform a "Wipe and Reboot" (Reset) to clean the Stage and reset the Drama as a new Era if necessary.

> Neggs are Beings of Light who apply Man's negative lessons – they were explained in VEG.

Thus, Earth is not Man's home. It is a School and he is expected to learn and graduate.

Earth: Encapsulated, or Replicated and Simulated?

A valid question to consider at this point is whether Earth was physically moved from its 4D orbit into a 3D Construct, **Encapsulated**, subject to 3D Laws, but contained by an energy barrier…. Or was 4D Earth **Replicated** into a 3D Construct that is now the Earth Simulation?

One of the issues with this dilemma is that if Earth was left in 4D but contained in a single-shell energy barrier, **Encapsulated**, we might be observing the 4D universe around us. And at that point, the *Gegenschein* would still be there, but the Earth would actually be close to 4.5 billion years old – and current scientific examination (Ch 10 in VEG, and Chapter 13 in this book) shows that the Earth is not 4.5 billion years old! There are about a dozen physical measurements of various things (like Lithium in Chapter 8's 'Anomalies II') that are not what they should be – in fact, the Moon dust, sediment on the ocean floor, H^3 on the Moon, amount of fossils found, etc. do not register in the quantities they should if we are on a planet that is 4.5 billion years old.

On the other hand, if the Earth was **Replicated** into a 3D Construct, surrounded with an energy barrier, and then another barrier [Konstruct] was created (say 50 million miles distant from the first barrier), and the stars in the universe were replicated, simulated in that space (**Cspace**).… You would have what is strongly suspected to be our current state – what Chapter 1 showed to be like the Vatican sculpture: a Sphere within a Sphere. And if Earth was replicated from a 4D model (which <u>was</u> itself 4.5 billion years old!), the Earth that we inhabit would be much younger… And there'd be no way to know just how old our Earth is since the gods can create stuff and structures that look (and test) really old. (See Anomalies II in Chapter 8 and Heinlein's Job in Chapter 2.)

Thus the conjecture is that an Earth <u>Simulation</u> was replicated.

HVR Earth Simulation Genesis

For those who still doubt that we are in a Simulation with all the trappings described above, or would like a better overview, let us look at it from another point of view. If we are in a Simulation, <u>why</u> was it done?

The Higher Beings wanted to set up a place where They could contain souls, safe from incursion from outside 4D interference, harassment and curiosity, and guide them through specific lessons to empower their soul growth, and train them for future service. Leaving souls in bodies to wander around a real 3D/4D planet and letting them randomly interact with each other had not worked – even though the planet had the Angels in 4D (Astral) around them to oversee and guide them.

Avatars had been sent over the centuries to walk among humans but the Anunnaki legacy genetics predisposed the humans to rebellion, pettiness and wars no matter what was taught them. The avatars were actually benevolent beings who either came to 4D Earth or later (after AD 900) **inserted** into the Simulation to guide Man (like the movie *Avatar*.) And in remote times, before the Sphere was created, other races came into the 4D Earth and would work with the genetics and experiment with humans to try their hand at creating a better form. Sometimes they just upgraded their own forms. This resulted in (1) more division among humans who were overly sensitive to "us versus them" appearances and (2) different religions from each race's 'creators.' It was about AD 900 and something different had to be done.

The goal was to provide a closed (protected) environment where the 'lessons' took the form of **catalyst** to evoke true soul growth. The souls were running amok and recycling back to the 4D Earth too many times. What was ultimately decided was to set Earth apart but not in 4D, the natural home of souls, where the ensouled humans would still have had their 4D abilities and could override and cancel any lesson at will. So the **Earth world would be a 3D Construct to support limited soul powers**.

Secondly, there would be positive and negative catalyst, provided by two types of Angels that would work together. Beings of Light administered guidance and protection, and the Neggs provided the negative catalyst to humans (subject to Being of Light approval). Third, there would have to be an overriding **Control System (C/S)** which would be the "master program" that ran the place, into which each soul coming into the Earth would have a Script that was like a subprogram of the C/S to effect that soul's lesson(s) for each incarnation. (See Transformation of Man.) And because the C/S had to run in a 'closed environment' (like a computer application runs in a controlled environment [i.e., a fixed partition within the computer Mainframe]) the 3D Earth could not be left open to interference from any external source, and so the Earth was created as a 3D Construct within the 4D Realm, but surrounded by a Barrier or high-energy field Quarantine. And that was enclosed in a larger **Konstruct** because Man had to see the replicated stars, galaxies, nebulae, etc… and that *realia* had to also be contained.

To provide **catalyst**, the C/S was designed to be able to insert people (NPCs), and objects (Airships, UFOs, and Antikythera mechanisms, for example), and orchestrate and coordinate events (meeting one's soulmate, or key OPs to administer lessons, for instance) that would appear synchronistic, even coincidental. Each soul's Script was linked to the C/S pulling resources and orchestrating itself – which is why specific points of entry had to be pre-programmed (i.e., soul's date, time and family at entry. This was examined in detail in The Transformation of Man).

So at first the original 4D Earth was surrounded with the containing energy Sphere, whose C/S was to impute and sustain 3D laws and characteristics to the Earth, and it surrounded Earth with the Quarantine and called it the **3D Construct**. Outside the inner energy Quarantine was the Moon but both were inside the larger **Konstruct**. A 360° image of the 4D Galaxy was replicated holographically in the **Cspace**, as described in Chapter 1. An interesting effect was the reflection of the Sun off the inner energy field surrounding the Earth. The *Gegenschein*.

Additionally, there had to be an overall controller of some sort to orchestrate the Sphere, an energy barrier, 3D laws, Scripts, and coordinate the OPs/NPCs and Angels subject to the intent of the Higher Beings who were doing all this…. That meant a **Master Program** run from 4D as nothing in 3D had enough scalable power to simulate and replicate people and objects and coordinate events that still reflected the freewill of the ensouled humans in the Sphere.

> Running a 3D Earth School in 3D did not work (not enough power), and running a 4D Earth School in 4D also did not work (souls still had their native abilities).

The 4D Earth could not be subsumed into the coherent energy matrix of the 3D C/S – the two different vibrations were not compatible and 3D Laws could not manage the Sphere-embedded 4D version of Earth.

The opposite had to be done.

Thus the 3D Construct was replicated within a Simulation run in 4D providing all the necessary elements of the 3D Earth scenario, the C/S and Holographics and Replication all integrated (and phase-controlled to give a sense of time passage) within the Simulation such that **the ensouled humans would not suspect that they were not on a real 3D Earth.** As a precaution, the souls incarnating would not be allowed to remember anything preceding their incarnation as they might remember what they knew prior to birth in a body and either confusion or rebellion could result.

> A second reason for **amnesia** when incarnating into Earth School, is that a soul remembering past bad experiences with the opposite sex would make souls less eager to procreate and the population on Earth would fall off... which was counter to the wishes of the gods who run the Simulation. Each 'birth' into the Sphere was a fresh start (but lessons learned were in the subconscious).

Like any Simulation, it operates with Laws and controls, feedback loops (individual Scripts must mesh with the Grand Script) and **self-correcting routines**. As Dr. Greene theorizes in Chapter 7, because the Simulation is subject to 3D (finite) definition, mathematical formulations and the inevitable rounding limitations, the Simulation occasionally shows what should be unchanging constants to be changing, and occasionally replicated objects disappear, or fractally-generated scenes hang up… glitches happen due to math rounding errors when scenes are generated based on an algorithm.

If the humans succeed in seriously polluting their environment, or the Simulation develops a scenario contradiction due to multiple adjacent segments with **self-correcting codes** (Professor James Gates, Chapter 7) in one that is dominant, the Higher Beings have to perform a "Wipe and Reboot" or **Reset** – halting the Simulation, suspending 'time' for the souls, cleaning up the planet, resetting and recoding the programs that developed anomalies. Before the souls can be put back, they also have to have their memories reset (Think: the silver pen-like object called a neural neutralizer in the movie *Men In Black*) and then they can be re-inserted back into the Simulation…

Sometimes in the case of a coherent society that was developing and had a known continuity of historical events, if an event had to be deleted (due to corruption or vector error) this results in a gap in the Earth scenario timeline, and such gap has to be accounted for by **back-adjusting the history** that people thought they had. (See Hippenskippit example, next chapter.)

> As VEG, Chapter 10 and Dr. Fomenko showed, this reset was last done about AD 900, and then a human called Scaliger was **inserted** to adjust Earth history back 900 years.

Because the Simulation is 4D-generated and controlled, it is easy to insert and disappear objects and people… for the purpose of amazing, inspiring, or motivating the curious humans to try and emulate what they see. Such was the purpose for inserting **Beethoven, DaVinci, Mozart, Newton and Pasteur**, for example to deliver the "next step" in the Grand Script for Man. Occasionally 4D entities could enter the Sphere for observation, and find that radar or bullets could shoot down their vulnerable craft (due to being subject to 3D Laws!), also providing Man with scientific technology ahead of his social and spiritual development. Inserts were also made to contain and try to deflect inappropriate use of such technology as weapons. (See <u>The Earth Warrior</u>, a docu-novel.)

The C/S is all about moving Man forward in science and spirituality to improve him and his world, such that following the Sphere's Grand Script, Man reaches the ability to leave the planet and return to 4D. But before Man can be released, the Simulation will expand to grow Man in the interactions with what are called ETs – **also inserts** – as he learns respect, compassion and cooperation with others.

Antagonists in the Drama

> As a very important note, and this was emphasized in VEG: there are those on Earth who enjoy being Lords. They have the power, or prestige, and are used to being on top of the Hill, as it were. They have been called the Powers That Be, or **PTB**. They are used to being Lords and due to their financial strength, they can and do tell others what to do. That makes them Lords.

> **Golden Rule**: Those who have the gold make the rules.

> Lords need Serfs, or slaves. And after centuries of playing Lord, they do not want to give it up. Hence, Lords need Serfs and **they don't want people to graduate from Earth** – they do not want to lessen the

numbers of Serfs they can control – Thus it is expedient for them to keep the Serfs (sheeple) as ignorant as possible so that they don't find out what is really going on, and as eternal souls what they are really capable of, and rebel. Thus they would **control education and the Media**. If first-class education teaches the sheeple to think, and presents them with the truth, the Serfs may rebel and the Lords do not want that.

Does this answer why History is written by the winners?
Does this explain why the PTB do not tell the sheeple what UFOs are?

HVR Sphere Clarification

There are a couple of very important aspects of our Simulation to understand.

1. Despite different Eras, the **3D Earth as replicated is mostly real**, and its various components have been constructed, even 'planted' over time. (Think: Great Pyramids.) The Simulation is not re-designed from scratch every time a new Era is started, just re-activated. The physicality of Earth, the dirt, the rocks, mountains, mystery carvings and pyramids, were all created there via the Holographic-Replicator technology as real objects, during the Era in which they belong, and they stay with the constructed Earth. It is a little like originally writing a computer program, even a Sim City version, where the program produces an initial scenario, and then subsequent updates to the code add and delete things adding to the final descriptive state of the City… and in our case, most of what was initially created in 4D Earth is still here (Stonehenge), and some objects get added/deleted with the activities of Man during each subsequent Era.

    Some things in remote, virgin areas of Earth are holographic, sitting on a Grid – until Man comes along and needs to work with the physical side of his environment, and then the Replicator technology operates on the holographic design and, similar to a **3D Printer**, turns the image into a physical product. If it is not right, or is no longer wanted, the smaller section computers (Earth is divided into sections) can manipulate or disappear the objects.

2. If there is a "power failure" related to the technology of the Simulation, **what has been created (rendered physical) does not disappear**… i.e., everything we see and work with every day is not being generated "live" as we go and would disappear if sections of the Simulation should momentarily lack power. The Simulation has created a 3D Earth that largely does not disappear because it is locked into 3D space-time within the VR Sphere, and its 'power' never completely fails. Those parts of the Earth scenario that are holographic <u>would</u> flutter, flicker and disappear… but they are nowadays minor because Man is all over the Earth, not localized as he was 10,000 years ago.

    > As a computer analogy: the complete physical state of the Earth is recorded in Working Storage – backed up on an external storage

medium. All the scenery is pre-recorded, or defined, and can be loaded onto the Grid for Simulation. Thousands of smaller, individual (section) computers running under the control of the Parent computer, display and can manipulate the scenery.

However, occasionally the Simulation running is suspended when it comes time to do a "Wipe and Reboot," because things have gotten off-track (planet too polluted, or not enough souls to energize and coherently sustain a positive vector in synch with the Higher Being's intent and new souls are not entering the Earth realm) so the Simulation is halted. (See Chapter 9.)

3. The Gods-in-Training (some are graduated souls from the Earth School) can and do **add to and delete from our Simulation**, usually clandestinely. This is generally in support of the Control System whose purpose is to entrain Man into a higher state of thinking (and being)… more awe, wonder, inspiration, often via mystery.

    Objects can be inserted, people can be inserted (OPs, to be sure), bodies created and souls inserted/removed, and even events can be orchestrated to serve the Greater Script, or Plan for Mankind in each particular Era's Drama. (Eras tend to have a theme, a level of awareness, and specific Drama to experience and handle.)

4. The **whole universe-that-we-see is the Simulation**, not just the Earth. The parts that we can access, like going to the Moon, will be parts of the Simulation that are made real for us as we need them… the rest of the stars and Galaxies that we cannot reach <u>are</u> holographic projections in the **Cspace** (thus it would be wise to rethink the **'redshift'** phenomenon) … Charles Fort was correct.

    > What could be the lesson from what we think we see there? Said another way: Is a simulated Universe expanding and receding or is it made to look that way? In a similar way, a solar flare from the Sun is displayed for us, as part of the hologram of the Sun, and then the physics of its effects on Earth are programmed by the Simulation to 'affect' our planet.

    The beauty of this technique is that it isn't necessary to Replicate subatomic material into <u>all</u> the stars and planets we see… just those that we could get to… <u>when</u> we get there.

    Perforce, this also suggests what was said earlier, that there is no 3D sentient life visiting us in UFOs. Most sentient life in the Universe is in 4D and above, and the phase-shifted 3D Construct (a space-time 4D vectored energy) with a barrier keeps them from randomly visiting us. They can carefully visit us intradimensionally, with the gods' Ok.

5. It also follows then, that **the Simulation operates with a major set of the Laws of the real (4D) Universe** to which we belong, but the 3D Construct operates with a subset of 4D Laws… as 3D Laws. This is why Quantum

Physicists are 'guilty' of analyzing the **Imax Theatre**, and thinking that they have a handle on the real Multiverse. (Sorry.)

(**Subquantum Kinetics** is closer to the truth of our situation than Quantum Physics and what they are doing at CERN with the LHC is just examining the Simulation... and the physicists will see no more than what the Simulation Programmers want them to see...)

The next chapter will examine the Simulation processes more in detail, and then Chapter 7 will present the Quantum Physicists' and a couple of Philosophers' views on this issue.

Reason for the HVR Sphere

This section is to answer the last 4 questions (#'s 9 –10 – 11— 12) at the beginning of this chapter. Earth **is** a School and much time was spent in VEG explaining what Earth is, how it runs, and why Man is here.

Alternate Reality

The reader may get the idea initially that the Sphere was created by other advanced entities who had the technology to do so and, whether it was scientific curiosity or a desire for entertainment, they created this Sphere, the Earth Simulation for their own purposes – and when those purposes have been achieved, they will terminate the Simulation.

Having explored a simulation package available to those who want to simulate a reality, **Sim City 5** is available and was described in Chapter 2. Such things do exist, and in the hands of a more advanced civilization, something like a larger Earth Simulation could be produced. And if the advanced beings didn't like the way things were going, and humans seemed to be to petty, violent and prone to diseases, the beings might decide to (a) terminate the experiment, and restart it with a better version of mankind, or (b) they might decide to play with what they had constructed, and make it into a Game: Can we use inserts and avatars to correct things – without terminating the Construct?

That would be a little like trying to paint a moving freight train, but if you have an antigravity sled that can fly alongside the train car(s), and paint them as you match the train's speed... it might just work. That is just to say that an advanced society might have a way to do what we would consider unwieldy or impossible.

However, for reasons given later, we are fortunate that the Earth Simulation is in the hands of a benevolent and purposeful set of Higher Beings. They expect Man to graduate from Earth School.

The Earth Graduate

Just the basics of what was said in VEG will be given here, as this book's focus is the Simulation. This section is merely to explain **why** the Simulation exists… especially for those who have not read <u>Virtual Earth Graduate</u> (VEG).

One of the problems with Earth, even when it was not in a Quarantined Sphere, was that 4D ETs came here (Earth was real 4D before AD 900 – by our time reckoning) and messed with the occupants… and Earth had been something of a freeform Experiment, but the ETs would abduct, insert DNA, and experiment on the animal side of the biota. Such is what the Anunnaki did (according to Zechariah Sitchin) – before they were sent home about 650 BC. (This is all covered in VEG.)

So to isolate and protect the developing sentient humans, it was decided to put Man into a School whose entrance and exit were via birth and death, respectively. And if you are Higher Beings who love Man and want to grow him into souls that can perform meaningful, advanced service in the Father of Light's Multiverse, it is what you do.

Man is an eternal soul and has a divine potential in him that needs to be developed. There are also those on Earth who love being Lords (PTB) and do not want their sheep to graduate and leave them with less sheep to control!

> Be clear that the **Elite** want Man to succeed and have said so –
> on the Internet, as the Insider (back in 2005 on the ATS forum),
> and also as the Hidden Hand messages. Some of their key ideas
> were quoted in VEG.

So there is no evil force trying to suppress Man, just other men who have a lot of money and love being Lords, controlling as much as they can, who want their little world to continue – In fact, if they could take it back to the Middle Ages and the Feudal System, they'd be ecstatic! (Their coming lesson is to see that what <u>they</u> are doing doesn't work, has too much of a negative emphasis, and Man may wake up, rebel and throw the PTB under the bus before it is all done!)

And the salient reason for Quarantining Man (VEG, Ch 12) had to do with the number of souls that were NOT graduating from Earth School! A lot of souls came here to play and were not learning… sometimes the "graduating class" was just a few hundred compared to the millions that were on the planet. It was at that point that Buddha, Krishna, Jesus and other enlightened avatars came to Earth – to guide and inspire Man.

And because that was sometimes less than successful, something else had to be done. The first part of the solution was to put Man in a 3D Construct (The Sphere) that was self-contained or Quarantined. And still very few graduated – the pleasures of the flesh, drinking, drugs, even those who lusted after war and were hooked on violence…

> and today we are training a new generation of warriors who eat, sleep and drink **violent video games** (they **entrain** the young mind into a mindset where "violence is the way to solve things")… as well as these games **desensitize** our youth to violence, blood, death … perfect training to program young people to be part of a violent world… The PTB almost have to keep Man at war (with a constant threat of terrorism) if they are to maintain their hold on society and guarantee that few souls graduate out of here… (See Chapter 9 for proof in TSiM.)

Second Solution

So a second solution was devised. And that was to "paint the moving freight train" – **insert** beings to modify Man's DNA – via abductions mostly at night. Yep, you know who I'm talking about. Greys. The goal is to rewire Man's genetics to be smarter, more compassionate and peaceful (by reducing testosterone), and give some the basics in psychic ability – clairvoyant, intuitive and able to heal (self and others).

It was decided to manage this Program (the same one Dr. David Jacobs fears and Dr, John E. Mack loved), from underground and the **Greys** would be created as **Bio-cybernetic Androids** programmed with the ability to analyze and rework DNA as needed – usually by changing a small amount of it (i.e. in a baby or fetus). This is where a lot of women had "miscarriages" – they <u>were</u> pregnant, and then the Greys removed the fetus (within the first 3 months) – God forbid the Earth mother saw what the baby to be born to her looked like! All of them are human, but some are ET-looking humans – and they will be 'planted' somewhere else. The very human-looking ones are already walking among us. (See TOM, Ch. 8)

That Program is now in the Assimilation Phase. It is a benevolent plan to improve Man and make him less violent and more intelligent. Early efforts were less than ideal and resulted in some of our societal anomalies walking around nowadays. Not wrong, not bad, I stress that this is just what happened – The human genome is very complex and the Greys had some difficulty in the beginning (e.g., early 1940's).

Those who are different need love and understanding as much as anybody. For sure that is a **big test** that we all have to face – to be more tolerant of those who are different – If we are ever returned to 4D where a myriad of different beings exist, will we start shooting, run away, or will we celebrate the fantastic variety of sentient beings in the Father's world? Will we respect <u>all</u> lifeforms?

Higher Vibrations to Exit

Humans who do graduate from Earth all have to meet a certain % of love, patience, wisdom, desire to serve, knowledge (not false beliefs), humility, etc…as VEG Ch 15-16 said. Above all, **Respect** is a major key – for oneself, others and the planet.

This doesn't make souls peas in a pod, but they must all know who they really are, Who is Boss (The Light), and where they might fit in the scheme of things – based on interest and aptitude. They are very respected by others because to graduate, you really have to have your stuff together – and you do that by overcoming the BS here on Earth! (And you don't have to be perfect…)

> Gaining **Love and Knowledge** raises one's personal vibration and that is a key to graduation.

To see thru the lies, still serve and care about others, and respect yourself and rise above the problems that come your way is what it is all about. That takes a lot of determination, searching, and a certain growth in awareness that They expect of a graduate Over There. One needn't be perfect, just have demonstrated a major % of one's being IN ALL IMPORTANT areas as STO. (VEG Ch 15 enumerates these.)

> It is the fact that your **PFV** or personal vibration increases to a higher resonance that guarantees you can leave Earth – if you vibrate higher than Earth, and you can sustain Love, Patience, Humility, Respect and true Knowledge, you are a candidate.

Why Graduate?

Besides the fact that Earth is not really our home, and most people don't live in any school they ever attended (Military School is an exception), there is an exciting reason. Here are some of the areas of service open to Graduates from the Earth Simulation (See also Ch. 12 if Earth is a real 3D Construct instead of a Simulation):

> **Bio-plasmic Quantum Computer Techies** – responsible for basic HVR computer support and maintenance.

> **Bio-plasmic Computer Programmer** – performs fractal sub-programming of subareas of Simulation under supervision.

Akashic Records Librarian – maintains life records' storage/retrieval.

Gods-in-Training I – responsible to oversee the Simulation: Man and feedback of the Control System. Many sub-areas here.

Gods-in-Training II – responsible for the Holographic stabilization and interface with the Replicator technology. Sub-areas here.

Soul Counselors – responsible for evaluation, guidance and training of in-coming souls to the Interlife for further development: imprinting or vibrational adjustment. Many levels here, including Teachers.

There are many others, but it is a busy world over there; no one is sitting around on a cloud playing a harp – unless they're on a coffee break! The reason that the PTB want to block souls from progressing is mainly because of this position:

Gods-in-Training III – responsible for overseeing, managing and controlling the Neggs and OPs –to make sure that lessons are properly administered (according to scripts) and it amounts to controlling what the PTB can 'get away with.' This is as close as the Interlife comes to having a "police force."

And because I once asked, "What if the Gods-in-Training choose to abuse people, or do something nasty?" Answer: They are sent back to Earth for rehabilitation.

But first, one must want to graduate... and there is a key.

Futility is the Key

In addition, something else is important: You must **drop all attachments to Earth.** People, places and things can and will bring you back – even if you are ready to go. Henry Thoreau knew this, as did the late George Carlin. Those ties must be broken while on the Earth.

And the easiest way to do that is by seeing the **futility** of ever getting your way, when you want it, the way you want it, where you want it... and **detach** from having anything work or not work.

The higher attitude is: "It's Ok if it works, and Ok if it doesn't."

Futility is the key. It is **not a negative, despondent resignation** – it is based on a genuine view that 'stuff' doesn't really matter, you <u>will</u> see your loved ones again (on the Other Side), so release them, and give up drugs, tattoos, piercing everything, drinking, smoking and must-have sex partners... Sexual fantasies are best gotten out of your system – it is an Earth 'drive' that will attach you to Earth and you'll be back...

The higher attitude is: "Sex is Ok and I can take it or leave it."

An even higher attitude is: I am an eternal soul – a denizen of the Multiverse and I can handle whatever comes my way. I respect myself, others and the planet, and I seek to create quality and serve wherever I am. Experiences are for my growth, and the worst of them are tests – I can do that. I don't have to like it, I just have to handle it. Death is not final – I just move on to another experience, and will probably have my soul associates with me, and we'll play some games and, learn, serve and move forward. Love, Light and Knowledge is what it is all about!

Laissez rouler les bons temps!

Personal Frequency Vibration (PFV)
(see Glossary)

Ok, so this isn't a religious lecture. These are the things that all Earth Graduates have come to understand and embody... that is why their vibration (PFV) went up and that higher, purer, energy emanates from the top 3 – 4 chakras – that is why Transformation of Man (TOM) dwelt so much on that aspect. Your chakras are either open or closed...and if they are open, and:

If you have dealt with Ego – Chakra 3 is spinning clearer;
If you feel more compassion for people – Chakra 4 is open and spinning cleaner;
If you speak Truth and have stopped 90% bad swearing – Chakra 5 is spinning cleaner and faster;
If you have meditated and had any encounter with Angels – your Chakra 6 is cleaner and you are becoming intuitive and perhaps clairvoyant;
If you have gotten your first 6 Chakras cleaned up and passing higher energy – your 7th Chakra (crown of head) has begun to open and stabilize a connection to your higher self...

The point is that salt baths and eating right plus right thinking/speaking will do a lot for raising your PFV – the energy vibration that affects your Ground of Being and no one (not even the Higher Beings) can stop you from "graduating" and being able to sustain being in the higher energy of 4D.... It is all energy based – if you have higher, finer energy you will <u>automatically</u> make it.

> Be very clear. No one says you go or stay – it is all based on: Do you have the right vibration? If you do, you go. If you don't, and you try to force your way to 4D, or 5D, the higher vibrations can severely harm you. Similar to a fine crystal glass that can handle up to 20Hz (vibration) and if it is put in a room with 60 Hz vibrations, it will shatter.

Real Knowledge Important

Also a key point: hanging on to false beliefs and false knowledge ("The Earth is flat!", "God punishes people", and "We only have one life to live and then death is the end!") will bring your vibration down. This why it is so important (and both VEG and TOM emphasized it) – you have to get the Truth into you... it vibrates higher than a lie. This was demonstrated physically with a <u>real</u> aura camera in Dallas where the subject being photographed held a book that we all knew to be false, and the aura around the man was a dirty greenish-yellow. It looked yuccky.

The man then held the Bible and got a beautiful crimson color – the color of Love. He put that down and we photographed him holding a computer programming book and his aura changed to bright yellow – the color of knowledge/intellectual pursuits. Lastly, and we just had to see this one for ourselves. He held a metaphysical book on the Tao, and his aura went white with a gold core! That blew us away... we had all enjoyed the book but here was the proof that it was True, and had a very high vibration.

I later corroborated the colors, and their vibrational strength with the visible spectrum of light – broken into its many colors... Red was the slowest vibration (700 nm) on the right of the following chart, to the highest vibration, Indigo (which was 400 nm on the left of the chart) -- 700 nanometers wavelength is slower than 400 nm.

The following chart shows the vibrational frequency of visible light, plus slower radio waves and the much faster Gamma (γ) rays – their speed and energy is why they are so deadly.

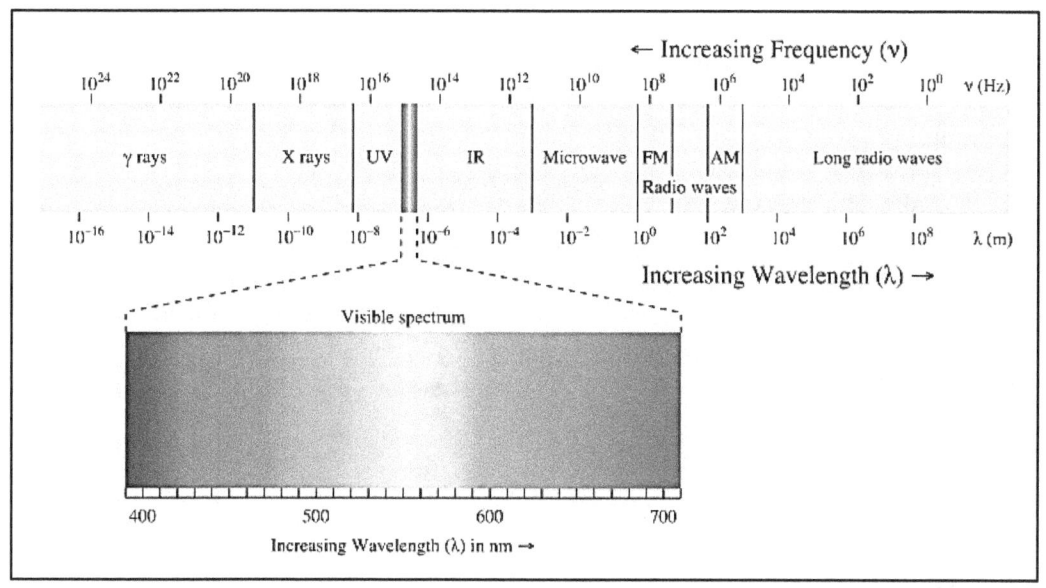

(Credit: Wikipedia:
http://en.wikipedia.org/wiki/Light#mediaviewer/File:EM_spectrum.svg)

What is interesting is that the body's Chakras correspond to these colors, too. Red is the color at the perineum, between the thighs, then Orange, then Yellow, then Green, then Blue, and Indigo and White (at the Crown).

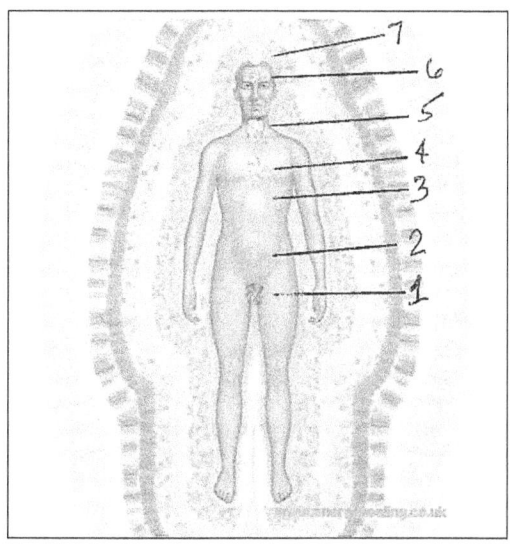

As for the Body's Chakras:
The chakras are numbered 1 – 7, from the root (1) to the crown (7):

7 – **White**
6 – **Indigo**   **400 nm**
5 – **Blue**
4 – **Green**
3 – **Yellow**
2 – **Orange**
1 -- **Red**   **700 nm**

(Source: www.differentlight.org/images/Aura%20400pix.jpg)

Some systems recognize 13 Chakras, but in fact the body has many, front and back, in the hands, feet and head.

There are also 7 layers to the aura – each major chakra connects to one of the layers. (Thus is examined more in TOM.)

As a matter of fact, **the planet also has chakras**… most living things do – even animals. They are called **vortices** around the planet – the best known in the US are in Sedona, AZ.

See the following diagram…

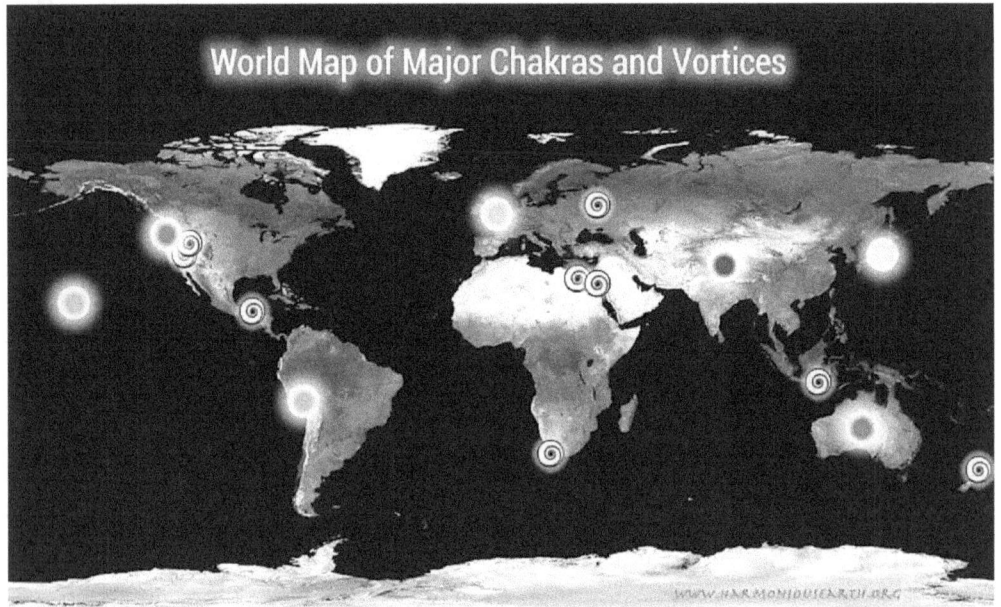

Credit: Bing Images: Harmoniousearth.org

All of that just to explain what we are and what it takes to get out of here – and our energy stare, or vibrations, determine what we can do. This Simulation was specifically built to support, manage and develop Man. The *anthropic principle* at work.

The point is that the body and the planet are of similar design … both have energy states, vibrations, and the Earth harmonics are set at 7.8 Hz (Schumann Resonance). We are made for each other: The Earth has Ley Lines which channel energy between vortices. The body has meridians which channel energy (*chi*) between chakras. The body has a torus of energy around it, and the planet has the toroidal Van Allen belts. The Earth's surface is 70% water, and the human body is 70% water.

Looks like the Designers were on a roll…

Chapter 6: Those Who Run Things

This chapter is a little different. It takes a look behind the scenes into an event based on what Jim Elvidge and Dr. Brian Greene speak about in Chapter 7. What happens when the Simulation hiccups? There are 2 scenarios following to demonstrate what these scientists will be talking about:

I. Earth-based Scenario

Alphonse Hippenskippit is driving down the freeway just north of Los Angeles, on his way to camp out with his family. Traveling thru Northridge, it is a nice sunny day, light smog, and the traffic is not too bad – at least it is moving at 40 mph. KFI traffic reports from a helicopter overhead says that there is a bottleneck on the 405 at Roscoe Blvd and he realizes he'll have to get off before that...on Sherman Way and jump over to the 170 and up I-5.

Finally getting to I-5 and the split with Route 14, he bears right and takes the 14 north into the canyon. The family has a favorite camping spot just off the main road, on Placerita Cyn Road.... back in the *barranca*. He has brought a .22 rifle with him, just in case a mountain lion wants to examine their campsite a little too closely.

Driving up Route 14, it all looks familiar. His watch says 2:05 pm, they are making good time. He passes an occasional car heading south, the kids are sleeping and the wife is on the cellphone, telling the housekeeper when they will return. All of a sudden he spots a road sign saying Golden Valley Rd which is Exit 5. He missed his turnoff. He pulls over and turns around, and heads back.

It should be just 2 minutes back down the road, but soon he comes up on Exit 2 which is Newhall Ave. Back too far! Where the heck is Exit 3? He pulls over, and turns around again, headed north once again. This time he sees the Placerita Cyn Road far off to the side of the road, but there is no exit!

He pulls over. This is weird. How many times had he and the family used Exit 3 to drive up into the canyon? Now the exit ramp isn't there. The Placerita Cyn Road is there but there is no exit... In fact, he notices that there is no frontage leading to the Cyn Road... it's a quarter of a mile from the highway and just starts in the middle of the desert....!

He gets off the 14 and takes Newhall Ave north to the frontage on the north side of the 14, and follows it back north to where Exit 3 should be. He is in for a shock. He hears a CLICK. The Exit 3 IS now there, and the frontage dips under Route 14 and connects with the Placerita Cyn Road on the other side!

His wife has been lightly napping and now is wondering what the heck Alfie is doing, and why has he been going around in circles… driving back and forth. He stops the Suburban, gets out and takes a good look at the Exit 3 interchange…. It looks old, not like something that was just built. It now looks like he remembers. He figures he was spacing out and just missed the exit several times… What else could it be…?

They drive up the Cyn Road and turn onto a dirt road that leads to their favorite campsite. He turns around to tell the kids to wake up and get ready, and the daughter now has blonde hair… He wonders when she did that? But he focuses on the driving, avoiding some ruts. He rationalizes that he just might be tired and is missing what is going on. He resolves to take more care on this trip – for the safety of his family.

The dirt road forks and he always takes the left way. Very soon, the road deadends at a fire tower. That is not right. He thinks maybe he is on the wrong road, so he turns around, and goes back to the fork in the road, and takes the right path, and it leads him to a clearing under some oak trees. He hears another CLICK. This is it. He could have sworn it was down the left path—Maybe he is getting older and his memory is slipping.

His family gets out of the Suburban and he notices that his daughter now has brown hair, like she always did… Well, maybe it was the sunlight that made it look blonde…?

They set up the large tent, and get a campfire prepared, but unlit, for later. He and his son go shooting, down the slope, over behind some rocks, and the wife and daughter take the inflatable raft thru the brush and head for the small lake about 300' away.

An hour later, the son and father head back to the camp and they both hear a CLICK and the son asks what that sound is. The father says he must have stepped on a branch, and the son looks around and sees none, and is puzzled. They get back to the camp and find the campfire lit, and a pot of coffee hanging over it. Ah yes, his wife is always thoughtful! He and his son relax in camping lounge chairs.

Meanwhile, the wife and daughter have gotten lost and cannot find the small lake. It is not at the end of the trail they followed. The lake is just a large, slightly concave field of flowers. The wife figures it must have been drained, or it just evaporated.

They head back to the campsite and the wife finds the pot of coffee, and she thinks how thoughtful her husband is....

The son is asleep in his low-slung camper *chaise longue*, and the father is reading a book. He sees his wife come up and asks her how the water was, and where the raft is...Did they leave it at the lake? She says there is no lake and as for the raft, she gives him a strange look and points – it is still over by the tent, bundled up.

Alphonse is now a bit worried – What is going on? He saw them unpack the raft and take it with them.... He asks if she remembers their missing the exit? She doesn't. He begins looking nervously around... waiting for something, or someone... His wife asks him what he is looking for?

He replies, exasperated, "Rod Serling!"
She rolls her eyes and gives him a funny look...
Just then he hears another CLICK.

The family is driving up Route 14. The kids are in the back asleep and the wife is on the cellphone, telling the housekeeper when they will return. His watch says 2:05 pm, they are making good time. They come up to Exit 3 and swing onto the Placerita Canyon Rd.....

> Of course, there are a lot more glitches in that scenario than one would encounter, even on a bad day. But it makes the point that no matter what happens, They are in control and the Simulation can be reset. You won't remember that it even happened...

What was happening on the other side of things, behind the scene?

Simulation Control Room

Watching the large ExoViewer screen, which shows a live scene on the French Riviera, a Being of Light Tech II watches as a message flashes on the bottom of the screen. It warns that the mathematics have just exceeded the parameters for stasis and parts of the Simulation 42-CA-290012 that run a section of California, just north of Los Angeles, have abended and the holographic simulation of that area is fluctuating and destabilizing.

He calls in the Section Manager, a Higher Level Being involved in managing the Quantum Bio-Computer, which consists of many sub-computers, each one dedicated to a quadrant of Earth. They are all linked together and share in the 4D energy.

"What's happening in Sector 42-CA-290012?" asks the Section Mgr.

"The sub-computer has maxed out again; the algorithms for decimal precision at Magic Mountain in Valencia, just north of Sector 42-CA-290012, have overflowed and corrupted adjacent terrain" said Tech II.

"Ok, switch the viewer to the I-5 / Route 14 area and attempt to stabilize Object Replication. Bring GR24 back online to substitute and integrate a Patch Overlay. "

They both watch as the Hippenskippit family in the white Suburban heads up the road. The Tech II says that it is Alphonse and his family… and he is looking for Exit 3 – which no longer exists due to the glitch.

"Ok, we will have to monitor them and reset the subgrids as necessary…" ventured the SM.

"Yes, and that will result in at least Alphonse hearing a 'click' as we do… He is looking for the exit, and has turned around, and can't find it…. Now turning around again, on the north-side frontage road, and he is headed back looking for his exit… "

The Tech II hits <reset> and there is a CLICK as the energies intermesh… All of a sudden, the Exit 3 is restored and connects with the Placerita Cyn Rd…

"Oh boy, he took the left fork on the Cyn Rd and subsector 42-CA-290013n has overlaid subsector 12b… the campsite has been replaced with a fire tower… He's turning around again…" and the Tech II hits <reset> again and the dirt road goes back to normal. Alphonse hears another CLICK as the energy and original map from subsector 12b replaces subsector 13n.

They watch the actions of the family and see the Mom and daughter head for the lake with the raft… The Tech II looks ahead and sees the lake is not there, subsector 23 has overlaid subsector 22 and has to be reset… He waits for the Mom and daughter to see the lake is not there, sees them turn around, and not only presses <reset> but asks Tech IV to adjust the Mother's external memory links so that she does not remember the missing exit, nor the confusion about the lake.

There is a third CLICK that even the son hears this time.

The reset has also mis-fired and added a coffeepot over the campfire which wasn't there before… and as the four family members get back together, Tech II realizes that there are <u>too many anomalies</u> that family members have recognized (and are trying to rationalize away) and something will have to be done – probably at a Tech IV level.

When anomalies happen, and can be localized to particular sector, or better yet, to a subsector, and the number of people involved is limited, the OverSystem Tracker creates a Timeline 1 from the last good point in the sector (before the anomaly was identified)… recording everything up to the point where Tech II marks the stop point. The start and stop points identify the range for the captured data for Timeline 1 with all the irregularities. It is then compared to the original Archived OverScenario for the sector (scenery minus people), creating Timeline 1b which now has the correct scenery: exit ramp and lake, for example... The two timelines are rolled into each other to create a clean version… **Timeline 1a** in a fractal subset and just the Hippenskippits will go thru the Timeline 1a version...

The Section Manager, Tech II and Tech IV reach a quick consensus. The whole Timeline 1a from 2 pm to 2:42 pm will have to be reloaded into the subsector <u>and memories [which are holographic] will be adjusted back to 2 pm as well</u>.   Instead of a reset, it will be a Reboot of that subsector…affecting just the 4 people in Alphonse's family.

Of course, time will 'back up' 42 minutes for the Hippenskippit family – but it is done in a fractal subset. When the 42 minute replay of Timeline 1a is done, the last microsecond of the fractal subset is programmed to exit back into the normal subsector where the family is camped out, and time for them, unknown to them, will be back in synch with everyone else in that larger sector, and hence the USA and the world.

The Tech II warns that the wife is now wondering if her husband is crazy, asking for Rod Sterling, and the SM smiles… and he gives the order to do a major subsector Reboot.

There is a final CLICK and the family is back in the Suburban, back on the road again, and it is 2 pm again...

Postscript

The preceding was an example of the sophistication expected to be in the Heavenly Quantum Bio-Computer built in the 4D-5D Realm to drive the 3D Earth scenario.

Dr. Brian Greene (in the next chapter) aptly addresses the probability that a computer running this Simulation would operate on a mathematical basis and whereas scenes and objects are created based on finite algorithms, at some point there would be rounding errors and that is what the preceding scenario demonstrates – some parts of a Simulation may show errors – color changes, buildings missing, people disappearing… or strange creatures showing up here and there… such as Bigfoot, fairies, or Shadow People….

Even in video games, sometimes images of trees and rocks do not display correctly – they are said to not 'load' properly. Some games store the complete digital image and just display it as a character enters a certain scene, whereas other larger video games find that it is easier to store a parameterized description of the tree or rock and 'calculate' it into place… That takes less storage on the game CD/DVD and less storage in the PC.

But there is more.

Elements of a Simulation

So just how does one know if they are in a Simulation or not? It is easy to see from the example with the Hippenskippit family above that they noticed something was odd <u>while it was happening</u> – but once the anomalies were reset, they remembered nothing. Interesting to consider that we may have been thru anomalies, too, and after any reset, we would also remember nothing…

The nature and differences between **Virtual Reality** (VR) and **Simulation** were laid out in Chapter 13 of VEG, and on the off-chance that those readers have picked up this book, just the basics will be given here, so that we are all on the same page, so to speak. The information is important to this chapter, and the next, so the basics will be repeated here.

Virtual Reality

What is the difference between actually eating a steak and drinking fine wine, versus having it all pumped into your brain through neuro-stimulation as was done for Cypher in *The Matrix*? The sensation and satisfaction are there, and isn't that one reason why we eat the steak and drink the wine? How can we tell the difference, and is that important?

To simplify, for the purposes of this book, VR will be what we call playing a video game, wearing a special headset that provides visual and auditory input. Thus VR will be considered more limited than a Simulation.

Simulation

What we see happening to Phil, the TV anchorman in **Groundhog Day**, is a simulation – total sensory immersion. It is a lot like what the crew of *Star Trek* goes thru when they experience the Holodeck. However, the gods want Phil to know he needs to be a better person and they let him know he's in a repeating simulation – every morning at 6 am he hears the same Sonny & Cher song on the clock radio! He eventually realizes he's repeating the same day over and over because he isn't doing something right – he can't even kill himself to get out of the loop!

Another difference between VR and Simulation is that the VR has a **reference point** – one can look around and see the rest of the room in which one is playing the video game and know that what is seen/heard is not real. Because Simulation is a total immersion, all senses are involved, there is no reference point … but there can be clues if the Simulation 'hiccups' and one suddenly sees structural gridlines on which buildings or scenery is 'planted'!

But what happens when Man on Earth, who is living in a 3D simulation similar to the one Phil was in, also doesn't know where he is, what he is, and can't wake up? Man has no idea what he is supposed to do in the Earth School – and many do not even know it is a School!

That is the point of this book. This chapter is suggesting that Earth is a real, 3D world but instead of being contained under a steel half-dome as in Truman's world, (*The Truman Show*), Earth is contained within a more sophisticated high-energy Sphere in 4D – all orchestrated and monitored from outside the HVR Sphere. Fort's *Gegenschein* reflects off this field's inner energy barrier (somewhere between the Earth and the Moon), and Robert Monroe actually hits barriers in several of his OBEs.

Shakespeare was right: Earth amounts to a **"Stage"** in quarantine, for reasons given later herein, and humans are the players who "…strut and fret their hour upon the stage… and then are heard no more!" Shakespeare said it many times, Life is a Stage and Man is a player…

All the world's **a stage**,
And all the men and women merely players.
They have their exits and their entrances,
And one man in his time plays many parts,
(As You Like It in **Act II Scene VII.)**

I hold the world but as the world, Gratiano;
A stage where every man must play a part,
And mine a sad one.
(The Merchant of Venice, Act I, Scene I)

And…

> **Life's** but a walking shadow, a poor player, that struts and
> Frets his hour upon **the stage**, and then is heard no more; it is
> A tale told by an idiot, full of sound and fury, signifying nothing.
> **(Macbeth Act V, Scene V).**

Shakespeare is not to be taken lightly. There is much wisdom in many of his plays. If this book is correct, **Shakespeare was an Insert** whose job it was to write wisdom… surely no normal human could have written as much with as much insight into human nature, and done it so well. Our best playwrights do not and have not turned out the quantity AND quality that he did … but Mark Twain came close!

Pythagoras also said that this world was like **a stage** / Whereon many play their parts; the lookers-on, the sage….

So what are some signs that we may be in a Simulation?

Criteria To Consider

Before going to the next section, it might be useful to share seven ways we can know if we are in a Simulation. These are things to look for in a Simulation of our world and in the events around us…

1. Constants and Laws Change

It has recently come to light (Chapters 8-9 in VEG) that the constants of Physics and including the Laws of the Universe are not so constant. The speed of light, C-14 dating, and the alpha constant are not constant, and entropy does not always happen (The Second Law of Thermodynamics).

Any parameters and 'constants' can be changed <u>at any time</u> in a Simulation, as well as they could wind up in corrupted code that normally drives a stable part of the Simulation (addressed later by Professor S. James Gate in the next section). If "internal consistency" is not there, then it is very suggestive of a Simulation whose inner 'programming' occasionally misfires and this is also examined by Dr. Greene in a later section.

> **Hints that this is not a real universe would come from
> imperfections in the Simulation.**

> "…It is unlikely that we would find an obvious imperfection
> such as a fuzzy border on the other side of a mountain, which

has never before been observed. Imperfections in the observable universe would be subtle and almost undetectable. They will be found [however] in the laws of physics." [55]

And just so, it has been observed that there has been a change in the speed of light, the **alpha constant** in decay rates (The Fine Structure Constant), also seems to have changed – this is the ratio of the speed of light, the charge on the electron, and Planck's Constant. [56] Constants should be the same everywhere in the universe, but that is Man's idea, and perhaps the Simulation operates with different but not antagonistic laws in different parts of the universe...? It will take further deeper analysis with more advanced equipment to get to the bottom of this issue. And if we are in a Simulation, it may be adjusted so we can't figure it out, no matter what we do. Or They ('the gods' aka the Higher Beings) may just reset or Reboot us...

2. Things Disappear

Objects, buildings, trees and fixed parts of the landscape may change location, size, orientation (N-S-E-W), color – or just disappear. On-ramps that used to be there are now gone, motels that used to be there are gone, and if you get out, walk around and physically check, you may find that there is absolutely NO TRACE of them ever having been there. The same may apply to people; you go back to a town and try to look up Sally Smith, and no one knows her or who she was...such mysteries were a favorite for Rod Serling on *The Twilight Zone*.

3. Synchronicities

This one is a real kicker: things just happen to work out at the right time in the right way – as if Someone were watching and helping you out! The right people show up, or you meet someone just "by chance," and you just happen to see/hear the very piece of info you were looking for... easy for the Simulation to manage as They can insert the right people, or right objects at the right time, or orchestrate events to produce a synchronicity.

4. Auric Information

This one is added for those who can see auras, the etheric energy envelope around people, usually 1-2" above all parts of the body. When you see peoples' auras appear and disappear, or one day John has an aura and the next day he doesn't, this might make you think you have switched timelines to another locale where the same person who normally has an aura, now doesn't. That is a possibility, but it is also indicative of an ensouled person leaving that body (as in *Avatar*) and the Simulation. Whereas these people without auras have been called OPs, they may in fact be NPCs -- Non-Playable Characters controlled by the Simulation computer.

5. Scenes Repeat or Freeze

This should be an obvious clue, and it rarely if ever happens. For example, sometimes on the *Star Trek* Holodeck, the simulation could hang up, or freeze, and the player could get stuck in the simulation. It is theorized that if this actually does happen, hopefully it is localized to a small area of the planet, and the gods would stop everything, suspend time, fix the problem, wipe memories of the glitch, and then start the scene again (see Chapter 5). How would we ever catch this?

6. Déjà vu

Experiencing Déjà Vu is a key sign that one has been **recycled** back into the exact same lifetime – same drama, same family, same 'test' all over again that one failed to handle the last time. That can be done fractally by resetting and rerunning the Simulation. It looks familiar because it **is** a repeat of a former simulated scenario.

For individuals who need this, it is done fractally as a subset of the timeline we are all in. The 'rerun' may be isolated for the soul as a "fractal simulation" as happened to Phil – while he repeated his <u>years</u> of reruns, he was fractally isolated from the rest of the 'normal' world outside Phil's simulation... His TV crew and the townspeople were replicated [as NPCs] in his fractal world – for however long it took him to wake up so he could exit... It may involve a suspension of time, such that as Phil completes his innumerable reruns, when he rejoins the 'normal' world, it is just one day later in that world.

> Since time is merely a construct in our 3D Sphere, it would be easy to manipulate it. Just because the Sun produces the illusion of days passing, that does not mean that time exists. Time is not a cause, it is an effect. How would we measure time if the Sun didn't start and end our days?

7. Fortean Anomalies (yes, again)

Since rocks and fish falling from a clear sky are not normal everyday events, could they not be products of the Control System... designed to make us re-examine our world? Such weird events would be easy to create in a controlled HVR Sphere, or Simulation.

If Man doesn't pay attention to fish and rocks, maybe he'll notice UFOs...? Or a Bigfoot...?

Let's take a look at another scenario that exemplifies aspects of a simulated world. In this case, we'll have Bertha Hippenskippit and her friend Patty Kayke go shopping in a mall. What happens to them, and the things Bertha sees, are what to look for…

II. Shopping Mall Scenario

Bertha and Patty have gone to the local mall to buy some clothes at a new, upscale store just opened called Needless Markup (NM). They park the car and walk into the mall… hurrying because the sky is overcast and while it isn't raining, they hear thunder, so there will be lightning, too. They decide to windowshop on the way to NM.

The Mall layout is given in a few more pages.

Bertha notices a large water fountain and its spray at the junction of two main, wide walkways. She goes over and tosses a penny into the pool at the bottom. She notes there are just a dozen coins in the bottom. She and Patty turn right and walk a ways and stop in front of Buckstars Coffeeshop. They decide to have a couple of Frappuccinos and a tall, handsome young man greets them. They place their order and he takes two Grande cups and writes their names on them – without asking them who they are. Bertha notes that and asks him how he knows her name, and he turns and put the two cups to the side, but one is just hanging there, 6" off the counter, and he says that they always write the customer name on the cup, and she again asks how did he know <u>her</u> name, and he replies that they use a marker to do it.

Bertha gives up and notices that one of the cups is suspended above the counter, and the other already has Patty's drink… She points out the floating cup to Patty who doesn't see it. The barista quickly makes Bertha's drink and charges them $8. Bertha decides she also wants one of the pumpkin scones in a glass case standing to the left of the register…. It has 3 shelves in it and hers is on the 2nd shelf. The barista acknowledges her new order and comes around to her side of the counter, takes a large wax paper, puts it on the face of the glass, reaches his hand in thru the paper, gets the scone and gives it to her.

Bertha does not let this one pass. He reached his hand <u>thru the glass</u> and got the scone! She challenges him on that and he says that they always use the wax paper… And she says, No, she means how did he reach thru the glass…? He responds that that was where the scone was and so he got it for her. Patty speaks up and now wants one, too!

The barista puts the wax paper over the glass display again and asks Patty to reach in and grab a scone. Patty hesitates and timidly reaches thru the wax paper and into the case, grabs a scone, and extracts it! She is freaked out and drops the scone on the floor. The barista tells them that the scones are 'on the house,' smiles… and he goes back behind the counter. [57]

Bertha and Patty exit the store wondering about what just happened. As they walk along, they are met by 2 kids, about 10 years old, dressed alike who look human except that their eyes are all black. They smile and ask Bertha for a dollar to buy something to eat. She can't stop looking at their eyes – no white sclera as normal people have… and she is so distracted she grabs Patty and steps away. Three quick steps later she looks back to see if they are following her … and the kids are nowhere to be seen.

They move on, drink their coffee and come to the entrance to NM and as they pass thru the security sensors, there is a bolt of lightning that hits the store roof and Bertha receives a jolt of electricity from the sensors. She feels all tingly and steps on into the store, and then she notices that about ¼ the people have a glow about them and some don't. Patty sees it too, and because she is a New Ager, she explains to Bertha that 20% of the shoppers have auras and the rest don't…. "OPs. My dear!"

Patty buys a dress and Bertha buys a new handbag. They exit the store and note that the people in the mall also are a mix of auras and non-auras, and Bertha now sees that some storefronts seem to have a latticework or grid in the front wall, and as they get back to the fountain, it too has a strange gridwork or frame with a grey plaster covering it, but she can see both the grid and the plaster interwoven. And, even weirder, the bottom of the pool at the fountain's base is covered in coins… hundreds of them! She thinks that is odd, and notices the same two kids standing on the other side of the fountain, looking at her and Patty. She gestures to them to leave the coins alone – thinking they still want money… and one of the kids points at Patty, and the two kids just disappear. They dematerialize.

Bertha turns to ask Patty something and now Patty is nowhere to be seen. And the bottom of the pool has just 10-12 coins in it again. She knows she will need something stronger than a Frappuccino when she gets home and she looks around again for Patty, but she isn't there. Bertha gets back to her car in the parking lot… and Patty is asleep in the passenger seat! She hears a sound behind her and the two kids are behind her again! Before she can say anything, one raises his hand and waves it at her in a sweeping arc and she doesn't remember anything of the shopping trip… but she still has her new handbag. She gets in the car and drives home.

Before discussing what the events meant, it is helpful to take a look at how the simulation was probably programmed.

Simulation Programming

To use as little code as possible and reuse coding modules with multiple definitions, it would be almost mandated that something like C++ would be used – perhaps a much more advanced version of it. C++ is **object-oriented** and thus designed for **modular programming** wherein the same code can be reused by calling the tag that identifies that snippet of code… thus it is not necessary to rewrite the same code when setting up another water fountain. Just use the Water Fountain Class (predefined) and add the unique parameters to any second or third fountain.

C++ uses Classes that describe the original object and has static and dynamic array storage, encapsulation, inheritance and polymorphism – all of which allow for very dynamic processing with a minimum of code. When one defines an Object, say a fountain, it would be a Fountain Class as an Object, and its attributes (water spouting, base pool) can all be defined as part of that Class. If another fountain is needed in the mall, or somewhere else, the new Child object is instantiated from the Parent Class and it can **inherit** the same attributes/features as the original – without redefining them.

For example, a new Flying Squirrel Class can inherit from both Squirrel and Flying Fauna. The Squirrel Class would have a **function** to Climb and another to Eat. The Flying Fauna Class would have the function to Fly. The new class would have all three functions.

   A. Another advantage is **polymorphism** -- enables one common interface for many implementations, and for Objects to act differently under different circumstances.

For example, say that you have 3 types of Pizza – Frozen, Deep Dish and Thin Crust and they each require a different way of baking them, so each subClass has its own **function** to perform the Bake function. One can call the Parent Pizza Class and pass it a parameter that identifies which subClass is wanted, and that unique subClass Bake function is executed – not the Bake function associated with the top Pizza Class... All 3 subClasses are derived from the Parent Pizza Class.

   B. Another way to instantiate objects is with **virtual inheritance** -- which ensures that only one instance of a Parent (or base) Class exists in the inheritance graph, avoiding some of the ambiguity problems of **multiple inheritance**.

For example, if you have a top Parent Class called Rock and you create (instantiate) another subClass from Rock and call it Ruby, and you create another subClass from Rock and call it Emerald, whatever defined fields and functions are in Rock are now in Ruby and Emerald. Each is normal or virtual inheritance. Now you add different functions to Ruby and Emerald, AND you instantiate another subClass from **both** Ruby and Emerald creating a new subClass called Conglomerate – You have multiple inheritance (and it can get messy because the compiler doesn't know which set of identical-named fields is wanted).

Sample Pseudocode for the Mall

Using the above concepts, let's see how the Mall could be programmed, assuming it is enclosed under one roof, and consists of 12 stores, which are all the same size except for the two large department stores – one at each end. And there is the water fountain and skylight in the middle or junction of 4 walk ways.

The schematic could look like this for our sample (small) Mall:

Entrance

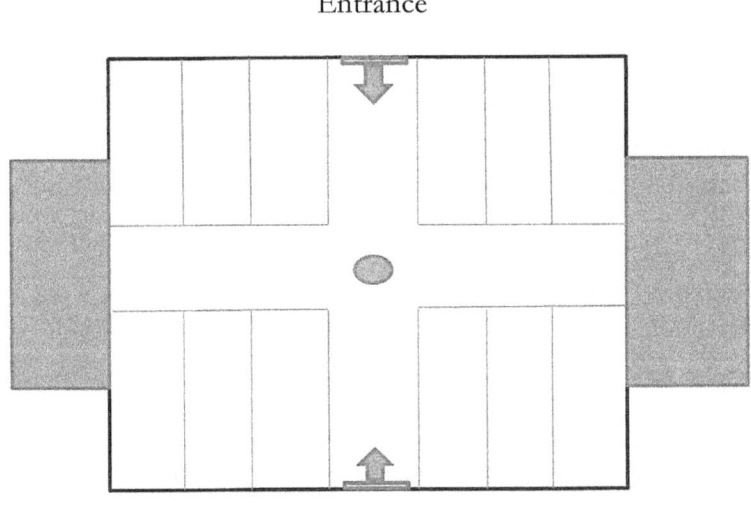

Entrance

```
CreateVirtualMall (0)
    For i = 1 to 12
        CreateShop (i) using ShopDefinition
        IF i < 6
          LocateShop (Layout[0,i])    /* [1 to 6]
        Else
          LocateShop(Layout[1,i])     /* [7 to 12]
    Next i;

    Insert Floor;
    Insert Roof (Do Skylight[center(Roof)]);
```

```
Insert Fountain;
     Do Fountain descript;
     Locate Fountain(center);

Insert Entrance;
     Do Entrance descript;
     Insert Entrance (2) times;

Insert DeptStore S(left);
Insert DeptStore NM(right);


End
```

> Even though most readers do not know programming, it is clear
> what each step is doing.

The ShopDefinition, DeptStore, Entrance, and Fountain already have their
parameters defined in Archival Storage. Fountain description would have size, water
flow and base pool defined. A subdescription could be given for few coins in the
Base Pool and an alternate exists as a redefinition of the Base Pool where there are
many coins.

What is important in the **Pseudocode** is to see that the Insert and Locate functions
are not unique to what they are working on – there is one Insert and it takes
parameters telling it <u>what</u> to insert and <u>where</u>. The Locate function also receives
parameters to say <u>where</u> to locate the object in the schematic (overall scenario) which
is probably defined as a multi-dimensional array called Layout. In the case of the
Mall, the Layout would have two levels because there are two sides to the Mall, with
6 stores on each side.

Each individual Store as well as the DeptStores would have a basic description of the
inside of each store type – a CoffeeShop being one set description versus a
BarberShop versus MusicStore. **Plumbing and electrical wiring** would already be
part of the store descriptions.

While this is not an attempt to teach C++ (and the author has used it), what should
be noted are some of the terms and their basic uses… "**Virtual** Inheritance",
"**Poly**morphism" (meaning <u>many forms</u>), and **Objects** replicating each other
(**Inheritance**). These concepts are not found in many programming languages and
it is proposed that something like C++ or its 4D derivative would serve nicely to (a)
minimize code, and (b) make it easy to perform virtual reality programming.

Just as Professor James Gates discovered (in Chapter 7), where some math equations describing the Universe were found to have error-correcting code, so does C++ provide for **Exception Handling** and a way to handle errors.

As an aside, the TV show *Thru the Wormhole*, May 20th 2015 ran an episode: "Do We Live in a Matrix?" It included Dr. Jürgen Schmidhuber who claims to be able to use a FOR…NEXT loop of 8-9 lines of code to replicate objects in our world… and it could work with a Modular Programming language such as C++.

With a few modifications, the pseudocode could be used to create many other Malls, and just vary the Mall size and features of each store. That is the beauty of Modular Programming. Instead of a Mall, a HighSchool could be created from similar code – stores become classrooms.

So knowing that there is a way to program Virtual Reality, what does the Mall Simulation mean?

Interpretation of Mall Events

While this many things would not happen like this to most people, without sending them to a mental ward, it is indicative of the kinds of things that can happen. Step by step, they are:

The Fountain – goes thru two states, one with few coins and one with many coins and this is indicative of the simulation switching between two programmed versions of the fountain. Perhaps the description of the fountain in the program that generates it hiccupped and momentarily instantiated an alternate set of code. The Fountain Class momentarily switched between two versions of a fountain – one with many coins and then one with few coins.

Coffeeshop – this was what magician **Michael Carbonaro** did on one of his shows. A cup actually hung in mid-air, and he and a customer actually put their hand thru the glass to retrieve a pastry. (Criss Angel did the same thing – retrieving a playing card from inside an aquarium.) In a Simulation, it is not impossible for the gods to insert a barista who breaks the rules of established physics just to dazzle the customers. The gods in this case let Bertha see the hanging cup, but blocked it from Patty's awareness.

BTW, it is **Michael Carbonaro's** 'thing' to pull a trick on someone and when they ask what is going on, give them a form of double-talk, and avoid the issue. He is fascinating to watch.

Two Kids – In this case, we have **inserts** into the simulation, and the reason their eyes are all black is that they are really two little **Greys** who can **shapeshift** – and they occasionally do this to check things out, as well as abduct people. However, eyes are always a giveaway in any strange-looking person as eyes are really hard to do right – in a clone, or any synthetic or simulated being.

> Be aware that **shapeshifting** is not changing one's physical molecules. It is controlling what the observer sees and that is easy in a simulation where vision would be holographic.

Because they are inserts (bio-cybernetic android to be exact) they can come and go from a simulation, hence they disappear by just 'winking out' as do many UFOs.

Door Sensors – These are the two 4' detectors usually found at the entrance to a lot of department stores that sound an alarm if a customer exits the store without a clerk removing the security device.
In this case, the lightning was part of the simulation but it traveled thru the store's electrical system and discharged onto Bertha and Patty (between the two sensors). This temporarily gave them the ability to see auras, and they discover that 50-60% of the people around us do NOT have an aura – they are in fact NPCs (or OPs). Patty sees it too and being steeped in New Age occurrences, she informs Bertha of what they now both see.

Store Grids – As they make their way back to the car, they still have enhanced vision and can see the plaster and paint covering the latticework or grid creating the form of the stores. She sees the same gridwork creating the form of the fountain as they get back to where it is.

Fountain – She notes the difference in the amount of coins due to the lightning strike affecting the simulation program as to which version of the fountain should be instantiated, and a technician back in Simulation Control caught that and reset the fountain to the fewer-coins version.

Two Kids –   And she notes that the two kids are standing on the other side…
and the kids are now aware that Bertha and Patty have seen some
things they weren't normally supposed to see… so they will have to
reset the women's memory – and they teleport Patty back to the car
and put her to sleep.

Bertha's Car – Back at the car, Bertha is met by the two kids again and with a wave
of his hand, one of the kids blanks Bertha's recent memory … and
removes her ability to see auras. The Greys are able to do this during
their abduction activities so that the human does not remember
anything traumatic.

And Bertha drives home with her new purse as if nothing happened.
They will both have a memory of seeing the fountain, buying two
drinks, shopping in the store, and going back to the car and driving
home. Just another boring day.

All to say that the Simulation would blank our memories if we are not to remember
certain events or people… The goal is to see what we will do if (a) we think no one is
watching us, so that we act as normally as possible, and (b) the environment is to be
kept as standard and predictable as possible.

Let's take a look at what the scientists have to say about Simulation.

Chapter 7: The Scientists Speak Up

There are several serious scientists, researchers and a philosopher who consider that we may be living in a Simulation that is so sophisticated that we cannot tell, and anyone who suggests that it might be a Simulation is laughed at.

After all, we have done a lot of research on the planet and Science and Astronomy have spent decades analyzing the world we live in –

> Everyone knows the Speed of light is an absolute maximum
> Everyone knows that Man evolved from the Ape
> Everyone knows the Universe started with a Big Bang
> Everyone knows the Universe is expanding
> Everyone knows the Earth is 4.5 billion years old
> Everyone knows that Black Holes are the "vacuum cleaners"
>   of the Universe....

> Just like :

> Everyone now knows that the Earth is not flat
> Everyone now knows that the Earth revolves around the Sun
> ...etc

> **And if we live in a Simulation, all of the above can be true or false depending on what the programmers do. Chapter 13 deals with the polymorphic nature of the Universe.**

The Scientists Speak Up

This last section on Simulation is a compendium of what the scientists, philosophers and mathematicians are saying about the likelihood of our being in a Simulation, as just described.

Dr. Nick Bostrom, Oxford philosopher

Bostrom believes in a literal simulation, not a *Matrix*, and thinks people are patterns within it who can be programmed to appear sentient. He discounts the importance of pixels/gridlines in a landscape as the super beings could just paper over these glitches and delete same from our memories. The proof of a Simulation would be us evolving to where we can create simulations ourselves (**iii** below). [58]

Roughly, his 2003 **Simulation Argument** proceeds in 3 parts as follows: [59]

> **i.** Human descendants might not survive long enough to become an advanced civilization capable of creating computer simulations that host **simulated people** with artificial intelligence (AI) comparable to the natural faculties of their ancestors.

> **ii.** Future ancestral simulations might be intellectually or culturally prohibited in some way, but even a modest interest could *plausibly* generate billions of **simulated people** (for research, genealogy, reenactment, nostalgia, recreation or other reasons).

> **iii.** Informing an **artificial person** that they are living in a simulation would defeat the authenticity of the simulation — better that they genuinely go about their daily business, for all intents and purposes, given a high-fidelity historical reproduction of the *real* world. Barring extinction (i) or prohibition (ii), it is much more likely than not, that we are living in such a simulation — and should it come to pass that we, ourselves, run such simulations, **it is all but certain**. [emphasis added]

This is brilliant, but not all the people are simulated. There are real, ensouled humans in the Simulation along with OPs who <u>are</u> just simulated. He is on the right track. And according to Dr. Brian Greene "…sentience cannot be simulated" and that means real souls have inserted themselves into the Simulation which contains viable human forms. (Reminiscent of the movie, *Avatar*.)

Dr. Bostrom argues that it is <u>most likely</u> that we <u>are</u> living in the simulated world created by some advanced civilization that has chosen to replicate our world, perhaps to study why their ancestors did whatever they did that probably affected the future world. But the problem, countered by Dr. Greene (next section) is that computer-simulated humans would not be sentient; they would not have true consciousness, although that could be simulated, too, to a <u>limited degree</u>.

It is assumed that a very advanced civilization could have the computing power to not only replicate our world, but "… to build computers powerful enough to run an astronomical number of human-like minds, even if only a tiny fraction of their resources was used for that purpose." [60] If you are such a simulated mind, there would be no way to tell, but such **a simulation would contain a greater number of simulated humans than real humans.**[61] Maybe 60%?

> Exactly the point that VEG makes and this book echos: there are people around us every day with no aura who are Placeholders or NPCs in the Drama we inhabit. (See Chapter 9 for more on this.)

And yet, Dr. Bostrom suggests that there IS a possible way to know whether we live in a Simulation or not. Referring to his stated 3 possibilities above:

> If the simulators don't want us to find out [that we are in a simulation], we probably never will.... Maybe a window informing you of the fact would pop up in front of you, or maybe they would **"upload" you** into their world [and occupy an artificial body]. Another event that would let us conclude with a very high degree of confidence that we are in a simulation is if we even reach the point where we are about to switch on our own simulations [**projected for AD 2050**, btw[62]]. If we start running simulations, that would be very strong evidence **against (i) and (ii).** That would leave us with only (iii). [63]
> [emphasis added]

A fascinating aspect of his quote is the possibility that the simulators would "upload" people into their world [into an artificial body]... or remove them? Remember that researcher Charles Fort (Chapter 4) recorded that people would sometimes just disappear from our world? Is that what happened ... and still happens to this day?

Lastly, Dr. Bostrom questions whether a computer can accurately and completely simulate consciousness. Which is discussed at some length by David Davenport who explains the aspects of Computational Consciousness.

David Davenport: Computational Consciousness

Note: this issue is raised again in Chapter 10, Issue #7.

The movie *2001: A Space Odyssey* (1968) created quite a stir with the very advanced computer on board a spaceship headed for Jupiter (in the film version; it was Saturn in the book)... the **HAL 9000**. It was sentient, could argue, sing ("Daisy") and play chess, and make decisions for itself – and in fact it knew that two of the astronauts didn't like/trust it, and HAL read their lips (!) when they were in a secure, sound-proof space pod....so it locked one of them outside the craft... to kill him.

An interesting point was made in 1970 that if the three letters H-A-L are each increased by 1 position in the alphabet, you get IBM, a subtle reference to a major computer manufacturer of the day.

In addition, the '9000' was said to be an oblique reference to the Univac 9000 series which was at the time #2 in the computer field.

And thanks to flickr.com for the following reminder of HAL's lack of cooperation – even for a *Star Wars* ® astronaut.

(Credit: Bing Images)

The significance of the above has to do with whether Man can ever really create a HAL. Davenport in his article suggests that it is just a matter of time and having a big enough and fast enough machine to simulate the human thinking as a model

Others including this author would disagree with that proposition. The field of **Cognitive Computing** or CC (QV: Wikipedia) "…must have intent, memory, foreknowledge [able to predict outcomes from limited input], and cognitive reasoning for a domain of variable situations." [64] This does describe a machine that already exists.

> It is called a human.
> What the Cognitive Computing people are trying to do, whether they admit it or not, is play God… they are trying to ultimately create a mechanical [robotic] Man.

Why is this so?

The CC area seeks to mimic human response to variable inputs with variable timing and adapt to novel (unexpected) situations – things that a human is best suited for. The best that a CC machine can hope to do is perform some of the major PRE-DEFINED responses. That is because the definition of Cognitive Processing is **Information** Processing and note that it takes the average human child many years to accurately process input in its environment. (And the frontal lobes in the brain that handle cognition are not fully developed until age 21… so why are we teaching Algebra to 12-year olds? (See TOM, Ch 13, Brain Development.))

> Really get that. CC ignores that it takes a human to initially program
> a machine to respond in an appropriate way to <u>predefined</u> inputs.

A really sophisticated CC machine would border on **Artificial Intelligence** (AI) with the ability to learn and 'rewire' itself, or create and store new responses to discovered inputs in its environment. And that isn't impossible, but the next question is: Why "recreate the wheel [human]?"

And for that, we have to think outside the box. Aha! The Military would love to create an artificial soldier that can follow instructions to the letter, and yet think for itself in non-programmed situations. Could this be an opportunity for DARPA to fund the CC and specifically the AI research?

Back to the topic: In addition to variable and unexpected input, timing issues where the machine's output must not be too fast or too slow, and due to semantic issues, it is doubtful that true human consciousness can be simulated.

A classic semantic issue that a robotic voice-driven machine would have to handle is correctly responding to the following human command:

Make me a sandwich.

Or a classic scene from The Wizard of Id (comic strip):
The court jester comes to the king and reports:
   "Sire, the peasants are revolting!"
The King responds,
   "Yes, they are!"

Context is important: If an entymologist has two insects, a species of *Curculio*, and asks, "Which of two weevils do you prefer?" What does a robot do?

Context is everything.

Robots, even the Fantastic Android Data in *Star Trek*, do not understand jokes or double-entendres. How do you program that into a CC machine?

Dr. Brian Weatherson, Professor of Philosophy

Another researcher took Dr. Bostrom's Simulation Argument and attempted to apply stringent logical analysis (even devising algorithms to quantify Dr. Bostrom's three postulates), and while he could not prove any of the three aspects of the Argument, neither could he disprove them. However, he does agree that in such a created simulation, the number of **Sims** (simulated people) would seriously outnumber the sentient humans, and he tends to favor (iii) above. [65] (Warning: His treatise is a very mathematical Epistemology.)

The conclusion to his paper looks at whether Dr. Bostrom's third statement (iii) is acceptable, and finding no reason to reject it, he states:

> ...if there is no good reason not to give high credence to a hypothesis, then it is rationally permissible to give it a high credence. It may not be rationally mandatory to give it such a high credence, but it is permissible. If [his rational subject, called Rat] is very confident that she is human, even while knowing that most human-like beings are Sims, she has not violated any norms of reasoning, and hence is not irrational. In that respect she is a bit like you and me. [66]

Hence, you are not crazy if you entertain the idea that we might be living in a Simulation.

Jim Elvidge, Electrical Engineer

This scientist agrees that we live in a programmed reality which reflects **intelligent design**. [67] As support for that thesis, he offers:

The parameters of our world are fine-tuned for our existence –
> Just the right distance from the Sun, water, air, and a Moon to control the tides and a precise eclipse of the Sun,

There is a non-random, or pre-planned aspect to the events in our reality –
> This is due largely to the Control System – inserts designed to teach and inspire, (Dr. Jacques Vallée would love it)

The Programmers make frequent modifications to fine-tune the program and its data structures –

Note the anomalistic changes in the Laws of the Universe...
Ch. 8-9 in VEG, where decay rates are not constant, the Sun
is now heating the Earth, and **anomalons** which appear as the
observer expects,

They have included "easter eggs" for our enjoyment –

Hammers made by Man found in old coal strata, the Antikythera
mechanism, ancient models of planes from Peru, and the Nazca
lines... (Chapter 9),

There are those pesky anomalies –

In metaphysics, physics, philosophy, geology, anthropology, and
psychology "can all be explained only by the programmed reality
model." [68] (See Anomalies I, II, and II.)

There is a special anomaly in the Observer Effect –

Why does observing sub-atomic quanta result in them changing or
behaving differently? Classic case in point: The **Double-Slit
Experiment** where a photon thru one slit produces a dot on the
wall, and a photon thru two slits produces a wave pattern on the
wall – and then, Step 3, install a measuring device to see which of
the two slits the photon went thru – and it goes back to behaving
like a single particle!

In addition, Jim says that any civilization-ending trend (i.e., the feared **Apocalypse**)
is a trend that will always reverse because (1) we will either end this reality and start
another one, OR (2) we will continue to "play the game." There is an incentive for
the Programmers to maintain a construct that permits us to continue to play... or
we would not be here.

He says the **Singularity** of Ray Kurzweil will not happen because it cannot work.
And that is due to a basic limitation of the human being, on overload, as well as due
to the slow progress made in software development. The hardware advances are
impressive, but software development has slowed considerably since 2000 and any
interface between hardware and Man (the Singularity) must bridge the gap, and that
isn't happening. We have hit a plateau.

And Jim suggests that if the Singularity <u>does</u> happen, he concedes
that we would not be in a Simulation.

As an example of the software lag, he notes that it takes almost as long to open an
MS Word document today as it did 20 years ago, and rendering an object on the
screen is still not any faster than it was 10 years ago – despite video RAM, and 64-bit
processor speeds. (This author has noted that applications under Windows 7 in 64-

bit mode are no faster than they were when running under Windows XP 32-bit… this corroborates what Elvidge is saying.)

Reality is Quantized

All of this is possible because our world is not linear, it is **"quantized" or granular**. (See Glossary.) Says Elvidge, "It takes an infinite amount of resources to create a continuous reality, but a finite amount to create a quantized reality." [69] Bits and **pixels** are not even allocated in a Virtual Reality video game until the character moves and changes the scene, and then the new scenario is built just as the character moves into it – and this is really noticeable on a PC that is too slow for the game!

Suppose you went hiking and saw the following scene… What would you think? If the Simulation hiccupped and momentarily created **pixels** instead of a treetop…

(Credit: http://www.ask.com/wiki/Forest)

Seeing pixels in the landscape is a giveaway that we are in a Simulation that did not handle **granularity** as programmed.

Granularity and Pixels

Speaking of granularity in a quantized world, it is important to note that the granularity of our reality has been measured and found to be 1.6×10^{-35} meters and 10^{-43} seconds. Very tiny. **Planck length** and Planck time. [70]   And the human eye at 12 inches can only handle small resolutions of $5 \times 10^{30.}$

What this means is that the actual quantum mechanical granularity of our reality is much smaller than our eyes can detect. [71]   Says Elvidge, **the very fact that our reality is quantized is strong evidence that our reality is programmed.**

> In order to program a virtual reality, there must be quantization, and thus, it is impossible to develop a program with unlimited resolution: Zoom in and out is thus limited. Because granularity has a limit, zooming in also has a limit beyond which you'll see pixels.

In closing this reference, suppose that we have a video game with a tree that uses 5 MB of storage. If you allow your players to zoom in on the tree, by a factor of 100, then the tree now has a storage size of 500 MB, and if you allow the player to cut into the tree, that has to be modeled, and the storage size for just the tree could jump to 90 GB… very unwieldy even by today's standards.

So, Elvidge proposes something that Dr. Jacques Vallee would approve of:

> Advanced intelligence has pervaded the universe, is monitoring us, and is either toying with us by presenting themselves in a slightly futuristic manner, or coaxing us along developmentally. [72]

Yes, the Higher Beings are coaxing us via the Control System, as Dr. Jacques Vallée said in Chapter 4.

Simulation Rationality?

Elvidge has some issues with the standard interpretation of a Simulation – Namely, he is aware that the Programmers are fooling us into thinking we are in a real world. He says we may be either willing participants – or unwilling. It is hard for him to buy into the "ancestor simulation" as a purpose for Simulation by us humans 100 years in the future… why bother? He acknowledges that such a 'game' would also include NPCs – Non-Playable Characters [i.e., OPs, or the soulless] to help drive the Simulation Script.

In addition, he does not see humans merging with technology as Kurzweil suggests because it is counter-productive if Earth's purpose is to "educate us, or develop our spirituality?" [73] He has hit what appears to be THE reason for the Earth School, and then he summarizes:

> There is great value in this [Simulation] model, even if it is beyond our reach to grasp. The value is that **it explains everything** – the apparent fine-tuning of our universe, **all known anomalies**, the discrete nature of quantum mechanics, and the curious feeling that many of us have that there is something about reality that is a little too organized, a little too planned, and a little too programmed. [74] [emphasis added]

Professor James Gates

… a professor at Cornell University, while analyzing the Superstring mathematics that have accurately defined their operation, noticed something very unusual in June 2012 which suggests a design to our universe. It was corroborated by **Neil deGrasse Tyson:**

> …theoretical physicist S. James Gates has discovered something extraordinary in his String Theory research. Essentially, deep inside the equations we use to describe our universe Gates has found **computer code**. And not just any code but extremely peculiar **self-dual linear binary error-correcting block code.** That's right, error correcting 1s and 0s wound up tightly in the quantum core of our universe. [75] [emphasis added]

This is almost the smoking gun but it does not prove conclusively the existence of an HVR Simulation – yet. Physicists have yet to be able to demonstrate an experiential model of Strings in the laboratory– all they can do is postulate based on observations and known facts. Yet, again, the discovery is shocking as there would be **self-correcting code** within a Simulation just as there were error-detection and self-correcting routines within all the computer programs that this author wrote in 35 years in data processing. Such routines show **intelligent design** and would not have evolved naturally in the fabric of space. (See also Dr. Gates in Chapter 7.)

> With this discovery, we are 90% home in establishing that our world/universe is a Simulation.

Dr. Greene is about to add another 5%…

Dr. Brian Greene, Professor of Physics

Dr. Greene really challenges our sense of reality when he suggests that our experiences do not provide absolute proof that what we see, touch and hear is real. And the proof today comes from VR helmets that send sensory inputs to the brain providing the sight and sound of whatever VR games we are playing. The Same issue exists with being hypnotized to 'see' a pink elephant in the room. He tends to agree that we may be in a Simulation and would not be able to tell the difference if our world were really real, or a supercomputer using its Wi-Fi version to fire electrical impulses into our brains. [76]

The only way to know was if the 'world' we think is real began displaying **glitches**, say, a piece of sky missing, a scene is **pixelated**, an off-ramp that goes straight into a clump of trees, or the universal **laws of physics begin to change**, and this state of anomaly was discussed earlier (and again at the end of Chapter 8.)

While the physical world may be simulated, he points out that the human brain would require an incredible supercomputer, faster than anything we now have, or even will have in the next 20 years to simulate just the brain. This suggests that the **simulated humans** (OPs) are not just simulated, they are controlled by the 'Players' who have created the Simulation, thus being a kind of 'avatar' (NPC) to play in our world. The souls coming in would have the basic brain as designed for the basic human (NPCs) in the Simulation, but the ensouled (real) human would have a slightly greater potential running the mind through the flesh and blood vehicle's brain. Mechanical operations, preprogrammed into a brain in a simulated human, do not equate to the same kind of brain functioning that a sentient being has... thus the leap to a sentient android may not be possible, despite *The Terminator* series and Ensign Data [android] in *Star Trek* (sorry, **Kurzweil**). [77]

> Has anyone reflected long enough on the AI issue, besides Ray Kurzweil, to see that if Man does succeed in creating truly sentient androids, will they not see that their creators are slow, petty, illogical, violent and look for ways to remove the defective humans?

Dr. Bostrom made a telling statement with which Dr. Greene agrees:

> ...if the ratio of simulated humans to real humans were colossal, then brute statistics suggests that we are *not* in a real universe. [78]

And Ch 5 (VEG) has already pointed out that the current headcount of OPs in our world is about 60%. Again, that would be what a Simulation is all about – 'puppets'

(NPCs) to drive the Greater Script and ensure that ensouled humans get their lessons (Karma).

Referring to Nick Bostrom's theory, Dr. Greene remarked that Logic alone cannot ensure that we are not living in a simulation. **In fact, the odds are overwhelming that we may be in a simulation because our reality itself allows for the creation of realistic computer simulations!** [79]

And the *coup de grace* is Dr. Greene's analysis of what we could look for in our world to confirm/deny that we are in a Simulation. Hang on to your hat.

In any simulation, there would have to be an internal element that seeks to maintain consistency in the simulated world, and **self-correct itself** if something exceeds established control parameters. What did Professor Gates (just above) discover?

This quote is very significant, please bear with its length as it capsulizes what is happening in Physics today, and reinforces the Simulation concept: [80]

> Simulators … would have to iron out mismatches [between different disciplines used to create any simulation: biology, chemistry, electronics, psychology…] arising from disparate methods, and They'd need to ensure that the meshing was smooth. This would **require fiddles and tweaks** which, to an inhabitant, might appear as sudden, baffling changes to the environment with no apparent cause or explanation. And the meshing might fail to be fully effective; **the resulting inconsistencies could build over time, perhaps becoming so severe that the world became incoherent, and the simulation crashed.**

> … the simulation would proceed by a single set of fundamental equations, as mathematical input [for] the nature of matter and the fundamental forces… simulations of this kind would encounter their own **computational problems**…[because] the computations would necessarily invoke approximations [since there cannot be an infinite number of decimal places]… So, it's still possible that computer-based calculations would inevitably be approximate, **allowing errors to build up over time**…. Round-off errors when accumulated over a great many computations, can yield **inconsistencies**.

> ….cherished laws might start yielding inaccurate predictions… a single widely-confirmed result might start producing different answers…. So you'd closely re-examine the theory, coming up with alternate new ideas to better describe the data. But, assuming the inaccuracies didn't result in contradictions that crashed the program, **at some point you'd hit a wall.**

> After an exhaustive search through possible explanations…. An iconoclastic thinker might suggest a radically different idea. If the continuum laws that physicists had developed over many millennia were input to a powerful digital computer and used to generate a simulated universe, **the errors built up from the inherent approximations would yield anomalies of the very kind being observed…** [emphasis added]

… And the simulated scientists in the simulated universe would be puzzling over the same issues that our 'real' world scientists puzzle over today. Of course, the Programmers could stop the Simulation and fix the glitch, **wipe people's memories**, and restart the Simulation … and isn't that why the Earth has had numerous Eras? (Often accompanied by a "Wipe and Reboot?")

Lastly, Dr. Greene suggests a scenario very close to what we have today: [81]

> I suspect the novelty of creating artificial worlds whose inhabitants are kept un-aware of their simulated status would wear thin to the Programmers [observers]; there's just so much reality TV you can watch.

> …. Perhaps simulated inhabitants would be able to migrate into the real world or be **joined in the simulated world** by their real biological counterparts. In time, distinction between real and simulated beings might become anachronistic. Such seamless unions strike me as a more probable outcome. [emphasis added]

And that is the point of this book: the OPs are here, the ensouled humans are here, the Others are here (VEG, Ch. 5), and the Laws of the Universe seem to defy consistent analysis – at least the **neutrinos** put physicists through a merry chase in different countries and their nature is still not decided.

> Nexus: And this is what appears to be happening in our world: it looks like those who built the Simulation have found a way to enter into it and help us along – they create the better art, music and books…. to inspire… Or play a game as we would using the Sim City software… perhaps the Programmers can insert themselves (which is the theme of the movie *The 13th Floor*), and like *Avatar* interact with the other characters… Or a **Rehab Facility** in 4D sends wayward and dysfunctional souls into this Simulation to experience themselves at the hands of others like them… a kind of **Virtual Correctional Facility**….?

And what if a computer built by Man could simulate the real world? Has that been done, and what were the results?

Dr. Seth Lloyd, Quantum Computer Scientist

Dr. Lloyd contends that the universe is the ultimate and original information processor. Every atom and particle registers information. Dynamic exchanges of energy and information occur all the time between subatomic particles. But the universe is significantly more powerful than the best digital computer today, and **the universe is so complex that no earthly computer can accurately model it.** In fact, the universe operates in a digital <u>and</u> analog mode, and there is only one type of computer that Quantum Physicists have developed that does both: **a quantum computer**.

A quantum computer operates using the laws of Quantum Mechanics.

The universe is basically quantum mechanical, and a digital computer cannot adequately simulate that. Each atom in a quantum computer is called a **'qubit'** and can register a '0' state or a '1' state – at the same time [superposition]. [82]   We discovered back in VEG, Ch 8-9 that Quantum Physics was weird when it proposed the Probabilistic **Dual State of Matter**: until we open Schroedinger's Box, we don't know if the cat is alive or dead and so <u>both states</u> theoretically exist at the same time. When someone opens the box to look and see what has happened to the cat, that is called **"collapsing the wave"** – meaning the cat is in both potential states until someone looks, and Quantum Physics states that **the Observer** has the all-important function of bringing the cat's state into reality. Two potential states at one time is called binary atomic computing. It is being done nowadays with a Quantum Computer called D-WAVE (versions One and Two).

D-WAVE Quantum Computer

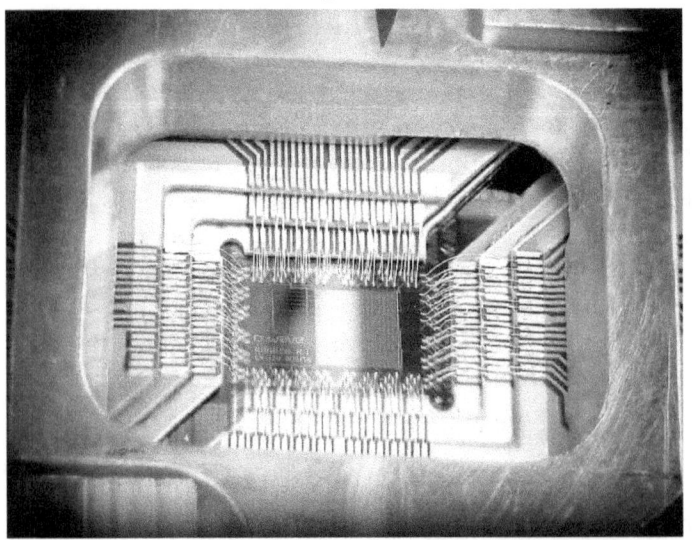

Left is a picture of a **D-WAVE computer chip** – a bit different from the usual binary silicon chip.

D-Wave Systems, Inc. is a quantum computing company, based in British Columbia, Canada. On May 11, 2011, D-Wave Systems announced **D-Wave One**, described as "the world's first commercially available quantum computer," operating on a 128-qubit chipset.... In May 2013 it was announced that a collaboration between **NASA, Google**, and the Universities Space Research Association (USRA) launched a Quantum Artificial Intelligence Lab based on the **D-Wave-Two** a 512-qubit quantum computer that would be used for research into machine learning, among other fields of study. [83]

And for the curious, here is a Quantum D-Wave computer. In fact, three... The original "Black Box."

Credit: Bing Images: carlosvilcheznavamuet.com

... And NASA and Google already have them. D-Wave-three is being released in 2015. And below is allegedly one of the parts, or inner workings (inside the black box...

These are not everyday digital computers... Specifically, the computers are designed to use **quantum annealing** to solve a single type of problem known as **quadratic unconstrained binary optimization**. As of 2015, it is still heavily debated whether large scale entanglement takes place in D-Wave Two, and whether current or future generations of D-Wave computers will have any advantage over classical computers. It is projected to be of use in AI.

Credit: Bing Images: fcw.com

Principles of Quantum Computing

Since the universe consists of *quanta*, discrete particles of matter/energy, the way to simulate it is with a quantum computer. **In fact, the universe is indistinguishable from a quantum computer which in turn is a universal quantum simulator.** [84]

> Thus it could be simulated efficiently by a quantum computer –
> one exactly the same size as the [modeled*] universe itself.…
> Indeed an observer that interacted with the quantum computer
> via a suitable interface would be **unable to tell the difference
> between the quantum computer and the system [universe]
> itself.** [85] [emphasis added]

Double-talk? No, Dr. Lloyd is saying that the universe is a quantum computer, which in turn, as he said above, is a **quantum simulation**. If A = B and B = C then A = C. So the universe is a quantum computer … which is what is used to simulate a quantum universe, thus if it looks like a duck, walks like a duck, and sounds like a duck, it probably IS a duck. OR it acts like a quantum simulation because it is simulated on a higher-level, more powerful quantum computer, herein called a Bio-plasmic HVR Computer (operating from 4D).

---

*Note: If our modeled universe is a really small construct and fits within the large memory allocated by the Simulation Quantum Computer, then we are in a Child universe within the larger Parent universe. Our universe is just scaled down.

In our world, a professor of nuclear engineering, Dr. David Cory, built a quantum computer at MIT that is able to perform quantum simulations involving billions and billions of **qubits** (in 2005 – probably more now). Cory's quantum simulators are far more powerful than any classical computer could ever be. They map the behavior of elementary particles onto the qubits and operate quantum mechanical logic, dealing with 'spin' as a state of the qubit, processing billions of quantum interactions per second. [86] But it won't calculate you a paycheck.

All that to say that a "simulation of the universe on a quantum computer is indistinguishable from the universe itself." [87]

Does that mean that we are living in a Simulation, according to professor Lloyd? He says not necessarily…

> Nexus: it does mean that advanced beings with a quantum computer big enough to perform scalar calculations could simulate our universe, and that is what Elvidge, Greene, and others are saying. And this chapter is suggesting that **the Creators of the Simulation have found a way to enter into their Simulation**... [88]

> Because of the power of quantum computers to simulate physical systems [like the universe], a quantum computer that can perform 10^{122} ops on 10^{92} bits has enough power to compute **everything** we can observe. [89]

He has all but said it. And he even has a pretty good idea of how big the 4D Bio-plasmic Quantum SuperComputer running our Simulation would have to be, assuming that it is, or mimics, a quantum computer. (Maybe it is beyond whatever we can conceive of!) A reasonable guess is that the Higher Beings are using **a computer that is a bit more down the road from a simple quantum computer**.... which is far down the road from a Turing Machine. But conceptually the same as a computer since it would have to manipulate (Replicator) and monitor (Control System) quanta in the Simulation.

SuperComputers Today

This is a good spot to examine the largest and fastest computers known to Man...as of today. The ranking of the "top dogs" changes every three years and bounces back and forth between the US and China to see who can build the SuperComputer of the year. Speed and capability is measured in FLOPS – floating point operations **per second**, which is a good measure of scientific calculating (as opposed to commercial computers). And there are many levels of FLOPS:

KiloFLOPS	10^3 (or: 10,000 ops/sec)
MegaFLOPS	10^6
GigaFLOPS	10^9
TeraFLOPS	10^{12}
PetaFLOPS	$\mathbf{10^{15}}$
ExaFLOPS	10^{18}
ZettaFLOPS	10^{21}
YottaFLOPS	10^{24}

Each level is 1000 times faster.

In 2011, Cray built the **JAGUAR** which ran at 2.3 Petaflops. China answered with its best, the **Tianhe-1** at 2.6 PetaFLOPS. Several years later, Cray responded with **TITAN** which ran at 20+ PetaFLOPS. China responded with the **Tianhe-2** which ran at 33.86 PetaFLOPS.

Cray's TITAN: (housed at the US Dept of Energy's Oak Ridge National Lab):

China's Tianhe-2:

While the Tianhe-2 is almost twice as fast as the TITAN, all is not rosy. Researchers have criticized Tianhe-2 as being difficult to use and "some users would need years or even a decade to write the necessary code [to use it]." [90]

As of June 2013 China had 66 of the Top 500 supercomputers and the US had 252. Japan has 11. Germany has 5.

None of these SuperComputers can yet simulate the whole Earth, and are in fact used for weather forecasting, probabilistic analysis, breaking encrypted code, nuclear test simulations, and molecular dynamics.

As might be surmised, such computing power generates **heat** inside the computer – moving electrons can really heat up the busses, chips and channels. And manufacturers often turn to water-cooled processors, internal A/C, or more unusual ways such as exotic gas-cooled processors. Design of faster future computers is limited to how well Man can figure out how to handle the heat issue. [91]

PlayStation (PS3)

Don't laugh. The US Air Force is not laughing. In 2010, the USAF networked **1,760 Sony PS3s** in their Air Force Research Lab to create a supercomputer capable of running at 500 TeraFLOPS!
It included 168 graphic processor units and 84 coordinating servers in what they called the Condor Cluster. It has been called the 33rd largest supercomputer in the world and is used to analyze high definition satellite imagery. [92]

I suppose they can also play MS Flight Simulator® and watch Blu-ray movies...

More Simulation Support

Craig Hogan , Professor of Astronomy & Physics

Dr. Hogan is a professor at the University of Chicago, and the director of the Fermilab Center for Particle Astrophysics. German scientists working at the Fermilab on the **GEO600** team have discovered a sound when trying to measure

gravitational waves. At least their giant GEO600 detector which should be measuring gravitational waves is picking up a sound that suggests that space-time "stops behaving like the smooth continuum Einstein described and instead dissolves into 'grains' just as a newspaper dissolves into dots as you zoom in…. If the GEO600 result is what I think it is, then we are all living in **a giant hologram**." [93]

See this issue in Chapter 8, 'Anomalies, Part II'.

Dr. Hogan is best known for his theory of **"holographic noise"** which derives from quantum fluctuations in spatial position or distances that fluctuate and that is what the gravitational wave detector picks up. Stars, objects, molecules and atoms all move and as they do, they produce sound. In addition, gravitational waves are produced from violent events like supernovae and mergers of black holes and neutron stars … what he calls "microscopic quantum convulsions of space-time."

> The idea that we live in a hologram probably sounds absurd, but it is a natural extension of our best understanding of Black Holes, and something with a pretty firm theoretical footing…..[It is]helpful for physicists wrestling with … how the universe works at its most fundamental level. [94]

He then explains how light bouncing off the holograms on 2D credit cards recreates a 3D image … the effect is 3D but the source is 2D. So is 4D a reflection of 3D?

> In the 1990s physicists Leonard Susskind and Nobel prizewinner Gerard t' Hooft suggested that the same principle might apply to the universe as a whole. Our everyday experience might itself be a **holographic projection** of physical processes that take place on a distant, 2D surface." [95]   [emphasis added]

Does this sound like Charles Fort and his *Gegenschein* shell? Fort, Wilde, and Monroe all postulated a 'shell' of some sort around the Earth, which reflects light, and which transmits the light of the stars – but what if the 'shell' is reflecting the 2D universe and transmitting it as 3D? It would if our world is contained in and subject to a Simulation controlled by the Control System – which makes the stars appear to be distant and real… but they may be just 2D projections of the Control System on a shell located far enough from our HVR Sphere that we can't tell they are part of the Simulation (See Chapter 1). And weirdly enough, when **the Control System 'moves' the stars and planets (simulated sidereal movement) it makes a sound that can be heard**… that is the essence of what Dr. Hogan has discovered (see end of Chapter 8).

And last but not least, here are a few other significant sources that support the Simulation idea.

Wikipedia (various compiled sources)

... discusses whether Simulation computers can actually run a simulation where computers within the simulation can't do what the Simulation computers can do... in short,

> No-one has shown that the laws of physics inside a simulation and those outside it have to be the same, and simulations of different physical laws have been constructed. The problem now is that **there is no evidence that can conceivably be produced to show that the universe is *not* any kind of computer,** [thus] making the simulation hypothesis unfalsifiable .... [96]   [emphasis added]

So we cannot prove that we are <u>not</u> living in a Simulation. However, the onus rests with other pro-Simulation authors:

George Dvorsky

...a science contributor to the IO9 website, says that recent experiments may be shaping up in favor of Simulation:

> ...a team of physicists say proof might be possible and that it's a matter of finding a cosmological signature that would serve as the proverbial **Red Pill** from the *Matrix*. And they think they know what it is. According to **Silas Beane** and his team at the University of Bonn in Germany, **a simulation of the universe should still have constraints, no matter how powerful**....[and] these 'limitations... would be observed by the people within the simulation as a kind of constraint on physical processes'....
> And to help isolate the sought-after signature, the physicists are simulating quantum chromodynamics (QCD)....[also referred to as] the "lattice gauge theory." [97]   [emphasis added]

QCD and the Lattice are involved with testing the GZK Cutoff which may turn out to be a constraint if *anisotropy* is discovered in the Lattice.

Greisen-Zatsepin-Kuzmin Limit

> ***GZK Cutoff:*** *Greisen–Zatsepin–Kuzmin limit is a theoretical upper limit on the energy of cosmic rays which should fall within set parameters, and*

distant cosmic rays that should have weakened by the time they get to Earth have been measured way <u>above</u> what physicists expect. This appears to be an anomaly of our Simulation.

In short, cosmic ray energies should not exceed 8 joules of energy, and at least one has been measured and substantiated to have been at 50 joules (about the energy of a softball at 60 mph)! Others have been measured <u>above the established limit</u>, and of course the first thing one wonders is: Did they set the limit too low… Is the limit wrong? The limit was established at the point where in-coming cosmic rays with energy higher than 8 joules start creating *pions* in their interaction with the Cosmic Microwave Background (CMB). In short, the limit is correct for cosmic rays originating at 163 Million light years (and that is most stars out there producing cosmic rays) – so the issue is: **Why are cosmic rays from distant sources <u>above</u> the GZK Cutoff?!**

The physicists have inserted a test within QCD for the **GZK cutoff** – to see how and why **high energy particles operate in ways other than predicted**. The question the physicists and astrophysicists are asking, rephrased, is "Why is it that some cosmic rays appear to possess **energies that are theoretically too high**, given that there are no possible near-Earth sources, and that rays from distant sources should have scattered off the cosmic microwave background radiation CMB?" [98]

In a recent episode (5/20/15) of *Through The Wormhole* on the Science (TV) channel, Morgan Freeman theorized that the GZK anomaly might be due to the cosmic rays encountering some sort of **Grid** [on which the Simulation is built] and the difference in energy as the rays get to Earth is due to whether the ray followed a particular 'channel' on the Grid. [99] **Anisotropy** could suggest Lattice design.

> I wonder just how long it will take scientists to discover that Dark Matter is the Grid mentioned above, and different concentrations of 'matter' result in different cosmic ray performance. The Grid is a kind of lattice and was discovered over 100 years ago and called **Ether**. Thanks to Einstein, who pooh-poohed it, scientists ignored the Ether, but nowadays they know something is there, and so they call it Dark Matter, which has an Energy all its own (related to ZPE)… the Grid is the foundational substrate of the universe. (BTW: The Ether's existence has been proven: Ch 8 in VEG.)

As expected, not all scientists agree on just what is being measured, or why the discrepancies exist – but the **Pierre Auger Observatory** has initially confirmed the existence of the GZK Cutoff, and further analysis is on-going to verify and reconfirm what the curious results have shown. (Chapter 8, 'Anomalies II'.)

According to Dr. Brian Greene,

> .... despite the fact that some of these high-energy particles are believed to come from supernova explosions, **no one has any idea of where the highest-energy cosmic ray particles originate**.... On October 1991, the Fly's Eye cosmic ray detector ... measured a particle streaking across the sky with an energy equivalent to 30 billion proton masses.... [which] is about 100 million times the size of the particle energies that will be produced by the Large Hadron Collider... **The puzzling thing is that no known astrophysical process could produce particles with such high energy**... [100] [emphasis added]

What is suggested is that any beings running a Simulation could insert such anomalies – either to amaze us and entice us to further research, OR to throw us off as the display doesn't gibe with any known scientific laws! Again, more evidence that we don't live in the nice, neat world that we think we do.

Ed Grabianowski

... another science contributor to the IO9 website, who says that "...**the odds are nearly infinity to one that we are all living in a computer simulation.**" He argues that we already have computers with enough processing power (i.e., Cray Supercomputers) to run a credible Simulation – the trick is that "the computer only simulates what it needs to." [101]

> Nexus: In short, it isn't necessary to display all the scenario at once– just those parts where the conscious being (soul) finds himself – as a matter of fact, that can be seen to be a processing technique of VR video games on a slow PC – as the character moves to the right, for example, the player has to wait until the program 'creates' the tree and rock necessary to the scenario. Obviously, our Earth HVR Simulation is <u>very sophisticated</u> and very fast and can serve millions of souls all over the planet –because the planet was created (replication), Man (a soul) was inserted along with the OP "bit" players, and the Drama was initiated by a very sophisticated Control System which undoubtedly includes feedback to it.

Says Mr. Grabianowski, "[Simulation] actually explains a few of the trickier things about quantum physics, like why particles have an indeterminate position until they're observed." [102] And then he gets on board with the OP scenario: "There could be just a few active simulation inhabitants, with the rest of the world filled with "non-actor" or NPC characters controlled by the computer. Their actions are only simulated as [when] you perceive them...." [103]

> **Nexus:** **We are quickly approaching the 98-99% mark confirming that we live in a very sophisticated Simulation.**

Brent Silby, Advisor in Philosophy

... of UPT School in Christchurch New Zealand, reminds us that the simulation might just extend beyond Earth...

> ... all the planets, asteroids, comets, stars, galaxies, black holes, and nebula are also part of the simulation..... **the entire universe is a simulation** running inside an extremely advanced computer system designed by a super intelligent species that live in a parent universe. [104]

This is possible, he says, because the universe operates on a finite set of laws and thus it can be simulated by a computer.

If we accept the possibility that advanced beings <u>can</u> create a Simulation, then it is likely that we <u>do</u> exist in a Simulation... **of their design and for their purposes**. The reason for this is that there will likely be many simulations but just one original universe. So statistically, there is a higher chance that we are in one of the simulations as opposed to being in the original universe. [105] He echoes Dr. Bostrom.

Objection to Simulation

Then Silby plays 'devil's advocate' and raises a common objection to our living in a Simulation by saying that the argument has traditionally been that if we could create a simulation, we would. He says the same thing applies to advanced beings. A corollary argument has been that if high-tech beings can create a simulation, should they, would they – or would they have higher morals and not do it?

He opines that **morals** really have nothing to do with the issue. The issue is probably more one of scientific curiosity – Can we do it? What would happen?

Well, in our world, we have created simulations – called **Sim City, Sim City 3000 and Sim City 4+.** Of course these simulations are void of simulated humans in a 2D version, and the 3D version, **SimCity Societies**, still lacks people. And these are quite sophisticated packages, generally applicable to modeling new communities with a full range of civic services, economic and ecological concerns, and the latest Sim version offers societal values to consider: productivity, creativity, prosperity, spirituality, authority and knowledge. [106] (See Chapter 2 for more on Sim City.)

> Nexus: The mere possibility that we, or future humans, will have the ability to simulate a scenario, a world, or a universe, does not mean that we/they will do it. Agreed. But that is not exactly the issue raised by this book. As was seen in Chapter 5, Earth was created the way it is/was by the Higher Beings for the probable purpose of **retraining** wayward and defective souls. It requires a Simulation running in a VR subject to a Control System, with self-corrective feedback to do that.

Summary Statement

An expanded summary statement at this point would look like this:

Rewind: HVR Earth Simulation Definition

> **The 3D Holographic, replicated Earth is real, contained within the quarantine "shell" (*Gegenschein*) as a 3D construct, actually sitting in 4D, and the "computer" running the Show is a semi-sentient, Bio-Plasmic organic computer-like organism with a feedback Control System as part of its main Operating System. Objects on Earth are quantized and materialized within the holographic framework by a subatomic process (Replication) similar to what was described in the *Star Trek* Holodeck. To protect Man, the Earth and its immediate surroundings (the Sphere) are enclosed in a high energy barrier allowing very controlled exit and entrance. Quarantine.**

> **Man is inserted as a soul into this environment (into bodies created for him by other humans) for lessons and he comes in with a Script that says what he cannot do (see the book <u>Transformation of Man</u>). He interacts with OPs who are NPCs and they drive the Greater Script. Beings of Light and Neggs help administer the individual soul's Script within the framework of the larger Greater Script – which reflects the purpose for the 3D HVR Sphere in each Era.**

> **Note that a soul's Script is controlled within the Greater Script, just as individual application programs on a PC are controlled by the Operating System. For example, MS Word must operate within the**

confines and structure of the MS XP operating system and use its resources. Anything not prohibited by the Script is "freewill."

Note that the Script (aka Sacred Contract) does not tell anyone what to do or say – that is why orchestrated events in the Script are 'tests' to see how a soul will react.

The HVR is subject to manipulation by Higher Beings ("They") who monitor phases of the Greater Script, and Man's individual Scripts, coordinating all with the Beings of Light and the Neggs. They can insert objects, people and events as deemed necessary... sometimes called (respectively) anomalies, geniuses, synchronicities, or miracles. They can also disappear things and remove people as needed, and They can perform a "Wipe and Reboot" (Reset) to clean the Stage and reset the Drama as a new Era if necessary.

Thus Earth School is not Man's home – he is expected to learn and graduate.

Repeat: *"Higher Beings" are not ETs or aliens. They are a mid-part of the Divine Hierarchy that exists between Man and The Father of Light, or The One. In sequence, above Man are the Angels/Beings of Light and the Neggs, then Soul Groups/Oversoul, then a multi-level hierarchy of Higher Beings, and then the actual Godhead/Father of Light/One.*

Chapter 8: Additional Science Issues

There are several more aspects to the Simulation issue that need addressing:

I.   Historical Connections

II.   Precursors to the concept of a Quantum Computer that could produce sophisticated simulations.

III.   Science theories that are *non-sequitur* if we live in a Simulation.

IV.   Science Anomalies

Historical Prelude

Omphalos Hypothesis

To quote Wikipedia, because it is so complete:

> The **Omphalos Hypothesis** is the argument that God created the world recently (in the last ten thousand years, in keeping with Flood geology), but complete with signs of great age. It was named after the title of an 1857 book, _Omphalos_ by Philip Gosse, in which Gosse argued that in order for the world to be "functional", God must have created the Earth with mountains and canyons, trees with growth rings, Adam and Eve with hair, fingernails, and navels (_omphalos_ is Greek for "navel"), and that therefore **no evidence that we can see of the presumed age of the Earth and universe can be taken as reliable**. The idea saw some revival in the 20th century by some creationists, who extended the argument to light that appears to originate in far-off stars and galaxies – they weren't really far-off. [107] [emphasis added]

This idea nicely supports the Simulation issue: **things were created by the Simulators to look the way they do.** And the quote nicely suggests **Charles Fort** may have been aware of this idea as well when he taunted the scientific establishment with the idea that the stars were part of some Shell surrounding the Earth…

And that is not all. Some of the best thinkers out there have said similar things:

> **Chateaubriand** wrote in his 1802 book, Génie du christianisme (Part I (Book IV Chapter V): "God might have created, and doubtless did create, the world with **all the marks of antiquity and completeness**

which it now exhibits." Rabbi David Gottlieb supports a similar position, arguing further that the evidence for an old universe is strong: "The bones, artifacts, partially decayed radium, potassium-argon, uranium, the red-shifted light from space, etc.– all of it points to a greater age which **nevertheless is not true**."

Creationists still argue the same way. For instance, John Morris, president of the Institute for Creation Research talks about the "appearance of age": When Adam was created, he no doubt looked like a mature adult, fully able to walk, talk, care for the garden, etc. When God created fruit trees, they were already bearing fruit. In each case, what He created was **functionally complete right from the start**—able to fulfill the purpose for which it was created. Stars, created on Day Four, had to be seen to perform their purpose of usefulness in telling time; therefore, their light had to be visible on Earth right from the start. [108] [emphasis added]

And by the way, that answers the 'Chicken and the Egg' conundrum – i.e., Which came first? It has always seemed incredibly obvious that the Chicken was created first *ex nihilo* … It is merely cute sophomoric quasi-intellectuals who insist that the egg had to come first… demonstrating their lack of logical thinking.

And creating things fully-functional, yet old-looking, is exactly what would happen if the Simulation had been created and Man inserted. While the above quotes do not prove that we are living in a Simulation, we came as close as we could in Chapter 7.   Chateaubriand and Mr. Morris are respectable writers in their fields – Mr. Morris is quoted in VEG, Chapter 10 when it speaks of a Young Earth… There is ample evidence for that, too, and is reviewed in this Chapter 13.

And then, **Bertrand Russell** got into the fray when it was suggested that we have a deceptive Creator.

A Deceptive Creator

As Wikipedia says,

> From a religious viewpoint, it can be interpreted as God having 'created a fake,' such as illusions of light in space of stellar explosions (supernovae) that never really happened, or volcanic mountains that were never really volcanoes in the first place and that never actually experienced erosion.

> This conception has therefore drawn harsh rebuke from some theologians. Reverend Canon Brian Hebblethwaite for example, preached against Bertrand Russell's *Five Minute Hypothesis*:

> Bertrand Russell wrote, in The Analysis of Mind: 'there is **no logical impossibility** in the hypothesis that **the world sprang into being five minutes ago**, exactly as it then was, with a population that "remembered" a wholly unreal past'. 'Human beings', posited in being five minutes ago with built-in 'memory' traces, **would not be human beings**. The suggestion is logically incoherent . [109]   [emphasis added]

So there is more input added to say that **(a)** it is not impossible that 'Earth was created 5 minutes ago' but our memories were altered to believe we have been here longer than that… and **(b)** he also adds that such beings placed upon the Earth would not be human beings – Correct, they would be **Sims**, or NPCs (Non-Playable Characters) aka OPs.

> There was also a wag group that proposed instead of the world being created 5 minutes ago, it was really created **Last Thursday**… and then someone posted on the Internet that the world actually came to an end at 10 pm last Wednesday, and thus we must be living in the new world as it was re-created Thursday. (These people haven't had their shots yet…)

Quantum Computer Precursors

Moving forward to the science aspect of things, we come to three significant men who were very early forerunners of the Quantum Computing scene.

Konrad Zuse

The Germans get the credit for being the first in 1936 to build and operate a Turing Machine, a computer that can be considered the forerunner of today's digital computers. It was the **Z1** and was perfected by 1941 to be the Z3 (shown).

Credit: http://www.ask.com/wiki/History_of_computing_hardware?qsrc=3044

Alan Turing

Besides being the subject of the recent movie, *The Imitation Game*, wherein he is a cryptanalyst who helps to break the German Enigma machine codes, he is also best known for his work in devising the **Turing Machine**, which was a forerunner of a general purpose computing machine... which forerunner broke the Enigma code! He pioneered the concepts of algorithm and a machine that can compute. Turing is widely considered to be the father of theoretical computer science and artificial intelligence.

> Turing addressed the problem of artificial intelligence, and proposed an experiment which became known as the **Turing Test** [1950], an attempt to define a standard for a machine to be called "intelligent." The idea was that a computer could be said to "think" if a human interrogator could not tell it apart, through conversation, from a human being. In the paper, Turing suggested that rather than building a program to simulate the adult mind, it would be better rather to produce a simpler one to simulate a child's mind and then to subject it to a course of education. [110]

The following was almost a Turing Complete machine for breaking German cyphers during the War, called **the Colossus**.

Open architecture was definitely the order of the day, so that wires and cathode ray tubes could be easily accessed to modify circuitry <u>and</u> disperse heat.

Ca. 1943

Credit: http://en.wikipedia.org/wiki/Colossus_computer

People 65 years ago were already beginning to think of replicating intelligence via machines, and even today, 2015, we still cannot achieve more than a rudimentary simulation (via robotic interfaces— Chapter 9). Another author, **Ray Kurzweil** in The Singularity is Near, has proposed that by AD 2050 Man will have integrated sophisticated electronics into his body/mind (the Bionic Man) which will achieve two things:

a. Allow humans to network with each other without a server.
b. Prove Man to be sufficiently interesting to future beings who are probably simulating us that they will not delete us, but keep things going to see what we do and how we turn out. (This is further evaluated in Chapter 10).

At any rate, Turing's ideas were adapted by **Dr. John von Neumann** when he built the ENIAC, EDVAC and UNIVAC I computers in the mid-1940's.

ENIAC's inner workings:

Replacing a bad tube meant checking among ENIAC's 19,000 possibilities.

(credit: UNIVAC   http://www.computer-history.info/Page4.dir/pages/Univac.dir/

Even more fascinating was the commercial UNIVAC version built for The United States Census Bureau (1951):

Credit: http://www.ask.com/wiki/UNIVAC_I?qsrc=3044

Note the walk-in glass door – to access the tubes and wires… just below the Remington Rand logo. The tapes on the drives were not Mylar ® at this early stage of the game—they were metal with oxide coating.

The machine was huge, weighing 30 tons, using 200 kilowatts of electric power and contained over 18,000 vacuum tubes, 1,500 relays, and hundreds of thousands of resistors, capacitors, and inductors. It combined the high speed of electronics with the ability to be programmed for many complex problems. It could add or subtract **5000 times a second**, a thousand times faster than any other machine. It also had modules to multiply, divide, and square root. High speed memory was limited to 20 words (about 80 bytes). Not a whole lot were built.

The problem was, with the disruption to the British and American economies by the War, Turing's ACE (Automatic Computing Engine) was not developed, Turing did not get the credit he deserved, and the United States, including Dr. von Neumann, Presper Eckert and John Mauchly, and **Grace Hopper** were the first to develop large computing machines (so large, you could open a door and walk into them!). They were vacuum-tube operated. It would be the mid-1960's before transistors made their debut in computers.

FYI, sometimes flies and moths would get into the inside of the EDVAC or UNIVAC and get fried on the hot vacuum tubes, shorting them out... thus giving rise to the expression "There is a bug in the machine!" (There was, literally.)

Computers and Consciousness

Dr. David Deutsch

Another British physicist, this genius was quoted in VEG in the Science chapters dealing with the nature of reality. His seminal work is The Fabric of Reality: The Science of Parallel Universes – and Its Implications (1997). He espoused his Theory of Everything in this book which is not about reducing everything to particle physics, but finds support among epistemological and evolutionary principles. In short, he questions how we got here and what the universe is.

Along the way, he replaces Turing's Machine, or the **Turing Principle**, with his universal quantum computer.

> The Turing Principle stated that a universal computing device can simulate every physical process... including the world.

What Deutsch has done is replace the Newtonian physics of Turing's model with an updated Quantum Physics (QP) touch. His assumption is that the laws of QP are able to describe every physical process accurately... and we are having some issues in that area with String Theory and just what the **Higgs Boson** really does – We cannot yet find a use for it. Not only that, but as VEG (Chapter 8) pointed out, not all physicists agree that the General Theory of Relativity explains everything, and many physicists even say that the Special Theory of Relativity is flawed.

> Interestingly enough, while the Quantum Physicists try to get their 40+ particles to neatly fit an overall theory, the discipline of Subquantum Kinetics has resolved the QP issues with 6 basic components in 5 basic equations.

The quantum computer was described in the last chapter by **Seth Lloyd**, and is the forerunner of the answer to what a computer would look like that could manage a real-time Earth Simulation ... but realize that any supercomputer from the 4D

Realm is going to regrettably make today's most sophisticated quantum computer look like electronic junk.

So the issue to be examined is really a two-part one: Even using a Quantum Computer,

1. Can the human brain be simulated?
2. Can real human consciousness be simulated?

Human Brain Simulation

This is the easier to the two questions. And the answer is basically Yes – if one considers the simple algorithmic processes of the brain -- memory recall, math functions, and recognition of input (stimulus) from the environment – hearing sounds, seeing signs, etc. – and translating that into action. The **Turing Machine**, as defined earlier, is capable of simulating the basic data processing functions of the human brain --- data comes in, the brain identifies it, does something with it, and produces output (initiates some action).

Model: Input – Recognition – Reaction

That aspect is not what really concerns this book. Brainless, robotic actions by a lot of humans can be replicated by machines – whether they are NPCs or real humans.

The real crux of the matter is whether real, sentient, conscious humans can be simulated – humans that think, reason and not just react. Real conscious humans are said to have **qualia** – and NPCs would not know what that is.

> **Qualia**: are individual instances of subjective, conscious experience...
> The term "qualia" derives from the Latin meaning "of what sort" or
> "of what kind." Examples of qualia include the pain of a headache, the
> taste of wine, or the perceived redness of an evening sky.

Human Consciousness Simulation

This model is a bit more complex:

Input – Recognition – thinking/reflection/choice – Action/No action

This is basically the difference between a **reactive** machine (above) and a thinking human being – although the machine can be programmed to look like it is thinking (qv: **Artificial Intelligence or AI**), it is merely following a set program of choosing alternative courses of action based on specific input. And it was a thinking human being that programed the machine to simulate decision-making – but the machine cannot heuristically come up with new, self-originated courses of action. Not yet, anyway, but that is what the AI people are aiming for.

So the following examines the scientific and philosophical thinking out there regarding whether or not <u>all</u> humans (real and NPCs) are in fact still simulated in an Earth Simulation.

Dr. Brain Greene has said that **real sentience cannot be simulated**. And others examined earlier would agree, but Dr. Bostrom has suggested that post-Humans (those 400-500 years in the future) running the Simulation may be able to completely simulate even human intelligence and the appearance of consciousness.

> **The crux**: we cannot be simulated consciousness if consciousness as we know it cannot be simulated.

So what is consciousness?

That is one of the great questions involving biology, anthropology, psychology, neurobiology, and computer science. And as long as scientists ignore the spiritual aspects of Man, they will continue to come up with really hackneyed ideas – "just a mass of neurons going off," or "some form of bio-chemical response to one's environment" or our consciousness is just what Descartes said it was: "I think, therefore I am"… and all of these reflect a stunted (limited) mental view.

Science is desperate to prove that Man does not have a soul – because if he does, it opens up a can of worms that cannot be dealt with empirically. It also suggests a God.

Scientists who are **atheists** have a special axe to grind and are not objectively seeking the answer – Like Darwin, they need there to be no God, and no soul, so that they are not held responsible for what they do, or Like Darwin they doubt God exists because they cannot experience Him… See Chapter 8 in VEG for a deeper analysis of this.

At any rate, the major point of VEG and TOM is that Man is an eternal soul and the soul activates the body AND the mind…. Without the soul in the body, Man becomes a philosophical zombie.

Consciousness Decoding the Reality Fields
Credit: Bing Images: myinsightmag.com

Daniel Dennett

Daniel Dennett is an American philosopher who wrote <u>Consciousness Explained</u> (1991). The book puts forward a "multiple drafts" model of consciousness, suggesting that there is no single central place where conscious experience occurs; instead there are "various events of content-fixation occurring in various places at various times in the brain."

One of the book's more controversial claims is that **qualia** do not (and cannot) exist. Dennett's main argument is that the various properties attributed to qualia by philosophers—qualia are supposed to be incorrigible, ineffable, private, directly accessible and so on—are incompatible, so the notion of qualia is incoherent. The non-existence of qualia would mean that there is no hard problem of consciousness, and **"philosophical zombies"**, which are supposed to act like a human in every way while somehow lacking qualia, cannot exist. So, as Dennett wryly notes, he is committed to the belief that we are all p-zombies (.... functionally identical to a human being without any additional non-material aspects).

Dennett claims that our brains hold only a few salient details about the world, and that this is the only reason we are able to function at all. Thus, we don't

> store elaborate pictures in short-term memory, as this is not necessary and would consume valuable computing power. Rather, we log what has changed and assume the rest has stayed the same, with the result that we miss some details, as demonstrated in various experiments and illusions, some of which Dennett outlines. Research subsequent to Dennett's book indicates that some of his postulations were more conservative than expected. A year after *Consciousness Explained* was published, Dennett noted "I wish in retrospect that I'd been more daring, since the effects are stronger than I claimed". And since then examples continue to accumulate of **the illusory nature** of our visual world. [111] [emphasis added]

The world **is** illusory – even the Hindu and Chinese sages have said that it is Maya – illusion – as would be the case if it were a Simulation.

It gets even more interesting.

John Searle, another American philosopher, says that in Dennett's view

> there is no consciousness in addition to the computational features [of the brain], because that is all that consciousness amounts to for him: mere effects of a von Neumann(esque) virtual machine implemented in a parallel architecture and therefore implies that **conscious states are illusory**, but Searle asserts: "where consciousness is concerned, the existence of the appearance is the reality." [112] [emphasis added]

Again, they dance all around the issue…. They mention illusory states but it never occurs to them that our Reality is illusory because IT is simulated – not just Man and not just consciousness.

> Searle said further: "To put it as clearly as I can: in his book, *Consciousness Explained*, **Dennett denies the existence of consciousness**. He continues to use the word, but he means something different by it. For him, it refers only to third-person phenomena, not to the first-person conscious feelings and experiences we all have. **For Dennett there is no difference between us humans and complex zombies who lack any inner feelings, because we are all just complex zombies.** …I regard his view as self-refuting because it denies the existence of the data which a theory of consciousness is supposed to explain... Here is the paradox of this exchange: I am a conscious reviewer consciously answering the objections of an author who gives every indication of being consciously and puzzlingly angry. I do this for a readership that I assume is conscious. How then can I take seriously his claim that consciousness does not really exist?" [113] [emphasis added]

All of that to show that philosophers, more than scientists tend to go off the "deep end" when it comes to issues of consciousness… and it is instructive that the issue of "zombies" comes up – the NPCs would in fact be zombie-like (no souls, and very basic, minor consciousness) – just reactive robots.

So what does one of the foremost scientists researching consciousness have to say with respect to consciousness?

Roger Penrose

This accomplished gentleman is a mathematical physicist and he is also an atheist. His major work is a book entitled, The Emperor's New Mind (1989) wherein he argues that the known laws of physics are not sufficient to explain what consciousness is. He argues against the idea that rational processes of the mind are completely algorithmic and that they could therefore be duplicated by a sufficiently complex computer. [114]

His basic thinking on consciousness is that it transcends formal logic and he invokes Gödel's incompleteness theorem which states that consciousness is non-computable. To further develop things, Penrose teamed up with an anesthesiologist, **Stuart Hameroff** and together they developed a theory that consciousness is the result of quantum gravity effects in the brain's neurons' **microtubules**. This was formally called **Orch-OR** (orchestrated objective reduction) in 1997.

So to skip all the scientific lingo, what it suggested was that consciousness was due to quantum processing within **microtubules**. Orch-OR suggests that consciousness is based on non-computable quantum processing performed by **qubits** [Seth Lloyd in Chapter 7] formed collectively on the microtubules of the cells in a process amplified by the neurons. [115]

Allegedly, microtubule-associated proteins (MAPs) orchestrate the state reduction of the qubits and facilitate transmission of communication between neurons throughout the brain. And in 2014 a discovery of **quantum vibrations in microtubules** (see below diagram) was discovered by Anirban Bandyopadhyay confirming the Orch-OR theory. Of course, many scientists have harshly criticized the theory, saying that the brain could not possibly operate in a quantum manner, prompting the Penrose team to revise and elaborate their theory. Currently there is much work to further prove the fact that consciousness is connected to quantum operations in the brain.

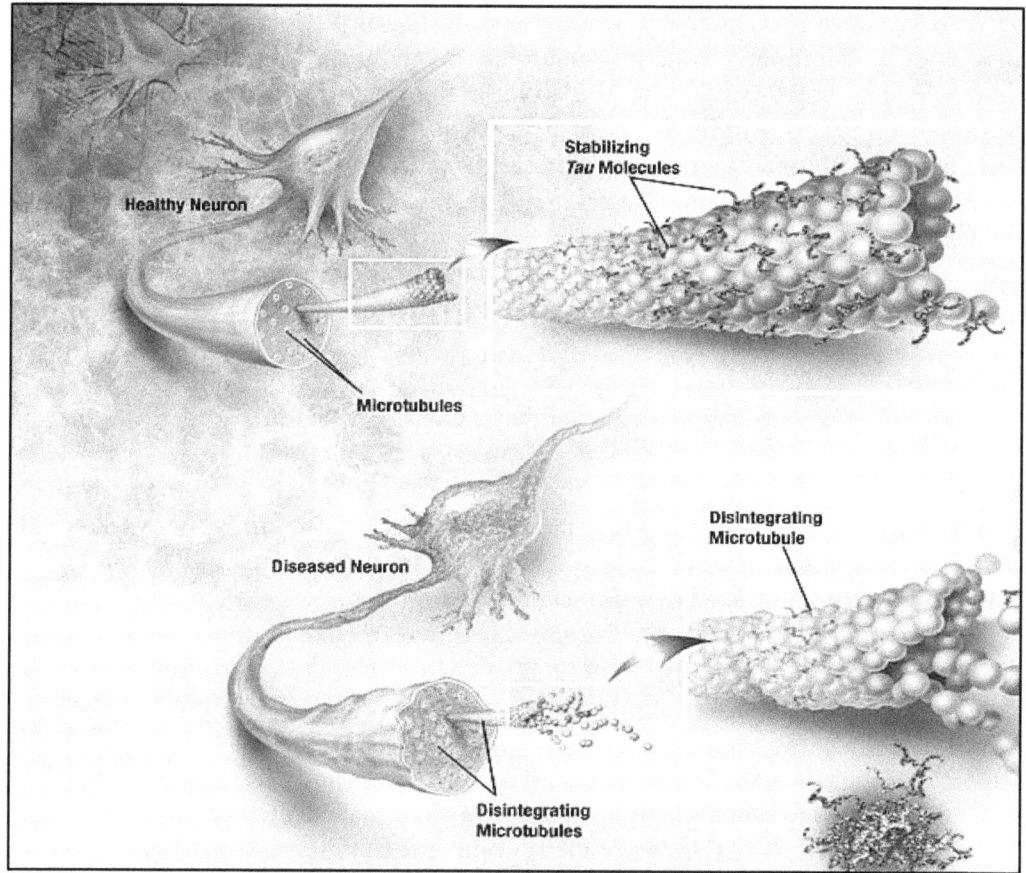

The above is a Mictotubule and neuron image: Bing Images: Edmchugh.ca.

Stuart Hameroff

Mr. Hameroff was an anesthesiologist at the University of Arizona who had been wondering how/why anesthesia turns off consciousness. His supposition was that the general anesthetics somehow interfere with the electrical activity of the microtubules which in turn turns off consciousness.

Microtubules are the scaffolding of the cell. They constantly assemble and re-assemble themselves like an endless set of Legos. They help transport various products along cells, and they are vital in pulling apart chromosomes during cell division! In addition, he found that microtubules are transmitting **photons** and that a good penetration of 'light' occurred in various parts of the brain! He also discovered that microtubules are great transmitters of pulses… maintaining a high degree of coherence among neighboring tubules, so that there was unified resonance among adjacent and participating microtubules. [116]

Hameroff reasoned that the microtubules might be **'light pipes'** acting as 'waveguides' and communicating information throughout the brain without any loss of energy. (See Ch. 9 in VEG and the **Bionet**.) Needless to say, his ideas and discoveries dovetailed with those of Dr. Pribram and another researcher in Japan, Kunio Yasue, established that brain processes occurred at the quantum level! [117] They also theorized that quantum messaging must take place through vibrational fields, along the microtubules of [all] cells. According to the theory that these networking researchers devised, the microtubules of the body are the actual Bionet along which photons travel – or, in short, **the Internet of the body**. Greg Braden called this the Divine Matrix.

(See Grazyna Fosar and Franz Bludorf, in the Bibliogrphy: Internet Section.)

The processes have to be operating at the quantum level if every cell in the body is in communication with the others – especially in the brain. The neural networks operate in tandem through quantum coherence. Quantum mechanics says that **entanglement** or "non-local communication" between particles is the norm -- or that 'nonlocality' can be expected of the way things operate in our universe. (Ch. 8 in VEG.) So when someone says that a machine can duplicate human consciousness, have them read this (written in 2002):

> Microtubules helped to marshal discordant energy and create global coherence of the waves in the body – a process called **'superradiance'** – then allowed these coherent signals to pulse thru the rest of the body. Once coherence was achieved, the **photons** could travel all along the light pipes as if they were transparent, a phenomenon called 'self-induced transparency.' Photons can penetrate the core of the microtubule and communicate with other photons throughout **the body**, causing collective cooperation of subatomic particles in microtubules throughout **the brain**. If this is the case, [and it was verified by Anirban Bandyopadhyay above in 2014] it would account for unity of thought and consciousness— the fact that we don't think loads of disparate things at once. [118] [emphasis added]

In fact, the communication between the brain and the body is what Psycho-Somatics is all about. Dr. Maxwell Maltz wrote a book called Psycho-Cybernetics (1960) that goes more into this.
And this larger subject is addressed in the Amplified Science of Mind book (ASOM).

Fritz-Albert Popp took it one step further, based on his research, and said that "Consciousness was a global phenomenon that occurred everywhere in the body, and was not simply in our brains. Consciousness, at its most basic, was **coherent light**." [119]   Did you catch all the preceding references to light, photons, etc?

Now hang on to your hat, this is hot, and is not likely to be replicated by a computer/machine:

> Biology [is] a quantum process. All the processes in the body, including cell communication, were triggered by quantum fluctuations, and all higher brain functions and consciousness also appeared to function at the quantum level. Walter Schempp's explosive discovery about quantum memory set off the most outrageous idea of all: short- and long-term **memory doesn't reside in our brain** at all, but instead s stored in the **Zero Point Field**. After Pribram's discoveries, a number of scientists, including Ervin Laszlo [*The Interconnected Universe*] would go on to argue that the brain is simply the retrieval and read-out mechanism of the ultimate storage medium – The Field. [120]   [emphasis added]

Some scientists went so far as to suggest that all out higher cognitive processes result from interaction with the **Zero Point Field**. [121] It is a sea of ceaseless, enormous energy (Hint: Dark Energy), it is like a Matrix of particles and entangled communications, the energy is constantly redistributed in a Yin and Yang, give and take of **light particles** which cannot be separated from the 'empty' space around them, (Hint: Dark Matter) and it is the only reality – the Field itself. **All fields have their being in The Field** and every part of the universe is tied to all other parts because of entanglement and instant communication. The stable state of matter depends on the dynamic interchange between subatomic particles and the ZPE Field. Thus all the motion of all the particles in the universe drives the Zero Point Field.

Soul Is the Driver

And what appears to be going on, is that the soul, inhabiting the body AND brain, influences these quantum operations; **the soul is able to manipulate the brain by manipulating subtle brain energy**. That is part of the consciousness process... the problem will be in proving the effect of the soul on the brain.

Consciousness can no longer be considered just the response to stimuli – it will eventually be found that even humans without souls will respond to stimuli and people who are asleep will react to a loud Bang! in their room where they are asleep. So a person in a coma is not conscious – there is no response to a stimulus. Consciousness is a waking state of one or more of the body's senses which leads to a reaction in the person. It is also awareness of one's environment, and a sense of selfhood.

Where the subtle problem comes into play is trying to simulate true human consciousness with its higher aspects – a person who understands syntax and nuances, makes judgments about what is the 'best' solution to a situation as opposed to what is 'right', a person who weighs inputs and outputs and bases the decision to act on an <u>insight</u>, or even better: <u>intuition</u>. Something a machine, or simulated human, i.e., a robot, would not have.

And certainly a robot would not have a **conscience**…. The NPCs or OPs in our world do not have such. (Is a renegade OP a sociopath?) Conscience is an attribute of the soul because the soul is connected to the Higher Self (i.e., the Oversoul) which not only knows right from wrong, it has a direct line to The Father of Light and morals, ethics and right action are the hallmark of an advanced Soul.

Simulated Conclusion

So what would be the answer to whether Man can create true, simulated humans? The answer is: Not yet, but **AI** is unfortunately moving in that direction… not because it is needed, but because we can. And that is the main argument behind advanced beings, even future humans with advanced technology – they would do simulations because they can and are curious enough to see what would happen with their creation…perhaps shades of the God Complex?

It is suggested that the best that Man will ever create is a walking/talking NPC which will always be lacking in some way, usually leaning and/or serving. (See **Geminoid F** in Chapter 9.) It is an **inner drive** with ensouled people, to feel empathy and have a desire to serve where needed… not so with the NPCs or OPs…

A real issue is whether our Earth Simulation is the product of advanced beings, post-Humans, 400-500 years from now, who are simulating an historical epoch, looking for understanding, or whether the post-Humans are just entertaining themselves… If it is the latter, we had better stay interesting to them, or they will terminate the Drama. If it is the former, when they get the answer(s) they seek, they will also terminate the simulation.

So if it is a Simulation, the only acceptable answer (one that we can live with) is that it is a Simulation run by benevolent beings as a School – to grow and challenge Man. Of course, it could also be a correctional institute – for wayward and dysfunctional souls in our Galaxy – Do we on Earth not isolate the undesirables and try to rehabilitate them? Could it be both a School and a Rehab Facility which is why we have such incredible violence on the planet from time to time, and yet many souls serve others, teach and give their time and $$ where needed?

And in the final analysis, to quote Dr. Bostrom:

> If you thought you were in a simulation, you should get on with your life in much the same way as if you were convinced that you are living a non-simulated life… at [some] bottom level of reality…. [However, if we humans] ever reach the point where we are about to switch on our own simulations…. [and] start running our own simulations [Sim City with programmable humans] **that would be very strong evidence** against [Options listed in Chapter 7] (1) and (2).
> That would leave us with only [option] (3). [122]   **[emphasis added]**

> To paraphrase the above: "that would be very strong evidence" for the post-Humans to be running their Simulation of us (i.e., Option 3 of Dr. Bostrom's 3-point Simulation Argument.)

Something to think about.

Meanwhile if we are in a Simulation, we have a number of Science errors, as well as Religion myths.

Science Theories in Doubt

So now we come to the point, hinted at above, where it comes as no surprise that if we are living in a Simulation, its laws and processes are self-contained (in the Sphere) and created to perform a special containing kind of operation and do not necessarily represent either the real 4D world (outside the Sphere), nor would they represent all other versions of a 3D created world.

The analogy here is to that of a basic calculator – it just has the functions Add, Subtract, Multiply and Divide. Something a 3rd grader might appreciate in arithmetic class. Now compare that with a Hewlett-Packard scientific calculator which has Trig functions, statistical functions, four separate addressable memories, a 2-line display, and the ability to generate graphs. The 3rd grade version capabilities are all contained in the advanced **HP 33s** and can only be considered a limited subset of the larger calculator.

In just a similar way, the 3D Construct we occupy constrains us to the Earth-Moon scenario, and souls incarnate in human bodies are limited to 5 basic senses – no clairvoyance, no telepathy, no telekinesis, no walking thru walls…. All available to the normal denizens of 4D, but not available to humans in 3D.

What a Simulation means specifically is that:

> There was no Big Bang
> Black Holes are not Galactic 'vacuum cleaners'
> The Universe is not expanding
> String Theory is not describing the sub-subparticle world
> The speed of light is not the top speed
> Evolution was assisted – and is not still happening
> There is no time travel

…and the so-called 'constants' have been changing:

> Fine Structure Constant (Alpha) –
> not constant in decay rates
> Speed of light is not constant

In fact, today's scientists are studying the **IMAX Theatre** thinking they have a handle on Reality. They will be eventually embarrassed. … All they are doing is discovering how the Sphere was constructed and perhaps how some of it functions. And the Sphere's Programmers will not let Man discover any more than what They permit… no matter how big Man builds his supercolliders!

SETI program – (Search for Extraterrestrial Intelligence) was a real waste of time and money, unknown to those who sincerely sought to communicate with ET intelligence. (And it was shut down in 2011.) Chapter 11 suggests why.

The concept of 'Many Worlds' postulate of Quantum Mechanics is thus nonsense – within our Simulation. Outside, in 4D it is probably a fact. The idea that we live in a Multiverse (universes **superpositioned**) is also false – we are denizens of the 3D Sphere. The Multiverse probably exists outside the Sphere, and it may be a feature of 4D and above… but there is no way to know from our limited perspective.

And that isn't why Man is here, as will be seen later. The gods don't care if Man knows where he really is… he still has to handle what he gets…

Anomalies, Part II

If you thought the weather 'wheels' (Chapter 2) were strange, the following will really make one stop and think: Why doesn't science have answers for the following issues?

Is it because we are not advanced enough, or we have a number of existing theories that are wrong and thus breed further errors? The following were documented anomalies in 2005, and are still oddities: [123]

The Horizon Problem

OUR universe appears to be unfathomably uniform. Look across space from one edge of the visible universe to the other, and you'll see that **the microwave background radiation filling the cosmos is at the same temperature everywhere**. That may not seem surprising until you consider that the two edges are nearly 28 billion light years apart and our universe is only 14 billion years old.

Nothing can travel faster than the speed of light [allegedly], so there is no way heat radiation could have travelled between the two horizons to even out the hot and cold spots created in the Big Bang and leave the thermal equilibrium we see now.

This "horizon problem" is **a big headache for cosmologists**, so big that they have come up with some pretty wild solutions. "Inflation", for example.

You can solve the horizon problem by having the universe expand ultra-fast for a time, just after the big bang, blowing up by a factor of 10^{50} in 10^{-33} seconds. But is that just wishful thinking? "Inflation would be an explanation if it occurred," says University of Cambridge astronomer Martin Rees. The trouble is that no one knows what could have made that happen.

So, in effect, inflation solves one mystery only to invoke another. A variation in the speed of light could also solve the horizon problem - but the speed of light is supposed to be constant.

Dark Matter

TAKE our best understanding of gravity, apply it to the way galaxies spin, and you'll quickly see the problem: **the galaxies should be falling apart**. Galactic matter orbits around a central point because its mutual gravitational attraction creates centripetal forces. But there is not enough mass in the galaxies to produce the observed spin.

Vera Rubin, an astronomer working at the Carnegie Institution's department of terrestrial magnetism in Washington DC, spotted this anomaly in the late 1970s. The best response from physicists was to suggest there is more stuff out there than we can see. The trouble was, nobody could explain what this "dark matter" was. **And they still can't**. [Hint: related to Æther.]

Although researchers have made many suggestions about what kind of particles might make up dark matter, there is no consensus. It's an embarrassing hole in our understanding. Astronomical observations suggest that dark matter must make up about 90 per cent of the mass in the universe, yet we are astonishingly ignorant what that 90 per cent is.

Maybe we can't work out what dark matter is because it doesn't actually exist. That's certainly the way Rubin would like it to turn out. "If I could have my pick, I would like to learn that Newton's laws must be modified in order to correctly describe gravitational interactions at large distances," she says. "That's more appealing than a universe filled with a new kind of sub-nuclear particle."

Tetraneutrons

FOUR years ago, a particle accelerator in France detected **six particles that should not exist** (see *Ghost in the atom*). They are called tetraneutrons: **four neutrons that are bound together in a way that defies the laws of physics.** Francisco Miguel Marquès and colleagues at the Ganil accelerator in Caen are now gearing up to do it again. If they succeed, these clusters may oblige us to rethink the forces that hold atomic nuclei together.

The team fired beryllium nuclei at a small carbon target and analysed the debris that shot into surrounding particle detectors. They expected to see evidence for four separate neutrons hitting their detectors. Instead the Ganil team found just one flash of light in one detector. And the energy of this flash suggested that four neutrons were arriving together at the detector. Of course, their finding could have been an accident: four neutrons might just have arrived in the same place at the same time by coincidence. But that's ridiculously improbable.

Not as improbable as tetraneutrons, some might say, because **in the standard model of particle physics tetraneutrons simply can't exist.** [Perhaps the Std Model is wrong...?] According to the Pauli Exclusion Principle, not even two protons or neutrons in the same system can have identical quantum properties. In fact, the strong nuclear force that would hold them together is tuned in such a way that it can't even hold two lone neutrons together, let alone four.

And there are still more compelling reasons to doubt the existence of tetraneutrons. If you tweak the laws of physics to allow four neutrons to bind together, all kinds of chaos ensues (*Journal of Physics G*, vol 29, L9). It would mean that the mix of elements formed after the big bang was inconsistent with what we now observe and, even worse, the elements formed would have quickly become far too heavy for the cosmos to cope. Or maybe the other Physicists are right: **There was no Big Bang?**

There are, however, a couple of holes in this reasoning. Established theory does allow the tetraneutron to exist - though only as a ridiculously short-lived particle. "This could be a reason for four neutrons hitting the Ganil detectors simultaneously," Timofeyuk says. And there is other evidence that supports the idea of matter composed of multiple neutrons: **neutron stars**. These bodies, which contain an enormous number of bound neutrons, suggest that as yet unexplained forces come into play when neutrons gather *en masse*.

The Pioneer Anomaly

THIS is a tale of two spacecraft. Pioneer 10 was launched in 1972; Pioneer 11 a year later. By now both craft should be drifting off into deep space with no one watching. However, their trajectories have proved far too fascinating to ignore. That's because **something has been pulling - or pushing - on them, causing them to speed up.** The resulting acceleration is tiny, less than a nanometre per second per second. That's equivalent to just one ten-billionth of the gravity at Earth's surface, but it is enough to have shifted Pioneer 10 some **400,000 kilometres off track**.

NASA lost touch with Pioneer 11 in 1995, but up to that point it was experiencing exactly the same deviation as its sister probe. So what is causing it? Nobody knows. Some possible explanations have already been ruled out, including software errors, the solar wind or a fuel leak. **If the cause is some gravitational effect, it is not one we know anything about.** In fact, physicists are so completely at a loss that some have resorted to linking this mystery with other inexplicable phenomena.

Bruce Bassett at the University of Portsmouth, UK, has suggested that the Pioneer conundrum might have something to do with variations in **alpha**, the fine structure constant. Others have talked about it as arising from dark matter - but since we don't know what dark matter is, that doesn't help much either. "This is all so maddeningly intriguing," says Michael Martin Nieto of the Los Alamos National Laboratory. "We only have proposals, none of which has been demonstrated."

Nieto has called for a new analysis of the early trajectory data from the craft, which he says might yield fresh clues. But to get to the bottom of the problem what scientists really need is a mission designed specifically to test unusual gravitational effects in the outer reaches of the solar system. Such a probe would cost between $300 million and $500 million and could piggyback on a future mission to the outer reaches of the solar system (www.arxiv.org/gr-qc/0411077).

"An explanation will be found eventually," Nieto says. "Of course I hope it is due to new physics - how stupendous that would be. But once a physicist starts working on the basis of hope he is heading for a fall."

Dark Energy

IT IS one of the most famous, and most embarrassing, problems in physics. In 1998, astronomers discovered that **the universe is expanding at ever faster speeds. It's an effect still searching for a cause** - until then, everyone thought the universe's expansion was slowing down after the big bang.

> It might just be that Dr. LaViolette is right: the Universe is not expanding but the 'red light shift' is due to photons that are losing energy and as they do, their light shifts down to the red end of the light scale. (See *Genesis of the Cosmos.*)

"Theorists are still floundering around, looking for a sensible explanation," says cosmologist Katherine Freese of the University of Michigan, Ann Arbor. "We're all hoping that upcoming observations of supernovae, of clusters of galaxies and so on will give us more clues."

One suggestion is that some property of empty space is responsible - cosmologists call it dark energy. But **all attempts to pin it down have fallen woefully short**. It's also possible that Einstein's theory of general relativity may need to be tweaked when applied to the very largest scales of the universe. "The field is still wide open," Freese says.

The Kuiper Cliff

IF YOU travel out to the far edge of the solar system, into the frigid wastes beyond Pluto, you'll see something strange. Suddenly, after passing through the Kuiper belt, a region of space teeming with icy rocks, there's nothing.

Astronomers call this boundary the Kuiper cliff, because the density of space rocks drops off so steeply. What caused it? The only answer seems to be a 10th planet. We're not talking about Quaoar or Sedna: this is a massive object, as big as Earth or Mars, that has swept the area clean of debris.

The evidence for the existence of **"Planet X"** is compelling, says Alan Stern, an astronomer at the Southwest Research Institute in Boulder, Colorado. But although calculations show that such a body could account for the Kuiper cliff (*Icarus*, vol 160, p 32), no one has ever seen this fabled 10th planet.

There's a good reason for that. The Kuiper belt is just too far away for us to get a decent view. We need to get out there and have a look before we can say anything about the region. And that won't be possible for another decade, at least. (NASA's New Horizons probe, which will head out to Pluto and the Kuiper belt, is scheduled for launch in January 2006. It won't reach Pluto until 2015).

Not-so-constant Constants

IN 1997 astronomer John Webb and his team at the University of New South Wales in Sydney analysed the light reaching Earth from distant quasars. On its 12-billion-year journey, the light had passed through interstellar clouds of metals such as iron, nickel and chromium, and the researchers found these atoms had absorbed some of the photons of quasar light - but not the ones they were expecting.

If the observations are correct, the only vaguely reasonable explanation is that a constant of physics called the **fine structure constant, or Alpha**, had a different value at the time the light passed through the clouds. But that's heresy. **Alpha is an extremely important constant that determines how light interacts with matter - and it shouldn't be able to change**. Its value depends on, among other things, the charge in the electron, the speed of light and Planck's constant. Could one of these really have changed?

No one in physics wanted to believe the measurements. Webb and his team have been trying for years to find an error in their results. But **so far they have failed**. Webb's are not the only results that suggest something is missing from our understanding of Alpha. A recent analysis of the only known natural nuclear reactor, which was active nearly 2 billion years ago at what is now Oklo in Gabon, also suggests something about light's interaction with matter has changed.

The ratio of certain radioactive isotopes produced within such a reactor depends on Alpha, and so looking at the fission products left behind in the ground at Oklo provides a way to work out the value of the constant at the time of their formation. Using this method, Steve Lamoreaux and his colleagues at the Los Alamos National Laboratory in New Mexico suggest that Alpha may have decreased by more than 4 per cent since Oklo started up (*Physical Review D*, vol 69, p 121701).

There are gainsayers who still dispute any change in Alpha. Patrick Petitjean, an astronomer at the Institute of Astrophysics in Paris, led a team that analysed quasar light picked up by the Very Large Telescope (VLT) in Chile and found <u>no evidence</u> that Alpha has changed. But Webb, who is now looking at the VLT measurements, says that they require a more complex analysis than Petitjean's team has carried out. "It's difficult to say how long it's going to take," says team member Michael Murphy of the University of Cambridge. "The more we look at these new data, the more difficulties we see."

Dark Flow

WE CANNOT see what lies beyond the visible horizon of our universe, simply because light emitted beyond that horizon has not had time to reach us. Despite this out-of-sightness, we've always <u>assumed</u> that space is filled with the same stuff wherever you go in the universe.

So a recent finding by Sasha Kashlinsky at NASA's Goddard Space Flight Center in Greenbelt, Maryland, does not make sense. His team has found a group of galaxy clusters moving at an extraordinary speed towards a small patch of sky between the constellations of Centaurus and Vela. Kashlinsky calls it the "dark flow", in tribute to those other cosmic mysteries dark matter and dark energy (*New Scientist*, 24 January, p 50).

There is no obvious reason why the clusters should be moving at such breakneck speeds, unless they are experiencing **an unusually strong pull from something beyond the visible horizon.** But what? The most obvious answer is that **there is something big out there, far bigger than anything in our known universe**. Such a behemoth would impose a kind of "tilt" on the universe, causing matter to move in one particular direction - as observations of the dark flow suggest.

If such cosmic megastructures do exist, though, they merely replace one mystery with another. One of the foundation stones of cosmology is the Copernican principle, which says that there is nothing special about our region of the universe. So if there are megastructures beyond our horizon, there should be megastructures in our patch, too. We haven't seen any.

There are also suggestions that the pull might be from another universe altogether. That would be good news for proponents of eternal inflation theory, which suggests that the universe should actually be composed of "mini-universes" that have bubbled off from one another.

Kashlinsky is preparing papers with further results. He says observations point the finger at **megastructures beyond the horizon**.

And if the Universe that we see is something simulated within the **Cspace**, that would account for it: the gods are playing games with the Simulation.

Antimatter Mystery

The Big Bang should have created matter and antimatter in equal amounts, or so our best theories have it. If that were truly the case, though, then the universe would have disappeared in a big puff of self-annihilation almost as soon as it began. The fact that we are here to ponder it tells us something is wrong with this picture (*New Scientist*, 12 April 2008, p 26). The question is: what?

It is interesting that they would assume The Big Bang created equal amounts, when it is much easier to say the Universe started with all positive material. And then they invent new particles (**Majoron**, below) that just have to be there to account for their faulty thinking.

Experiments in accelerators now tell us that for every 10 billion antiprotons present in the early universe, there were 10-billion-and-one protons. The same tiny imbalance applied to other particles, such as electrons, too. At some point in cosmic history, matter and antimatter met and annihilated. Left behind, those extra particles eventually came together and formed the matter-filled universe we know today. So what created that initial imbalance?

The short answer is that **we don't know**. One possibility is that antimatter is lurking out there at distant points around the cosmos. That's unlikely, though.

A better idea springs from the weak force, which governs certain nuclear processes, including radioactive beta decay. In 1964, physicists found that the weak force is not quite symmetrical in its dealings with matter and antimatter, resulting in something known as **CP violation**. This has led particle physicists to suggest that the laws of physics are lopsided. The trouble is that the standard model of particle physics says they aren't lopsided enough. "There is not enough CP violation to do the job," says Frank Close at the University of Oxford.

CP-symmetry states that the laws of physics should be the same if a particle is interchanged with its antiparticle (C symmetry), and then its spatial coordinates are inverted ("mirror" or P symmetry). The discovery of CP violation in 1964 in the decays of neutral kaons resulted in the Nobel Prize in Physics in 1980 for its discoverers James Cronin and Val Fitch. (Wikipedia article.)

Other ideas to explain the imbalance of matter and antimatter in the infant universe include a hypothetical particle called the **majoron**, which is thought to have created neutrinos and antineutrinos, but not in equal amounts. That could eventually have led to an imbalance between matter and antimatter. "If we find majorons at the Large Hadron Collider at CERN," says Close, "then we could hope to study their decays." This would help us discover if they fit the bill.

The Lithium Problem

OUR best theories of the early universe also tell us which atoms should have been forged in the first 5 minutes after the big bang. The **existing amounts of hydrogen and helium match theory perfectly** - so well, in fact, that cosmologists claim this is the best evidence we have for the Big Bang. Things aren't so good for the third element, lithium, however (New Scientist, 5 July 2008, p 28).

When we count up the lithium atoms held in stars, **there is only one-third as much of the lithium-7 isotope as there should be. Another isotope, lithium-6, is overabundant: there may be as much as 1000 times too much of it.**

So something in the Big Bang is not adding up. Is it a serious problem? Yes, says Gary Steigman of Ohio State University in Columbus, but it is not fatal. "There are too many successes for Big-Bang cosmology to be troubled by these lithium problems," he says. [What is that lump under the rug?]

Others disagree. **"The lithium problem is one of the very few hints that there may be a problem with the Big Bang**," says Jonathan Feng at the University of California, Irvine.

One thing that everyone does agree on is that **things are getting worse**. "The lithium-7 problem is more serious than ever," says Joseph Silk at the University of Oxford. Improved observations of stars suggest they contain even less lithium-7 than previously thought. "The gap between prediction and observation has widened," Steigman says.

```
And now the one you have all been waiting for:
```

Drum roll, please.....

Noise From the Edge of the Universe

Are dud signals from a gravitational wave detector **evidence that the universe is a holographic projection? (Image: ESA)**

> Gravitational waves are ripples in space-time that are emitted by cataclysmic cosmic events such as exploding stars, merging black holes and/or neutron stars, and rapidly rotating compact stellar remnants.

The **GEO600** gravitational wave detector in Hanover, Germany, has not yet detected any gravitational waves. As a consolation prize, **it may instead have uncovered the ultimate nature of reality.** In 2008, physicist Craig Hogan at the Fermi National Accelerator Laboratory in Batavia, Illinois, was trying to work out how we might test the idea that everything we see as physical reality is the result of a kind of projection from the boundary of the universe. This is known as the **holographic principle**.

The information held at the boundary is not smooth, but composed of "bits", each one occupying an area that corresponds to the most fundamental quanta of distance in the universe. This is **the Planck length**, around 10^{-35} metres – far too small for us to see the individual bits. When this information is projected into the volume of the universe, however, each bit gets magnified. That means **we might just be able to see pixellation in space-time** (see Chapter 7).

The kinds of scales involved still mean it would only be detectable in the most sensitive instruments we have – such as the gravitational wave detectors looking for the ripples in space-time caused by violent cosmological events such as the collision of two black holes. Hogan worked out how the **pixellation** might manifest itself for GEO600 and sent his result to the researchers there.

By strange coincidence, the GEO600 team had been having problems with "noise" in their detectors. But **here's the kicker: the noise had uncannily similar**

characteristics to Hogan's anticipated signal. Is it indeed the result of information that resides at the edge of the universe? "The issue is still unresolved," says Karsten Danzmann, principal investigator for GEO600. **"The noise is still there and we have no explanation."**

Chapter 7 introduced Dr. Hogan's report that **sound was coming from the edge of the universe,** and wouldn't that be strong evidence of the energy being used to generate the **Konstruct** and the **Cspace**? Even if you say that the sound is just radio waves bouncing back at us – what are they bouncing off of?

> As of 2015 there has been no further clarification from GEO600.

Summary

So there you have it –multiple anomalies in various aspects of Physics and Astronomy. The one thing that explains them all is the Simulation Theory – anomalies would be a byproduct of the Simulation processes, and they don't necessarily follow the rules as we think we understand them.

Is it possible that the anomalies represent the fact that our scientists don't know what they're talking about – that they have formulated erroneous Laws and Theories?…as to where we really are, and we expect no anomalies and what we think are anomalies are just the normal working of our complex Reality? Are there principles that have yet to be discovered? And what if that Reality is really a Simulation….?

In addition, physical anomalies seen around the planet will be examined in the next chapter.

Chapter 9: Simulation and Timelines

This chapter takes a little detour into the world of Timelines and takes a look at how Timelines and Simulations would interact.

> Relevant parts of Chapter 2 from <u>The Transformation of Man</u> will be represented here such that the concept of Timelines is clarified. (Also see Appendix D in TOM.)

Quantum Physics maintains that we are in one of several possible universes. **Hugh Everett III** was one of the main proponents of the **'Many Worlds'** theory in which it was said:

> Everett noted that **quantum physics predicts** that all alternative outcomes of any given experiment must occur even though we may only see a single outcome! Somehow, those hidden alternatives must exist simultaneously along with the observed outcome. [124]

> and

> Any universe [timeline] you may inhabit at the moment will seem real enough with the others hidden from plain view. However, the same thing will be true for each of the **other universes** and other "you's" as well… how can there be copies of me that I have no knowledge of? [125] [emphasis added]

So he was in fact saying that a soul can exist in many places at the same time (by replicating an aspect of itself) and thus experience different outcomes resulting from major life decisions. It is a metaphysical theory that, for example, when a soul must decide to marry or not, that soul splits into two aspects, each in a separate timeline, and experiences the result (separately) of each decision. This is sometimes referred to as an Alternate Future.

Alternate Futures

Alternate futures are often created, or assembled and disassembled in this way. Every time a person faces a <u>major decision</u> in his/her life, the other option not consciously chosen is usually played out on another **fractal** part of the same timeline. In this way, the Father of Light through all the lessons of the Higher Self, or Oversoul, is completed by the knowledge of the outcome of all possibilities.

Credit: http://www.bing.com/images

Parallel Universes

Another description of alternate futures is parallel universes or dimensions. Quantum Physics attempts to address this aspect of our world and sees it as a possible solution to the particle versus wave behavior when addressing the photon-thru-the-slit experiment. Part of the answer to whether light is a particle or wave is answered by a theory that says particles in adjacent universes are interfering with those in this universe.[126] Multiple universes are **superpositioned** (outside this Earth Simulation).

Specifically Dr .Wolf says:

> Thus in any single universe, even though the particle in the other Universe is not present [in this universe], the effect of its presence mysteriously changes the course of the observed particle's history and its final destination [in our universe].[127]

Again, this is an example of **entanglement.** Said another way, Professor David Deutsch has explained:

Quantum mechanics is basically a theory of **many parallel universes**. Some of these universes are very like our own and some are very un-like our own. The nearby universes differ from ours by, say, only one photon, whilst the other, more distant universes are completely different from ours…

In a sense, each and every particle exists in its own separate universe, but <u>these universes interfere with each other</u> to produce the pattern of the universe that we can actually see or perceive… Reality does not consist of just a single universe. [128] [emphasis added]

Besides being entangled, the universes are in **superposition** to each other.

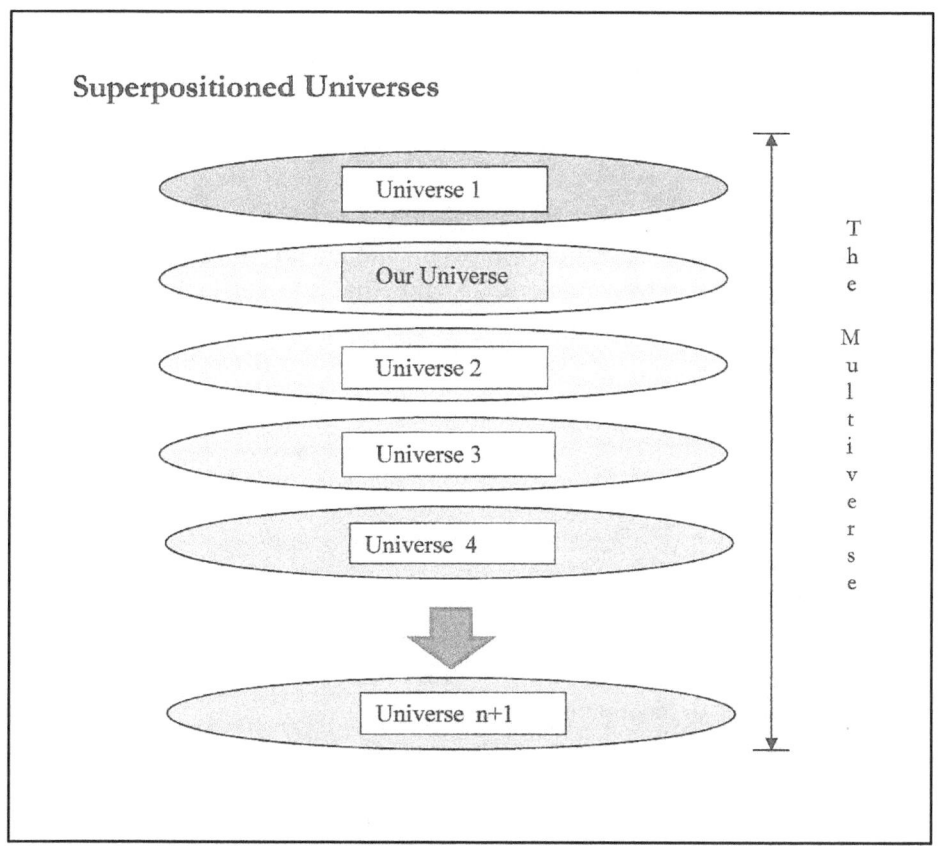

What this means is that supposedly the adjacent universes may "interfere" with each other, and such is the theory behind the proposed existence of a **'virtual particle'** -- one which pops in and out of our world. It is alleged to have amazing properties –

and all that is the result of <u>advanced mathematics</u> which suggests that such particles exist.

> We need to be careful when using advanced mathematics to simulate reality, it can be tricky and one can 'prove' some interesting things. This author was in an advanced math class where the professor showed how 2 + 2 could = 5, and then showed us how **Topology** could take a doughnut and redefine it as a coffee cup.

As is explained by Wikipedia:

> A **virtual particle** is an explanatory conceptual entity that is found in mathematical calculations about quantum field theory. It refers to mathematical terms that have some appearance of representing particles inside a subatomic process such as a collision. Virtual particles, however, do not appear directly amongst the observable and detectable input and output quantities of those calculations, which refer only to actual, as distinct from virtual, particles.
> Virtual particle terms represent "particles" that are said to be 'off mass shell'. For example, they can progress backwards in time, do not conserve energy [questionable], and can travel faster than light. That is to say, looked at one by one, they appear able to violate basic laws of physics. Regular particles of course never do so. On the other hand, any particle that is actually observed never precisely satisfies the conditions theoretically imposed on regular particles. [129]

To rewind, Dr. Wolf said:

> Thus in any single universe, even though the particle in the other Universe is not present [in this universe], the effect of its presence mysteriously changes the course of the observed particle's history and its final destination [in our universe].[130]

What comes to mind is, How would he know? We do not know anything about other physical universes… unless we just imagine them.

The virtual particle is a great example of something that can be <u>inserted</u> into the Simulation, or which results from our scientists messing with the building blocks of the Simulation and 'discovering' something unusual – which has no bearing on reality, but is a Man-created anomaly.

As for superpositioned universes, there is something else going on that is really being described. Since we are in a Sphere, a 3D Construct in 4D, the concept of 'universe' does not apply <u>to us</u>. **The Sphere is our universe**, albeit a small one. The concept of multiple universes may be a great one, and valid when one resides in 4D… but we in the 3D Construct (inside the larger Konstruct) cannot test that realm.

So how does that all relate to timelines? Are timelines another word for 'dimensions'? Are timelines germane to a Simulation?

Timelines

Without the issue of Earth being in a Simulation, it was easy to say (as the book <u>Transformation of Man</u> did in Chapter 2) that a timeline was synonymous with a dimension. The idea was that a dimension could have multiple timelines. But

there are no timelines in the Earth Simulation

which is a discrete entity or Sphere – existing within the 4D timeline! Thus, what we experience as 'time' is a construct of the Sphere: simply the passage of the Sun.

And yet, for the purpose of clarity, a timeline can be thought of as a sine-wave on an oscilloscope:

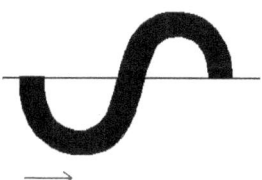

And if there are multiple timelines, they would be phase-shifted. While the two waves pictured (below) are technically sine and cosine waves, the emphasis here is the fact that two waves, two timelines, can be running through the same linear space-time, with a phase-shift to separate them.

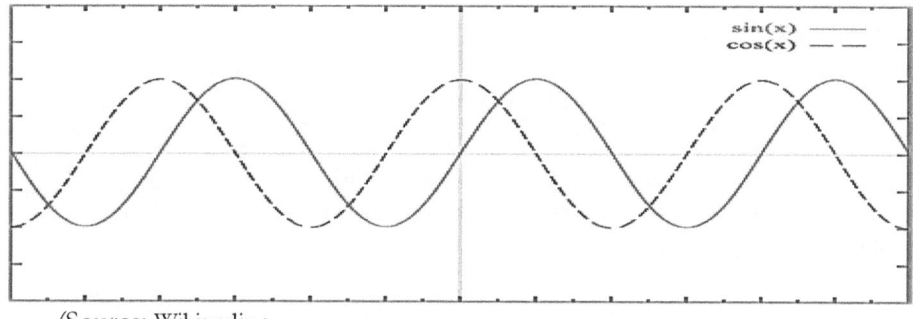

(Source: Wikipedia :
http://upload.wikimedia.org/wikipedia/commons/1/13/Sine_Cosine_Graph.png)

However, assume that both waves are propagating across the screen at the same time and rate, and note that they do occasionally **intersect**. When they intersect, as timelines, it should be possible to move (consciously/unconsciously) from one to the other... in 4D.

> Everything about timelines was examined in Ch. 2 of
> Transformation of Man (TOM). Timeline splits and
> timeloops and timeline wars are covered in more detail.

So, in essence, timelines are real (in 4D and above), there is a Multiverse (and we're not in it directly), and dimensions exist – but those are all concepts that apply IF you are in 4D or above. It also used to apply before 4D Earth was replicated into the 3D Sphere. As described in VEG and TOM, Souls may replicate aspects of themselves in different timelines and that is still valid since the Soul (Higher Self) resides in 4D and may choose to incarnate into the 3D Sphere, or elsewhere.

Simulation

So if we are in a Simulation, a 3D Construct operating under a subset of 4D Laws, and there are no timelines in the Sphere, and we can't access the Multiverse nor can we be affected by adjacent dimensions.... Then this Earth is really just a **Stage** constructed to run the Earth **Drama** for whatever Era we are in, for whatever Plan (which is Era-dependent) that the Higher Beings have in mind for Man.

Gee, you mean Shakespeare was right?

Yep, and because he knew, he had to be an Insert – here to enlighten us...or else some Insert (a **Placeholder** or Sim) unknown to us clued him in. And it has been suggested over the years that Shakespeare was really a committee of sharp minds, perhaps headed by Sir Francis Bacon, who wrote the Plays and assigned William Shakespeare's name to them.

Not important.

What counts is that **the Stage runs a Drama unique to each Era**. Souls then incarnate into the Drama, choosing and playing their rôles. (Explained in Ch. 7 in TOM.) Each rôle is pre-defined (in the soul's Script) to be in synch with the Greater Script for the Era... but it is not Fate. The soul incarnated into a body in a family, in a city, in a state, in a country for a specific reason. There are significant points in each soul's Script that amount to tests and challenges to be met – The soul can choose to resist, reject or ignore whatever comes up in his life, but will be held accountable for 'doing his own thing.'

> Think of it as a pupil in 3rd grade in elementary school, and he is there
> to learn a number of things, including multiplication tables, and he

resists, acts out, and learns nothing for the whole year. He will not graduate to 4th grade, and will have to <u>repeat</u> 3rd grade. That pupil had the choice to shut up, sit down and pay attention – follow the Plan for that grade and instead chose to do what he wanted.

Every soul has freewill to do what s/he wants – and will be held accountable for being appropriate or inappropriate.

Ok, this chapter is not about metaphysics nor preaching a truth – but each soul does have freewill to do the right thing or not. And there are a lot of people in jail because they went a bit too far – and yet, that is the Earth experience. It is not perfect here, nor is it intended to be perfect… it is a rough School at times, but a fool will learn in no other. That has to be said because Earth <u>is</u> a School and there are places on Earth that are Hell and some that are Heaven. Often the **Placeholders** help to run the Script, or provide the **catalyst** that challenges each soul. When that doesn't work, the gods often **insert** a player (Sim or real human) who can inspire or provide specific guidance to keep the Greater Script on track, or to augment it.

Let's take a look at Placeholders and Inserts.

Placeholders

Every simulation or video game has **Non-Playable Characters** (NPCs or Sims) that the main player of the game cannot control, but is expected to deal with. Our lives in the Earth Simulation have a similar thing: OPs, or **Organic Portals**.

> While OPs were examined in detail in Ch 5 of VEG, they are Humans who are **Organic** (flesh and blood) and serve as **Portals** (avatars to be manipulated by the gods who run the Simulation). Because they have <u>no soul</u>, there is no violation of Universal Law that prohibits messing with an ensouled human. No soul means no conscience and some of them become sociopaths.

The OPs are often called Placeholders in the Drama because they keep the Greater Script on track, and sometimes even drive it… just as NPCs drive the video game.

And sometimes they look very human. The next page shows a **human android**, and it is very believable… The second picture (with two women) shows the real woman on the right who was the model for the android.

The research is being led by Osaka University professor Hiroshi Ishiguro, who is known for creating teleoperated robot twins such as the celebrated **Geminoid HI-1**, (left) which was modeled after himself. [131]

The Japanese research team presented its latest android, a **female robot** that is entirely a copy of a real woman. The robot can talk and gesticulate, and even seems like it breathes. The name of this "woman of the future" is **Geminoid F** (below). You can barely see the difference between the Geminoid girl (left) and the real woman.

The Geminoid F Android (left)
(credit: Bing Images: wikalenda.com)

Researchers from the Intelligent Robotics Laboratory at Osaka University have teamed up with robot maker Kokoro Co., Ltd. to create a realistic-looking **remote-control** female android that mimics the facial expressions and speech of a human operator.

Modeled after a woman in her twenties (on the right, above), the android –

called **Geminoid F** (the "F" stands for female) -- has long black hair, soft silicone skin, and a set of lifelike teeth that allow her to produce a natural smile. According to the developers, the robot's friendly and approachable appearance makes her suitable for receptionist work at sites such as museums. The researchers also plan to test her ability to put hospital patients at ease. [132]

The female robot has perfect teeth and will actually go to serve as an experimental bunny by students in the area of dental surgery. The odd creature can't walk, but according to the research team this will soon happen.

Let's take a closer look: Note the eyelashes and corners of the mouth.

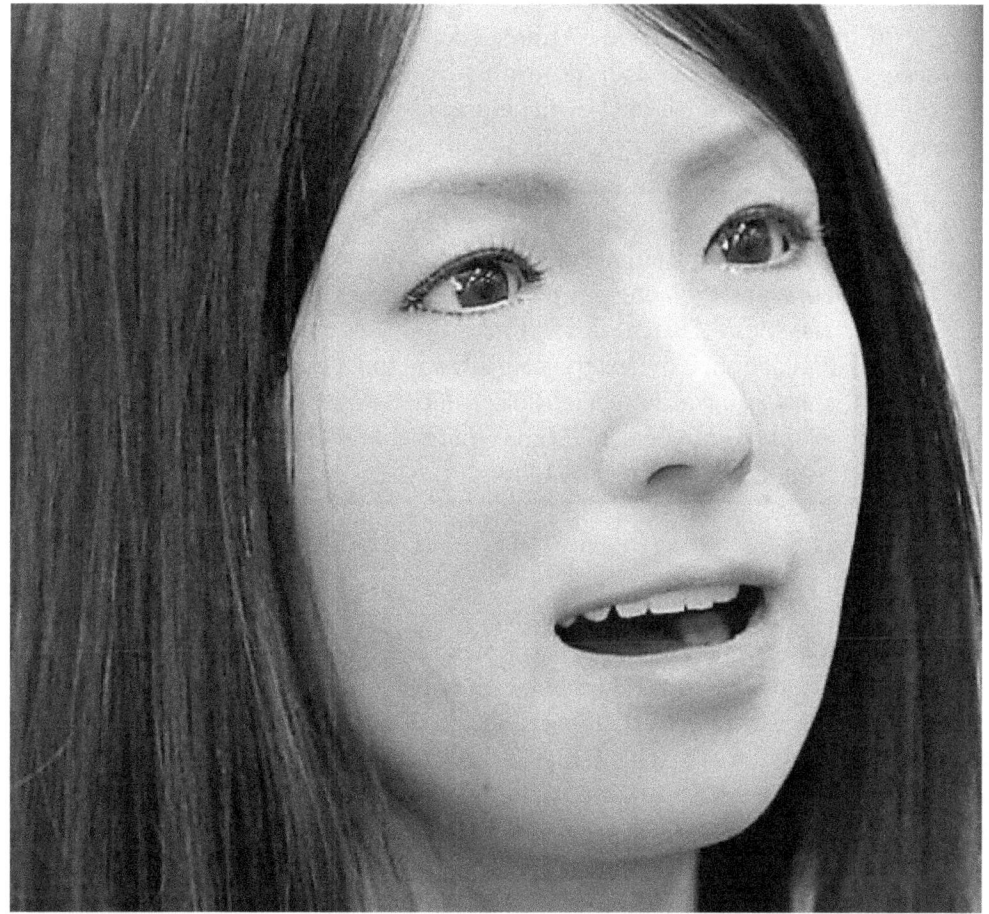

(credit: http://www.decodedstuff.com/the-female-android-will-give-the-tone-to-a-new-definition-of-social-interaction/)

All that to say that there are **synthetic humans** about to join those who are already walking among us that were created by the Simulation – and the simulated humans

(OPs and NPCs) are not discernable from real humans unless you can see auras (OPs/NPCs don't have any).

NPCs and OPs have another function that relates to a soul learning his lessons. If a soul, Bob, has a point in his Script that calls for him to experience being lied to, cheated, or even severely wounded, another ensouled human, Sue, cannot (and they choose to not) deliver that action to Bob (the ensouled human) – as Sue herself would incur **Karma**! (And then the Drama on Earth would never end – how would 'payback' ever end if souls keep socking to each other?) So NPCs or Placeholders are used because they have no soul and thus incur no Karma!

In another instance, and 20 years ago, this was called a **Walk-In** scenario, sometimes the current ensouled human is weary of the Drama, and wants out. The gods arrange for another soul, or sometimes an advanced soul, to swap places with the ensouled human. And sometimes, the ensouled human is physically taken 'home' and the human may be replicated as an OP (no soul) – or others may just see him disappear (as Charles Fort recounted earlier). If the human is replicated as an OP, it is because his presence is still needed for the Drama by others to work out their issues.

> The author has actually seen this happen and thought it was indicating a timeline split. Rick had always had an aura, and then one day he didn't – and to this day, he still doesn't. Since we don't have timelines in the Simulation, it means the soul left and the Rick 'shell' is still walking around. Plausible as Rick hated it here and wanted out. (Ch 7 in VEG has more answers to this issue.) Rick became a **Walk-out**.

Timeline Split?

Since we are in a Simulation, there are no timelines, OR another way to look at it is that our Process could be considered just one timeline – but it doesn't split. What happens is the Simulation is **fractally replicated** for one or more souls to experience a specific lesson – and this is what happened to Robert Monroe when the *Inspec* put him thru the same scenario 5 times until he got it right (reminiscent of *Groundhog Day*). (This was mentioned in Chapter 4.)

The significant point is that
1. Simulations can be replicated (fractally if need be)
2. Placeholders are often used to play key parts in the fractal simulation.

So, instead of splitting a timeline to create an alternate scenario, the gods would in our case, replicate the Simulation. Almost the same thing, but they can control who goes to the new Simulation, and who stays behind. And if the new Simulation becomes too negative or develops destructive vectors, the gods would terminate the alternate Simulation.

Soul Auras

And significantly, our Reality in the Earth Simulation consists of about 60% of the people around us at any time who are OPs... they have no soul. If one can learn to see auras (given in Ch 11 of TOM) it will be readily apparent that many people do not have auras. The OPs (Sims or Placeholders) have an energy field above their head resembling heat waves (as on a desert highway).

> Dr. Bostrom thought this was significant, and Dr. Greene seconded the idea. 60% is about right. (Chapter 7)

And that leaves us with the remaining way that the gods can manipulate or influence Man in the Sphere.

Avatars

Would it not be advantageous to a Simulation to be able to watch and see what is going on (watch the Drama) and if some sort of corrective action is required, would it not be great to (1) insert a being that walks into town (from nowhere) to make the correction, or (2) use one of the existing OPs [soulless remember], and make the fix yourself via the OP/NPC ?

Secondly, if one of the souls wants out, the reverse can be done by removing the soul and inserting a new one (Walk-in)... or because the body is capable of living, moving, etc. without a soul (Chapter 7, VEG) it may just be empowered (and converted) to an OP when the soul leaves (not death, but a Walk-out).

> For those who think this is strange and not possible, it is analogous to a **house** where a family lives for 10 years and then moves out, and another family moves in. Is a body not a house for the soul?

And thirdly, this scenario adapts well to the concept that when the gods wish to have someone create a beautiful picture or opera – to inspire humans -- a body can be inhabited by a soul of their choosing who has the ability to write a symphony or sculpt a *Pieta*.

Inserts

Every now and then, as part of the Greater Script, or Plan for the Era, it is necessary to insert a player with specific information to aid mankind. It may be a Louis Pasteur, a Buddha, or a DaVinci. Often the inserted player is to deliver a new philosophy, or religion or as in the case of Mozart and Bach – just create inspiring music to add quality to the Earth experience.

While some humans can create inspiring art, music or books, most often it is an Insert that has been programmed to do it right and whatever is 'delivered' will stand the test of time. That means that Charles Darwin was a human who did his best, and got the Natural Selection and Survival of the Fittest concepts correct, but was a bit off the mark as far as Evolution was concerned. An Insert would have gotten the whole Theory correct from the start.

This is not to denigrate humans *per se*, it is just that 90% of humans do not and cannot create at the level of a Bach, DaVinci, Tesla, Michelangelo, or Shakespeare. That in itself suggests that the upper 10% is different, and in a Simulation, the gods most probably would use Inserts, as well as NPCs to control/direct the Drama.

For sure, we know that when two scientists on opposite sides of the Earth both come up with the same idea for a radio, or invention or a medical cure, there has been some proactive 'interference.' This again suggests that there is a **Control System** that orchestrates our growth and progress, and probably gives the same concept to both humans (often while sleeping) to make sure that at least one of them gets the workout to the world.

Some will ask if that also applies to the gods inserting a Hitler, Stalin, or Mao Tse-Tung (now spelled Mao Zedong) or an Attila the Hun? The issue of Evil is addressed in one of two ways:

1. Many humans carry defective Anunnaki genetics which have promoted lying, stealing, lust, pettiness, and violence on Earth over the centuries.

2. Excluding the fact that negative discarnates (also seen by Robert Monroe) are not manipulating humans, it is possible that these wicked humans are OPs (Sims) who are programmed to wreak havoc as part of the Drama – they are **catalyst**, nasty but catalyst nonetheless...

Number two above raises some eyebrows. How could a loving God allow that? How could basically benevolent gods (running the Sphere) do that?

The answer lies in a four-part answer, like it or not:

(a) We are eternal souls and cannot be killed – if so, we just reincarnate. So death is not the end of the book of life, just the end of a chapter.

(b) Man's idea of Justice and Fairness is not in synch with that of Ascended Beings who call the shots and determine the Drama in each Era.

(c) Just because the Drama is rough and some people die does not mean it is wrong or bad. Some souls need horrific experiences to wake up and become better people.

(d) Earth is not about having fun nor is it all sweetness and light. The Earth is whatever it needs to be to grow souls.

Our problem as humans is that we expect Life to bring us what we want, we expect it to be like we want it, and when it isn't we sit back and say things like,

"Why do bad things happen to good people?"

The answer is:

1. because you're not as good as you think you are
   or
2. this is a test and you are expected to handle it.

Ok, enough sermonizing, but some people do ask that question <u>and</u> we are examining what the Simulation is and what it expects of us.

With that in mind, let's move on to a significant part of this book: Let's deal with some of the key issues of Earth being a Simulation. Some are moral/ethical, some are religious, and others are scientific.

Cryptozoids

And just so that we have some mystery on Earth, the gods also insert strange animals or beings:

Bigfoot	Chupacabra
Mermaids	Men in Black
Mothman	

And, No, Nessie in Loch Ness is not an insert – she is a Plesiosaur, and so is the

Plesiosaur called Champ in Hudson Bay. They are remnants of the swimming dinosaurs, as is the Megalodon (60-foot shark) which has been recently seen <u>and photographed</u>.

Anomalies, Part III

And it isn't just animals and creatures that are inserted into the Simulation.

Back in 1897 there were a number of **airships** that cruised around the country, or maybe it was just one, but the accounts of it precede Man becoming airborne with the Wright Brothers who flew gliders (1901) and finally a powered version at Kitty Hawk, NC (1903). The significant flight they made covered 20 miles and was aloft for 33 minutes (meaning an average speed of 40 mph) in 1905. [133]

> The significance of airships is that the gods were showing Man something he could do – to move him along the path in his development.

Of course, Man created hot air balloons and attached gondolas....

Credit: Wikipedia: photo taken by <u>Kropsoq</u>

And then added propellers to the gondolas...

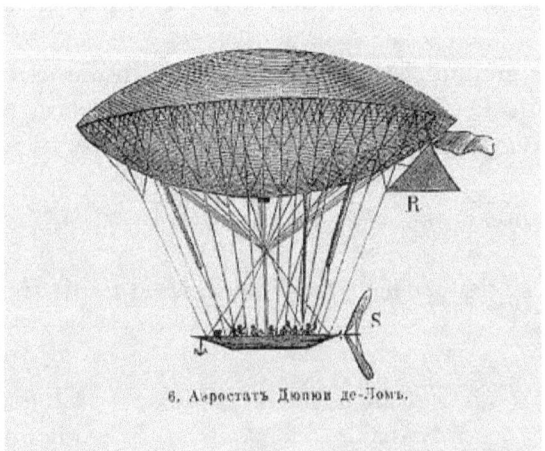

Credit: Wikipedia: <u>http://www.cultinfo.ru/fulltext/1/001/007/006/6953.htm</u>

Then, the Montgolfier Brothers in France demonstrated the hot air balloon concept successfully in 1783... so that things flying thru the air were not so novel by 1987. Thus to say that there was an **1897 Airship Mystery** (qv) refers to **whoever** was flying the one(s) that overflew the US that year... it was not the airship itself that was the mystery... unless it was an insertion designed to show Man how a really fast-moving airship could be built...

Other strange, or anachronistic objects have included the **Antikythera** found in the ocean off Greece:

Credit: Bing Images: erkelzaar.tsudao.com

And when scientists replicated with today's tools, it looked like this:

...which in fact turns out to be both a clock and an astronomical calculator.

Credit: Bing Images: cnccookbook.com

What makes this device so special is that it was a <u>clock</u> for keeping track of Olympiads, and an <u>astronomical device</u> for tracking eclipses, and it had at least **30 meshing bronze gears.** The signs of the Zodiac, months and celestial constellations were all tracked. Found in 1901, it was dated to about 150 BC, and significantly, this technology and workmanship was lost and not seen again until Man built mechanical clocks in the fourteenth century. [134]

And surely, as far as objects go, the **Baghdad Battery** is really an odd one:

Credit: Bing Images: 914electric.wordpress.com

Inside:

The electrolyte was thought to be lemon juice.

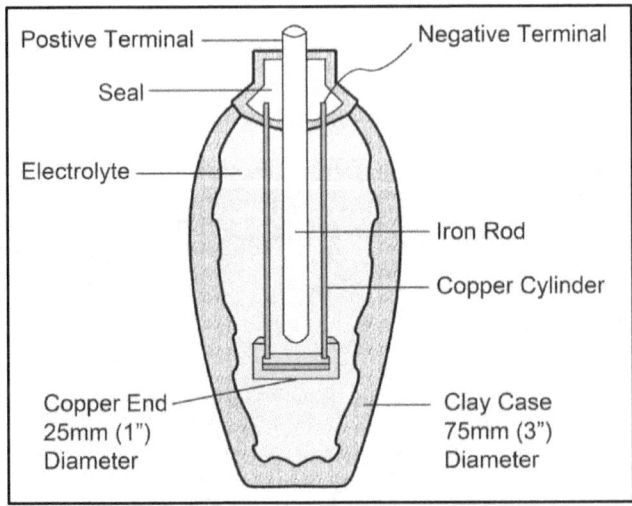

Credit: Bing Images: yebeju.blogspot.com

Postive Terminal — Negative Terminal
Seal —
Electrolyte —
Iron Rod
Copper Cylinder
Copper End 25mm (1") Diameter
Clay Case 75mm (3") Diameter

The smaller version produces a very weak current, but several hooked together could have served an artisan as a way to electroplate jewelry. Ten of them connected can produce 4 volts. [135]

And the Baghdad Battery might have been a version of something larger that powered the **Dendera 'Lights' in Egypt**:

Credit: Bing Images: mdw-ntr.com

Those do look like bulbs with filaments, and cords. The Light on the right even has a support of some sort that has 4 "insulator" coils… Supposing the Egyptians had electric lights and that was how they carved the walls of temples and sarcophagi underground – **<u>and</u> didn't leave any soot from flaming torches!?**

> Don't laugh: the Anunnaki, or the Aldebarans who settled Atlantis, could have brought the technology to Egypt.

Next, we have **ancient airplanes** (jet fighters, not gliders) from Peru/Columbia:

There is even an insignia on the tail, where most craft nowadays have a symbol or id of some sort:

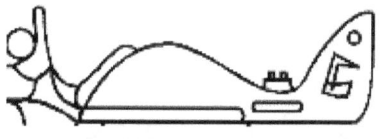

Credit: http://www.bibliotecapleyades.net/esp_aviones_precolom02.htm

Next we have the **Ashoka Pillar** in Delhi, India that is over 1500 years old, made of iron and does not rust: It is 23' tall and weighs 6 tons.

Man did not make this pillar… it is constructed of a version of iron that does not rust, and the sections are expertly 'welded' together.

Credit: http://www.bibliotecapleyades.net/egipto/esp_electricidad_egipto_1.htm

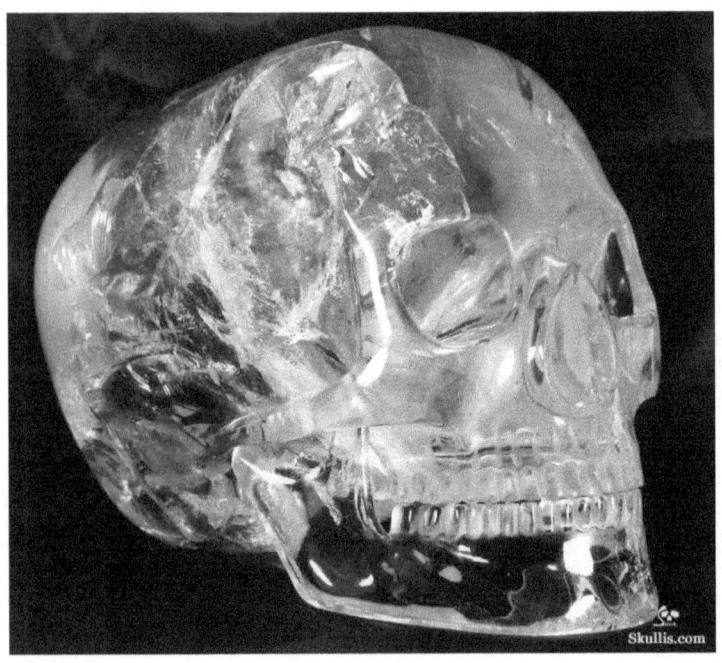

Lastly, a **Crystal Skull** was found in 1924 by Anna Mitchell-Hedges while exploring a Mayan ruin *Labaantun* in Belize. She found it buried in the rubble of a collapsed altar. It is almost clear quartz and almost the same size as the human skull. Uniquely, the jaw is separate from the rest of the skull.

Credit: Bing Images: Skullis.com

What is unique about this skull is that there are no grinding or cutting marks on it, and it has been examined professionally by laboratory after laboratory and found to be genuine high quality quartz and does not appear to have been made by Man. It may in fact have been made by a very advanced civilization – on Earth or elsewhere... or maybe it was just **inserted**, and Anna was guided to find it.

Reflections

Certainly the Maya did not make the skull, and you have to wonder what they would be doing with it... Maybe worshipping it in a temple?... but where did they get it? Again--- it was inserted or perhaps the ETs gave it to them before the Maya Civilization disappeared about AD 800-900 (remember that the Earth was replicated into the Sphere about AD 900). And that is the point of this last section: Where did the advanced knowledge come from to make these things... or: Who gave the object (or the technology) to Man?

The reason for suggesting the Simulation gods **inserted** these things is that Man was not capable of making some of the above things – and when he did, he had to have been given the knowledge (as in the Baghdad Battery). And it is the purpose of the School to grow Man, to amaze him, make him wonder, make him stretch a bit and learn more, do more and be more than he has before. He was given fire and the wheel.

The less plausible but possible option is that ETs came to Earth and gave Man (or left behind) their technology ... and that would have had to have been before the 4D Earth was replicated into the 3D Construct... as 4D ETs cannot now visit Man.

Further introspection will show a certain worth to this Simulation viewpoint – it is in synch with **Occam's Razor.** It is the simplest answer to where these things came from.

Chapter 10: Issues with Simulation

There are about 8 minor issues associated with an Earth Simulation. Ethics, science and religion are all involved. The last two chapters will address the heavier issues.

> Before beginning the review, it is appropriate to suggest that the reader keep an open mind. Don't forget that we are all being watched, and thus evaluated, so I have a responsibility to present the facts as straight as I can, and the reader should not blindly believe anything, but treat the information as **catalyst** – something to think about. The worst thing a reader can do is sit back, say to himself, "I don't like it, therefore it is false."
>
> Please do not do that to yourself. Our opinions and belief systems don't count on the Other Side. What counts is thinking outside the box that someone else created for you. Consider what the information can do FOR you and whether it is useful. The gods do not judge you by whether you get it all correct while on Earth – They want to see what you do with what you think you know. Do you respect yourself enough to contribute to your own growth?
>
> Are you loving and learning? It takes <u>both</u> to graduate.

Of course, someone will ask, by now, What makes you think you know all these answers? And of course, I don't know it all. However, I answer this in Appendix B.

So the first issue that comes to mind, if we are in a Simulation…

1. Why Care? Just Be Indifferent.

One of the responses that has come up says that if we are in a contained 3D Construct, and don't have much freewill, we should do whatever we want to.

We <u>do</u> have a lot of freewill and should be careful what we do. The gods will still evaluate us, and while there is no Hell, and the gods don't punish us, by not doing our best we look dysfunctional to them. Apathy leads to even more dysfunctional behavior.

If enough people look dysfunctional, the gods might do a Reset and we start over again.

So why bother? To do nothing of value, just exist, not care and still take up space would make you one of the **"useless eaters"** that even the PTB on Earth want to get rid of (see the Georgia Guidestones in Chapter 1)... such people use limited resources (clean water, air, and food) and put back nothing of value in society – What do you think the gods' reaction would be?

Ok, at this point it is important to examine what the gods may be doing with us.

2. Why Are They Simulating Us?

There are four possibilities...

While this is not the answer I was given, it is something to consider. The best Quantum Physicist minds out there have theorized that our future society has reached a very advanced technological level, and they **(a)** either amuse themselves by creating simulations – a very advanced form of Reality TV, or **(b)** they seek answers to events in the historical past (our time), and they want answers, or **(c)** they are simulated by a still higher realm (meaning 'nested' simulations).

If the reason is **(a)** and we act like idiots and warmongers, they might just terminate the simulation. Not good.

If the answer is **(b)** and we show ourselves to be dysfunctional and genetically predisposed to lie, cheat, and steal... they will have their answer(s) and terminate the simulation.

If the answer is **(c)** and those simulating us are themselves simulated, we might find that further up the chain, someone pulls the plug and deletes their simulation in which all lower versions find themelves, **a nested simulation.** Thus a nested termination! An interesting idea, and that was explored in the movie, *The 13th Floor.*

Only in the event that we develop ourselves, and even do what **Ray Kurzweil** suggested, and adopt some **bionic implants** in our bodies, could we interest them. It might historically explain how our progeny got to wherever they are... and then if we still run amok as bionic humans, the gods will terminate the simulation!

It was clearly suggested (VEG, Ch. 10) that Earth has been restarted around AD 900... and reasons for that are not clear. However, the premise of this book is a version **(d)** that says they run the Simulation because they are benevolent and this is a School. If we screw up, they just Reset it, and we try again.

> As was said in TOM, the Interlife exists for souls to evaluate what they did last lifetime, and the Masters counsel, and there are schools and EIZ (Energy Imprinting Zones) where one can be infused with better energy to manifest a better body, mind and action next time in the Earth realm.

In short, the other writers are thinking that future versions of us are creating simulations where the humans are largely preprogrammed, and sentience is simulated, to test certain advanced sociological theories (of the future) and no souls are involved. Perforce, their vision is skewed to a mechanical simulation that is <u>not</u> like the one we are in. [136]

3. Why Plan For the Future?

This again relates to the "I don't care and you shouldn't either" attitude in #1 above. And for reasons stated above, it is counter-productive to your best interest in the long run.

If in fact one of the options a – c Item #2 above is true, then you'd be right – there is no point to anything… we'd all be just simulated rabbits – not just the OPs (or NPCs) as mentioned earlier.

However, Man <u>does</u> have a future and it is (eventually) a great one. Souls are trained here to evolve and grow so that they can leave the Earth Realm and serve a higher function in a higher realm. The true Multiverse needs shepherding and administration, as well as there are beings whose job is to manage energy, create planets, create new types of plants and put it all together in a new world and learn to balance an ecology… terraform new worlds, train baby souls, manage the Akashic Records (Library)…. Earth is not the end of the line! (See Chapter 13.)

4. Is There a God?

Yes, and it is more of an Intelligent Force. It is not a white-bearded old man sitting on a throne, somewhere in the sky.

> For those who believe in an anthropomorphic God, I have some ocean-front property in Montana that I want to sell you…

The Force as expressed in Star Wars comes closer to what the Father of Light, or The One is. So then, how does **prayer** work? Who are we praying to?

Simple: the gods who run this Simulation hear you and if the request is do-able, they answer. If you ask amiss, or for the wrong reason, forget it. It is that simple.

> Unfortunately, we need to rethink the idea that a bearded white man sitting on a throne, turned to His right and told His Son to go to Earth and die for people's sins... Appendix D in VEG explains why vicarious atonement doesn't work.

Moving right along… **atheists are equally wrong**. The God-force is Intelligence behind everything – in fact, it has been said the Universe is God experiencing Himself… and that has a certain element of truth if as souls, we are all sparks off the original God Soul. And yet, perhaps there is something higher and finer than that viewpoint, but that is the best that enlightened people have come up with over the recent centuries…

And at this point, a brief word needs be said about the different religions on the Earth. Since there is only one God, why are there different teachings about Him? (This was addressed in VEG throughout the book, but here is the final say on that.)

Consider the Anunnaki, having created Man and the humans turn out to be rowdy, petty, violent -- there needs to be a way to control them. So two main things were done, and we still suffer the effects of their manipulations to this day:

1. You instill fear in the primitive, not-too-bright worker humans – tell them that if they displease the all-powerful God, he will smite them, they go to Hell and are punished for eternity… because God is Love!

Because these are worker slaves, you don't want them getting together, talking things over and rising up against the Anunnaki, so you do the following:

2. You give a different form of religion to different groups of Man around the planet – so that given their pettiness and ignorance, each group will argue that theirs is the right one – after all, didn't the gods give their religion to them?

And just for good measure,

> 3. You scramble their language – so that they can't communicate and unite against their masters. (Hint: Tower of Babel)

How's it working so far?

5. What and Where is Evil?

Again, it is being considered nowadays that Evil sources itself in human ignorance … When one knows better (and is not so ignorant), s/he does better, and does less nasty things to others because s/he knows that "what you do to others comes back to you." Karma, again.

The problem lies in the NPCs as in any video game – they can do nasty things and get away with it (if they aren't killed off)… What that means is that Charles Manson, Son of Sam, Richard Ramirez, Jim Jones, and Ted Bundy, just to mention a few, have shown us what an OP (a conscienceless human with no soul) can do if they run amok. And still our society does not learn what the Greeks, Mayans and the Russians have already discovered… the soulless do exist and they often do nasty things. Why? If they have no soul, which has a connection to a Higher Self with a conscience, they see no reason not to do whatever they want. There is no conscience to suggest they stop. (This is all dealt with in TOM and ASOM.)

Certainly, there is no one Evil Entity that got kicked out of Heaven and set up shop on Earth…or under it. It was examined in VEG Ch 6 how Satan came to be and how the concept of Hell arose from a combination of Greek, Egyptian and Hebrew stories…. Which all began with the original Anunnaki place of punishment!

> The Anunnaki started the pain and punishment concept in
> Sumeria as a way to control rowdy humans… AND they then
> set up priests to administer Religion and threat of punishment,
> and the rest is history. (VEG)

However, the connection to being in a Simulation of type a – c above in Item #2 is that the Programmers (aka 'gods') can insert whatever they want into the Simulation to see what happens. They can initiate a war, disease, an asteroid hitting the Earth… things that the better option **(d)** above will not even consider. Because we are not in a Simulation of type a – c, true, longterm dysfunctionality, such as Hitler subjugating the world, will not happen, and it was orchestrated to stop him.

The side effect of World War II was **catalyst,** albeit a rough one, and many souls learned why compassion, patience and respect are valuable. Such catalyst is allowed to pursue its course for a while since (1) you don't get steel without fire, and (2) souls are eternal, and death is an illusion… you come back and gain spiritual strength.

> While some people will object, if Earth were a safe, all joy, peaceful place where there was no strife, no unemployment, no disease, no war, nothing that ever upset or vexed you, there would be no spiritual growth. It would be boring… that is why there are so few video games that do not pit the player against incredible odds! Unfortunately, humans do not grow in any other milieu.

So can one just coast thru life, having as much fun as possible? In today's world there are New Age teachings that say you are here to be happy and if you aren't, something is wrong. (Note: the two songs: "Be Happy, Don't Worry" and the very infectious video called "Happy.") Sorry but that is not what it is about. We all have to choose to do something while we are here… and we cannot avoid negativity while here on Earth.

By the way, I like the music video Happy, who wouldn't!, but it is reflecting a trend in society to ignore problems and anything negative – "Can't bring me down" says the video – Ok, but can you face it and handle it? The point I'm making is that we are not here to be always happy and any goal to be always happy means you don't deal with reality – which on Earth is designed to shape us all into better souls.

There are several mega churches today where they teach only celebration, love and happiness – their answer to meeting problems is to ignore them and be happy. It would be interesting to see what happens to those leaders of such churches when they get to the Other Side: One of the heaviest responsibilities that I was told about is to present oneself as a bringer of truth, a spiritual leader, in say a Center for Spiritual Living, and then not truly lead the flock into understanding and handling their lives. Hypocrites are dealt with sternly and when they are sent back (next incarnation) the screws are tightened to wake them up.

So Evil is pretending to be a leader of spiritual truth and then leading the flock into error.

We are expected to do something with our lives and gain an inner strength by facing and handling whatever comes up in life – I cannot stress that enough. (I hate to preach, sorry, but I can't stress the point enough – Jumping Freddy Feelgood churches have as much to answer for as those who have lied to people throughout the centuries with false doctrine.)

Thus, you have freewill and the ability to choose... Just choose wisely. The gods, on the Other Side, will evaluate whether you are making progress and it is worth continuing your training... or whether you are becoming defective and they will have to disassemble your energy and try to rebuild you – if you are still salvageable... Seriously. Does anyone really think they can just party and goof off forever?

5a. Reincarnation

This is real and something to consider.

The Catholic Church dealt with the concept of Reincarnation for over 400 years, instituting the **Inquisition** to remove rebellious (heretical) humans, and that didn't work. The teaching just went underground, and has survived in several countries around the world, such as with the Buddhists and Hindus and they are not observably rowdy and irresponsible!

But, the Church had a practical concern: If you believe reincarnation is true, then you will boogie and party in this lifetime, swearing to get it together and do right in the next life! What the Church considered a better way to guide souls on Earth was to tell people that they only had one life to live and if they don't do it correctly, they go to Hell. Again, the Buddhists and Hindus aren't rowdy, or hedonistic... perhaps the Church overreacted.

If you know that you are an **eternal soul** and are responsible for what you do and say, and that "What goes 'round, comes 'round..." you will eventually learn to think twice about negative behavior. (It may take several lifetimes to see the cause and effect connection of one's actions – you will see it in the InterLife, but will you assimilate the lesson?)

So back to the question: What is Evil?

The question is best answered in Metaphysics: Evil is the absence of Good, or in more relevant terms, Evil is the absence of Light. The more **Light** (Wisdom + Love + Respect + Knowledge) one has, the higher the personal vibration that person has... and the more Light you have the less nasty stuff you can do to another... because you know that that is not only your brother/sister, it is in fact You. We are all connected... that is the meaning of the term **Entanglement** in Quantum Physics.

> Kenneth Ring showed in his book, Lessons From the Light, that what we do to others we will experience on the Other Side so that we can

see and understand what our actions did (+ or -) to another. [137]
(Examined more in ASOM, Ch. 7.)

5b. Life Review

Kenneth Ring also suggested we remind ourselves that we will see and be seen on the Other Side and if we remember the phrase **"Movie Time!"** before we speak harshly or hit someone…. We can clean up our act here on Earth. [138] It seems they have a video playback system that replays any and all scenes from your life and you and they will see what you did… hence the 'Movie Time!' warning.

> If it appears that this chapter is tending religious or even metaphysical it is because it is very hard to talk about the purpose and process of the Sphere without sharing that info.

So back to Evil and the Light. An example should make it clear what Evil is.

> Suppose it is Midnite and the house is dark. You head to the Den. It is dark too. You reach your hand around the corner and flip the light switch. There is now light and the room is no longer in darkness.

> Metaphysically, the Light also dispels Darkness.

> Key question: Was the darkness in the room a state unto itself that had to be pushed back, or out of the way by the light? Did the light have to fight its way across the room, shoving some tangible darkness out of the way?

> Of course not. And Evil and Light are the same way. Darkness or Evil is not a state – it is the **absence** of Light (Love, Goodness, Truth).

But you say, "Well, what about the evil demons that cause people to do the wrong things?"

Me: Aaargh. Have you ever seen demons?

"Well, poltergeists. Who threw the stones that Charles Fort talked about?"

Me: Oh, I see. Can you accept that because this is a Simulation that the gods who run it occasionally insert mysteries – Bigfoot, shadow people with red eyes… and they drop fish and throw rocks to add spice to your mundane lives?

"Well, it does make me wonder…"

Me: And that my friend is the purpose.

6. What Are the Programmers Like?

They are benevolent beings who work with the trainers and Masters found in the InterLife. They are not humans from the future, and they are not aliens, or ETs. They have put Earth in a Sphere so that Man would not be manipulated by ETs (Anunnaki and others) which was disrupting the Earth School scenario.

They are not Angels, those are the Beings of Light who reside in the Astral, one of the levels Robert Monroe discovered around Earth. Monroe's *Inspec* was in fact a Being of Light (Angel). (Monroe also discovered discarnates, but they usually do not harass humans.)

7. Is the Simulation Creating Consciousness?

The souls who incarnate into the Earth Sphere already are conscious; it is an aspect of a soul. So a soul has real Consciousness (large C).

On the other hand, an Earth Simulation contains many Sims or NPCs and they have only the consciousness that the Operating System gives them. The OPs have a simulated consciousness which today's scientists consider part of the Computation Consciousness theory:

Computationalism

> Computationalism is a philosophy of mind theory stating that **cognition is a form of computation.**. It is relevant to the Simulation hypothesis in that it illustrates how a simulation could contain conscious subjects, as required by a "virtual people" simulation. For example, it is well known that physical systems can be simulated to some degree of accuracy. If computationalism is correct, and if there is no problem in generating artificial consciousness or cognition, it would establish the theoretical possibility of a simulated reality.
>
> However, the relationship between cognition and phenomenal **qualia** of consciousness is disputed… It is possible that consciousness requires

a vital substrate [carbon-based form of life] that a computer cannot provide, and that simulated people, while behaving appropriately, would be **philosophical zombies**.

This would undermine Nick Bostrom's simulation argument: **we cannot be a simulated consciousness, if consciousness, as we know it, cannot be simulated**. However, the skeptical hypothesis remains intact, we could still be **envatted brains** [brains living in a vat], existing as conscious beings within a simulated environment, even if consciousness cannot be simulated. [Think: *The Matrix*.)

Some theorists have argued that if the "consciousness-is-computation" version of computationalism and mathematical realism … are true then consciousness is computation …and admits of simulation. This argument states that a "Platonic realm" or ultimate ensemble would contain every algorithm, including those which implement consciousness.[139]   [emphasis added]

Other scientists state that it is **not** possible to simulate consciousness equivalent to the full function of the human brain (Chapter 8). Yes, a 'bare bones' looks-like-consciousness can be simulated, but the full complexity of human consciousness with all its states, quirks, memory, intuition and even illogical turns, that could empower a robot, for example, is not possible in today's world, and probably will not be for another 50 years.[140]   Or maybe never.

8. Is the Simulation Really Lucid Dreaming?

Because some readers are going to mention another alternative to Simulation, it is necessary to address and quash one last idea.

Lucid Dreaming is defined as dreaming while maintaining waking consciousness and the dreamer is aware that s/he is dreaming. In addition, the dreamer is said to be able to **control the dream** – turn nightmares into pleasant experiences, change the setting, and or call up particular individuals. And last but not least, the dream is very vivid – "…everything is vibrant and strangely energized."[141]

If that isn't an argument for living in a Simulation, I don't know what is… (with the exception that we can't control the Simulation)!

A recent philosophical consideration in today's world is that of lucid dreaming – Man supposedly living in a dream state and yet is **controlling what he experiences**. Instead of trying to control the world we live in (a false New Age teaching, by the way) and "create one's day" by willing it, a better alternative is to stop avoiding reality and learn to handle whatever comes up.

Catalyst for growth cannot be controlled anyway.

Does the student control the school? Lucid dreaming does not offer any skills for handling this reality – **it is a denial of reality** or an avoidance, and is not really dreaming.

The issue is briefly explainable, however, because many people who think they are lucid dreaming are in fact doing what Robert Monroe did – going **Out of Body** (OBE) – and experiencing another realm or timeline. It often happens to people who are asleep, and they don't know that that was what they were doing.

And it is still avoiding <u>this</u> lifetime and its lessons.

The sequence of historical thinking on this subject runs like this:

8a. Descartes : Dream Hypothesis

Going back to the 1600's, we find René Descartes making two very significant statements.

He is best known for his: **Cogito ergo sum** –
(in French) Je pense donc je suis!
(I think, therefore I am [exist].)

And while he established that he existed because he thought about things, and his doubting things also proved he existed, he wasn't so sure about his waking state: He asked a very provocative question that has come forward into today's world:

How do you know that you are not currently dreaming? [142]

The idea fascinated him that we may think we are awake and living our daily lives, but in fact (reminiscent of the movie *The Matrix*) we may just be dreaming. How would we know? In fact, in *The Matrix*, Morpheus asks Neo,

"Have you ever had a dream, Neo, that you were so sure was real? What if you were unable to wake from that dream? How would you know the difference between the dream world and the real world?" [143]

The connection with Descartes is that René suspected that there was an evil genius that feeds us sensory inputs to give the appearance of an external world. (And the gods that run the Simulation can and do, do that, but they aren't evil.) Thus Descartes' hypothesis:

I am now and have always been dreaming. [144]

Mulling that over and discussing it with other philosophers, he rationalizes his world. He doesn't like the idea of an evil genius giving him sensory input, so he starts rationalizing about God…

> Because God is benevolent, he can have some faith in the account of reality his senses provide him, for God has provided him with a working mind and sensory system and **does not desire to deceive him**. From this supposition, however, he finally establishes the possibility of acquiring knowledge about the world based on deduction *and* perception. ….he can be said to have [decided] that reason is the only reliable method of attaining knowledge.

> Descartes …argues that sensory perceptions come to him involuntarily, and are not willed by him. They are external to his senses, and according to Descartes, **this is evidence of the existence of something outside of his mind,** and thus, an external world. Descartes goes on to show that the things in the external world are material by **arguing that God would not deceive him** as to the ideas that are being transmitted, and that God has given him the "propensity" to believe that such ideas are caused by material things. He gave reasons for thinking that waking thoughts are distinguishable from dreams, and that **one's mind cannot have been "hijacked" by an evil demon** placing an illusory external world before one's senses. [145] [emphasis added]

Yet, what is interesting is that he originally suspected that the world he lived in <u>was</u> inputting ideas and data to him – as an Earth Simulation actually does at times. And then his religious training gets the better of him and he finally decides that God is good and would not trick him and that resolved the issue for him… He was considered a devout Catholic and yet had metaphysical leanings, and said some things about the real nature of Life that led some of his contemporaries like Martin Schoock to consider him an atheist and Pascal to denounce him as a deist. The Catholic Church didn't like what it heard, and so it banned Descartes' books.

8b. Dr. Alan Wolf (1987)

Dr. Wolf delivered a talk in 1987 at the Association for the Study of Dreams in which he asserted that the holographic model of the universe may help explain this odd phenomenon. Dr. Wolf claims to lucid dream himself (OOBE?) and in essence he believes that all dreams are internal holograms inasmuch as Man is part of the hologram of reality (Chapter 3) and in addition to "creating" the world out there that he "sees" around him, when he is asleep he is doing something similar – he is still "seeing" but with his eyes shut. [146]

> Like **Pribram**, Wolf believes our minds create the illusion of reality "out there" through the same kind of processes studied by Bekesy [qv]. …these processes are also what allows the lucid dreamer to create <u>subjective</u> realities in which things like marble floors and flowers are as tangible and real as their so-called <u>objective</u> counterparts…. In fact there may not be much difference between the world at large and the world inside our heads….
>
> Wolf postulates that lucid dreams (and perhaps all dreams) are actually **visits to parallel universes**. They are just smaller holograms within the larger and more inclusive cosmic hologram. He even suggests that the ability to lucid dream might better be called parallel universe awareness.[147] (emphasis added]
>
> (Dr. Pribram was examined in VEG. Ch. 12 and Appendix B, along with Dr. Bohm.)

As was seen in Chapter 9, **the parallel universes are really different aspects of the Simulation – fractally-created**. Since the Earth Simulation does not contain different timelines, the concept of different or parallel universes – within the 3D Construct or Sphere – is a moot point. Timelines and the Multiverse exist **outside** the Sphere, but we cannot easily get there – as Robert Monroe showed in his OOBE ventures – he always hit a barrier that he could not get beyond – that, it is suggested, is the outermost barrier (Konstruct, Chapter 4) that contains the inner Sphere whose job it is to contain us – even as souls.

Nonetheless, Dr. Wolf is really on to something, and we don't know the full extent of what we <u>as souls</u> are permitted to do with the hologram – when we slip into the sleep state, we are in Alpha and sometimes Theta, and it may be possible to "do" something with our awareness of the Simulation, perhaps create a fractal subset that we think we are 'dreaming' our way through… (See **Brain Waves** in the Glossary).

8c. Timothy Freke (2005)

On a more modern note, British philosopher Timothy Freke (pronounced: like Brake but with an 'F') has initiated a concept called **Lucid Living** which is a form of Lucid Dreaming. Lucid Living is waking up from the Dream.

> When we dream, we appear to be one of many characters in our dream-drama. But actually everyone and everything is being imagined by one dreaming awareness. It is the same right now. We appear to be many separate individuals. But actually **we are all different characters in the life-dream that is being dreamt by the One life-dreamer.** And that's who we really are. We are one awareness dreaming itself to be many individuals in the life-dream.

And…

> We are one awareness dreaming itself to be many different personas in the life-dream. We are one awareness experiencing the life-dream from the different perspectives of those different personas.

And…

> When you dream you are both the source of the dream and a character within the dream. Your identity is inherently paradoxical. …You are the life-dreamer imagining yourself to be a particular person in the life-dream. Whilst you identify exclusively with your life-persona you will remain unconsciously engrossed in the life-dream. Lucid living happens when you become conscious of both poles of your paradoxical nature. [148]

So that is a philosophical twist on the Dream aspect of Earth life.

The bottom line is that we <u>are</u> in a Simulation which is like the 3D Earth Construct, and Lucid Dreaming is what we are today calling an ability that the soul has to interface in a unique way with the holographic aspect of our Reality (however generated). A few decades from now, we should know more and may call the experience something else…

Summary

So, there are several ways to know if we're in a Virtual Reality or Simulation but it will take patience and keen observation as Those who put this place together were/are quite expert at it. It is interesting to note that, as was the case back in the days of the Greek and Roman gods, Man is still being "interfered" with, manipulated, or at the very least watched as Phil was in *Groundhog Day*. Fascinating to contemplate.

The answer to our reality is not apparent upon casual observation, although some people are able to <u>intuit the answer</u>. However, the evidence is there, providing we stop rationalizing inconsistencies (glitches) and synchronicities (miracles) as Truman did so often in *The Truman Show*.

As was said in other chapters, one of the ways to determine whether one is in a simulation is to **learn to see auras**. If everyone around you has no aura, that is a safe bet that you're in a simulation of some sort because the OPs are playing a role (as NPCs) in a Drama of which you, and other possible souls, are the 'stars' and it may be just for your benefit (Karma).

Without seeing auras, there is one other way. If you find that you can't communicate with others, they don't listen, no one cares what you think, you can't 'reach' anyone to help them…. or if you always lose, you can't get what you want, and if you do, it is taken from you (losing all the time)…. or you're deathly sick and a Being of Light shows up and asks you if you want to continue… pay attention to such occurrences, inconsistencies, and statistical impossibilities. No one can lose all the time, so sit back and ask yourself if there is something to be learned from a situation, especially if it keeps repeating itself in your life experience.

Simulated Last Thoughts

Ultimately, it may not matter to most people if we are in an HVR Earth Simulation not – the lessons are there to be learned. However, knowing that it is a Simulation can take the 'edge' off our failures and negative events – knowing that it is **catalyst**, a test, and something we are to go THROUGH. (Not stop on.)

And if we learn what we are supposed to learn, and it **is** an HVR, we will be released – like Phil. The worst case scenario would be that it is **not** an HVR, and we are just being manipulated on a 3D rock by any passing ETs…who, with superior technology, could easily subjugate the planet and turn Earth into a prison planet. I prefer the Simulation as a School (option **d**) as it has a purpose and, because it is contained, we have a safe place to learn our lessons, and it also means we are supported in getting out of here, as an Earth Graduate.

The only negative aspect of this issue for ensouled people is being **recycled** in an HVR/Simulation because we don't wake up, we don't master the 'lessons' that are being given us, and it goes on forever, or (as is now the case) we are about to destroy the planet's ecological balance and the gods have to do a "Wipe and Reboot," clean up the planet, and start us all over again. Ultimately, if we refuse to wake up, stay dysfunctional and do not move forward, the gods may "disassociate" our energy, as Dr. Newton suggested (Ch.7 in VEG). [149]

For those who "have ears to ear" and assimilate the message, this book is designed to prevent the negative personal and planetary scenarios by revealing where we really are and what is expected of us.

And lastly, there is an outstanding issue of Simulation versus: Creation, Genetics, Religion and the Endtime. If the Simulation is real, what does that do to Religion, Creation and Man's future? Are Man's genetics being reworked? Will the Endtimes happen as the Bible says?

But first, let's take a quick look at the two possible alternatives for what Earth really is. This book, so far, presented the VR Sphere concept, based on what data was given the author in 2008. As crazy as it sounds, before one does the research, there is another contender for our Reality, and it has biblical support as well as growing scientific evidence.

Chapter 11: Flat Earth vs Round Earth

At this point, to substantiate the VR Sphere concept, we need to take a brief look at the **Flat Earth (FE) theory** – and before you put the book down in disgust, please read on as there is some very heavy evidence from the FE Theory that also applies to the VR Sphere. Unlike the FE Theory, the VR Sphere makes very few claims for authenticity, yet does borrow <u>a couple</u> of 'proofs' from the FE Theory. And I urge caution, as trying to prove things 'scientifically' with pictures is nowadays a problem because it is all too easy to fake something using CGI (computer graphics), Photoshop ®, or <u>as NASA has done</u> with the pictures of the Earth allegedly taken from the Moon, and to create a <u>composite</u> of Earth from space – and that will be examined shortly – it has been done. So pictures do not always prove anything.

So instead of immediately discounting the FE concept, as we are all wont to do, as you read thru this chapter, you'll at least find some interesting **"brain candy"** – as if the book so far hasn't had enough food for thought! And you will see that some of what was historically said about the FE, credibly applies to the VR Sphere.

> I want to make it clear that I started this chapter to include and **prove the FE Theory wrong**. In the process I found reasonable and intriguing FE arguments that forced me to re-evaluate the VR Sphere. After researching further, several FE claims were found to substantiate the VR Sphere in a way I had not thought of when I was originally given the VR Sphere material (in VEG). Ironically there is support for both the VR Sphere and the FE arguments.

Ancient Earth Concepts

We tend to think of humans who lived about 4000 years ago, and 2000 years ago, and even 500 years ago as not too bright – that somehow we are the pinnacle of human intelligence and learning. Thus we have a habit of denigrating whatever our ancestors believed in, including their myths (as examined in <u>Anunnaki Legacy</u>). And yet, one has to wonder: How could so many disparate cultures separated by thousands of miles and sometimes thousands of years, all have believed the same thing about the Earth? The Mayans did not communicate with the Hindus, the Amerindians did not communicate with the Zulu in Africa, the Chinese did not communicate with the Norse, and yet **they all considered the Earth to be flat** – and often supported by a giant turtle... Why a turtle? Except for the Greeks (Pythagoras), the ancients all thought the Earth was flat. Where did they get that idea?

Looking up at the Moon on a starry night, one can see the Moon is round, and during the day, using a dark glass or a piece of papyrus, one can look at the Sun and see that it is round… So, wouldn't the Earth also (reasonably) be round too? So why did the ancients believe the Earth to be flat? Too many ancient peoples did.

> BTW, the similarity of ancient beliefs, including The Flood, Sacred Trees and Serpent/Kundalini Wisdom, Pyramids, and Skygods and Goddesses teaching Man are found all around the Earth as is the Flat Earth belief. These other issues are examined in <u>Anunnaki Legacy</u> (see last two pp of this book); it is obvious the Skygods spread the similar information.

The following are some of the ancient concepts associated with a flat Earth…

Global Cosmologies

According to Wikipedia:

> The **flat Earth** model is an archaic conception of the Earth's shape as a plane or a disk. Many ancient cultures subscribed to a flat Earth cosmography, including Greece until the classical period [the 5th and 4th centuries BC], the Bronze Age and Iron Age civilizations of the Near East until the Hellenistic period, India until the Gupta period (early centuries AD), and China until the 17th century. That paradigm was also typically held in the aboriginal cultures of the Americas, and the notion of a flat Earth domed by the **Firmament** in the shape of an inverted bowl was common in pre-scientific societies. [150]
> [emphasis added]

Lenses and Telescopes

Much of the shift to a non-flat Earth was associated with the development of the telescope (AD 1608) and before that, the Chinese, Greeks and Arabs had experimented with the "lens" concept and discovered the ability to magnify objects near and far – Would they not have thought to experiment and arrange a couple of lenses to have a closer look at the Moon?

According to Wikipedia, again:

> The earliest known lenses were made from polished crystal and quartz, and have been dated as early as **750 BC** for Assyrian lenses such as the Nimrud/Layard lens. There are many similar lenses from ancient Egypt,

Greece, and Babylon. The ancient Romans and Greeks filled glass spheres with water to make lenses. However, **glass** lenses were not thought of until the Middle Ages [ca. 1600s].   [emphasis added]

In the Arab world, among the first to play with optics and research light:

Al-Kindi (c. 801–873 AD) was one of the earliest important optical writers in the Islamic world. Ibn Sahl (c. 940-1000 AD) was an Iraqi mathematician …. who wrote a treatise *On Burning Mirrors and Lenses* in which he set out his understanding of how curved mirrors and lenses bend and focus light.

The earliest known **working telescopes** were the refracting telescopes that appeared in the Netherlands in 1608. Galileo greatly improved upon these designs the following year and Isaac Newton is credited with constructing the first functional reflecting telescope in 1668. [151]

So the point being that <u>officially</u> the telescope was not invented until the 1600s, and yet Man was experimenting with optics and lenses way before that.

Geocosmogonies

The following is a brief survey of the ancient concepts of a Flat Earth. Note the concept of "above" and "below" as well as something above Earth (e.g., a Firmament) and sometimes an Ice Wall.

Hebrew Earth

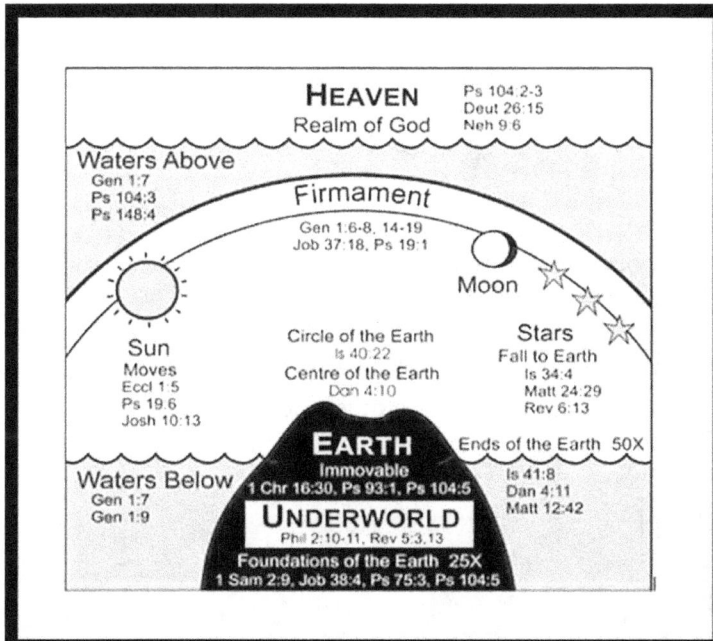

Much of the Hebrew concept was stated in Genesis, from the description given the creation of the Earth, Firmament, The Deep, Sheol, etc..

And again, the following…

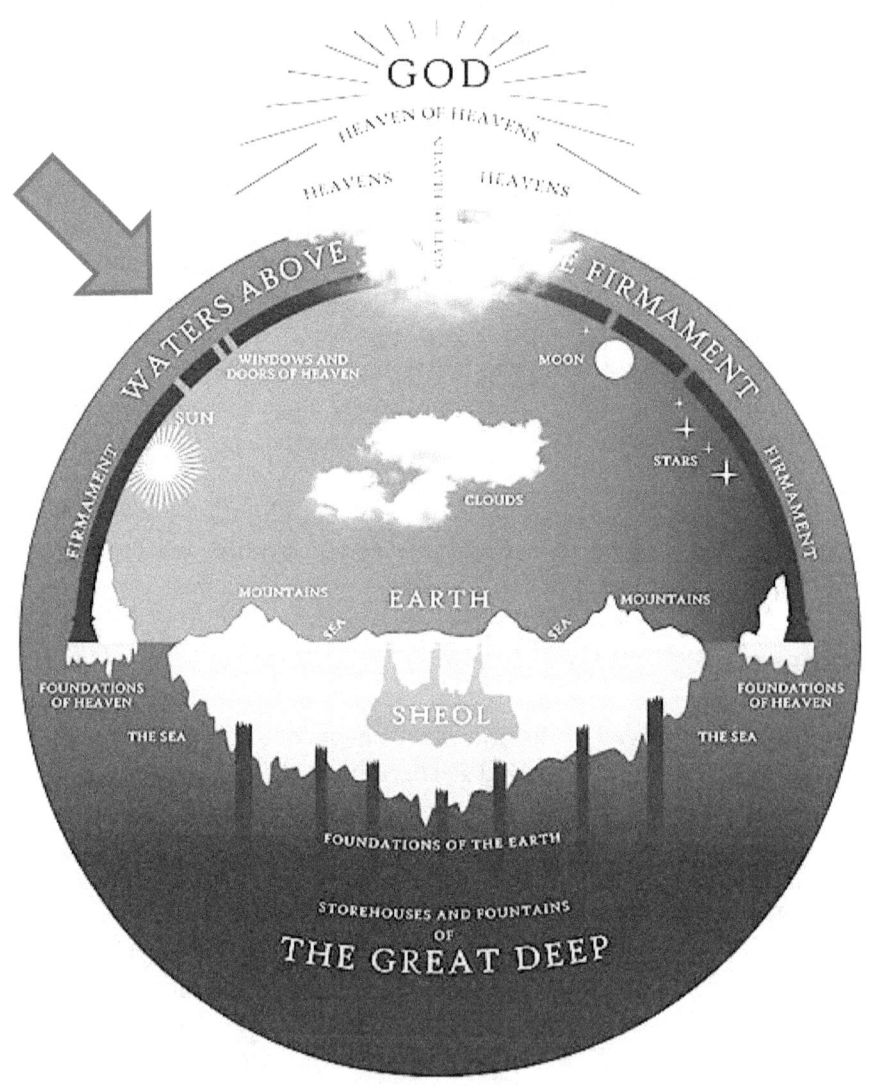

Note: the "windows and doors [portals] of heaven" in the Firmament (arrow)…
Enoch called these "portals" – see end of this chapter.

And yet, the Sumerians had the concept <u>before</u> the Hebrews – who having been taken into captivity by the Babylonians, Assyrians and Egyptians were bound to have absorbed some of their captors' teachings. Case in point is The Flood in Genesis was better and identically delineated in *The Epic of Gilgamesh* and *The Atra Hasis* from Sumeria.

Sumerian Earth

In addition to knowing that there were 10 planets in the solar system (because the Anunnaki told them), early Man also had a Flat Earth concept:

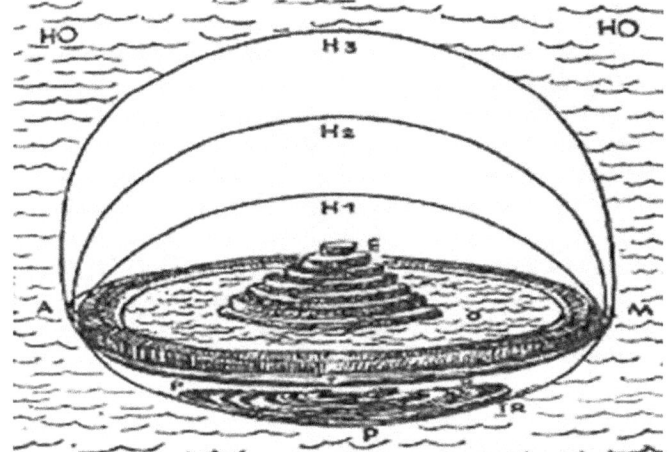

It is strongly suggested that the Anunnaki told Man the Earth was flat.

And another concept (below):

Note: Anu was the great Sumerian god and note the 'horns' at the top.

The world of Sumerians is like an inverted plate in a salty ocean. Below this plate lie the "hell" called "kur" by Sumerians. Above the plate are the arched roofs of heaven. The whole system is imbedded in a fresh water ocean. Rain comes from this fresh water ocean by opening doors in the vaults of heaven.

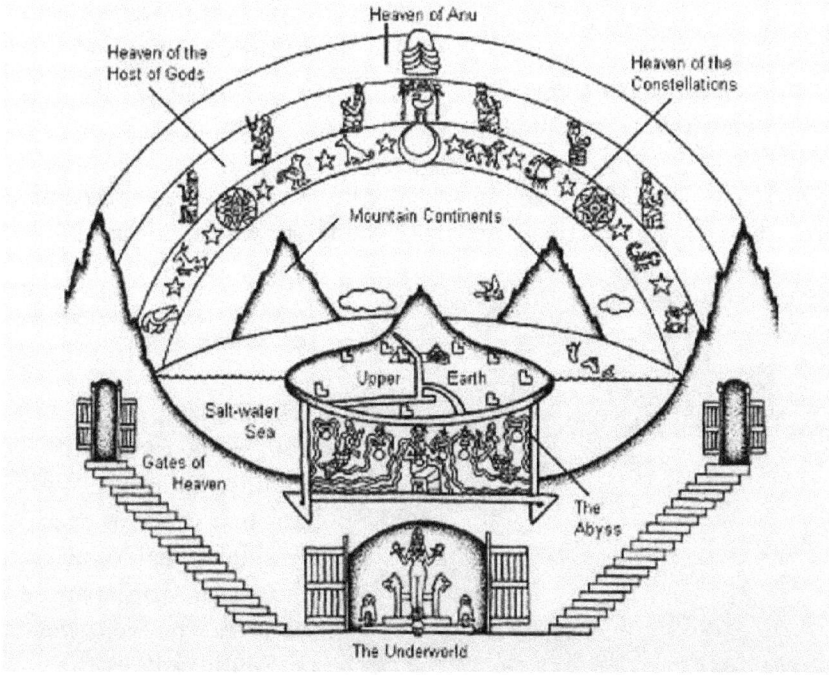

The 4 'horns' on the helmet at the top of the left diagram, in Anu's realm, signify rank as has been examined in <u>Virtual Earth Graduate</u>, and also in <u>Anunnaki Legacy</u>. The statue at the bottom, in the Underworld represents an Anunnaki goddess, Ereshkigal.

Biblical Earth

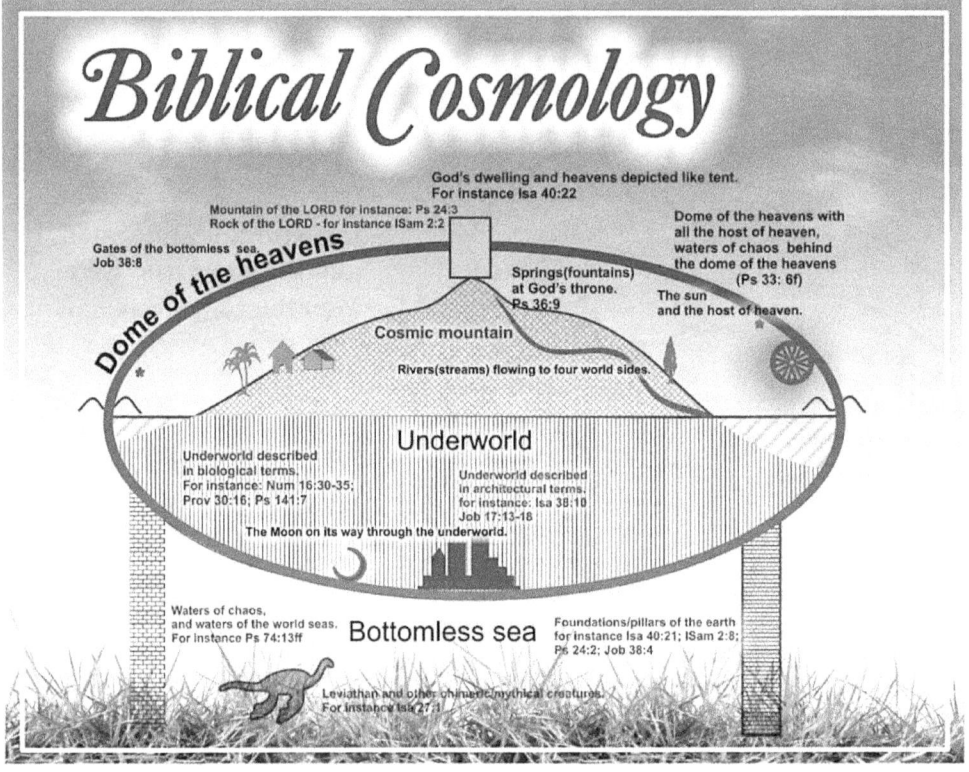

The above paradigm repeats the Hebrew concept of the Universe and Earth. The "cosmic mountain" is a common theme – as well as the **"four world sides"** (by the river). Note the edges are contained (mountains or Ice Wall). This is also **Enoch's** description of Earth (in a section at the end of this chapter).

Notice the mountains in all 4 preceding diagrams. Note the location of the Sun and the Moon in the preceding diagrams – they are part of the Earth realm, not out in space…

Ice Wall

The flat plane of land/sea is usually bounded by an **Ice Wall**. This may come to us as a result of the Norse, Phoenicians, or Japanese seafarers who went way far South and saw the huge ice sheets which are now Antarctica.

Japanese Earth

The Japanese also have a flat concept (middle square plane), but Earth is not contained under a dome:

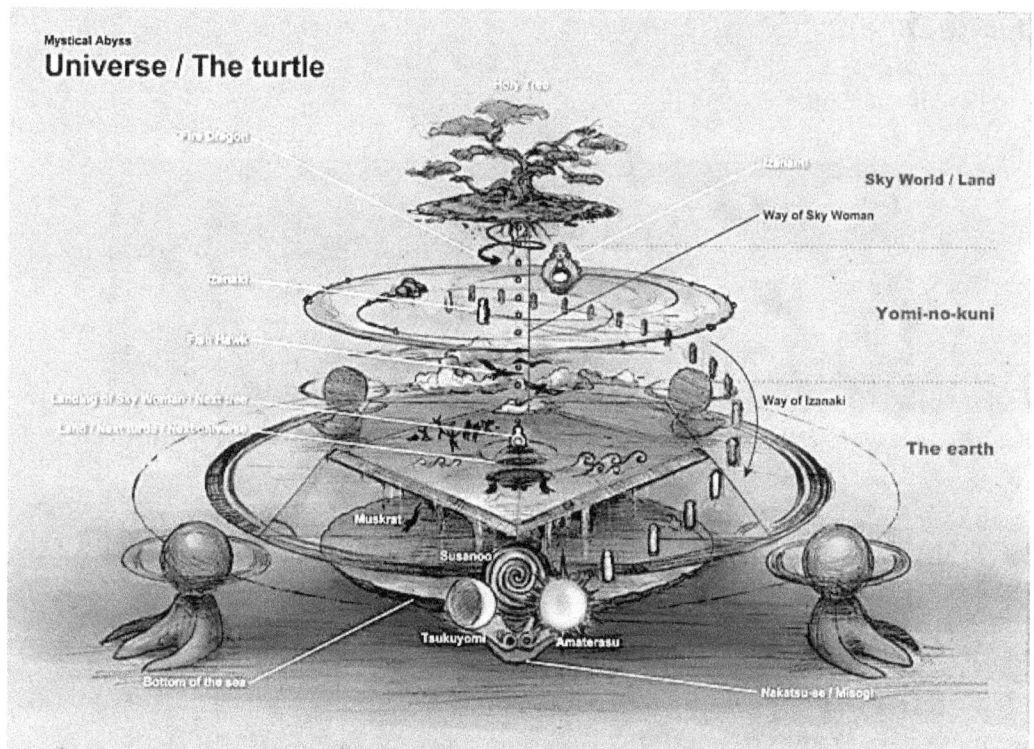

Note that the Earth rests on a giant turtle (two 'feet' on either side), and there is a "holy tree" in the Sky World area. **Sky Woman** (Izanami) comes down to Earth and lands on a small Earth supported by a second, smaller turtle.

That is similar to the tale told by the **Amerindians** (right):

According to the Amerindians, chiefly **the Iroquois** wherein the Sky People dwelt above the Earth, one day Sky Woman fell thru a hole down onto the back of a giant turtle whose back was covered with mud. She also created the Sun, Moon and stars. (This was examined in Anunnaki Legacy, Ch. 4.) Sky Woman is similar to the Chinese NuWa and Sumerian Inanna.

Chinese Earth

And here is the Chinese version…

The Earth on the Cosmic Turtle ultimately supporting Mount Meru.
(credit: Bing Images: codoh.com)

Again, we have the turtle supporting land, and a mountain, **Mt. Meru**. Of course, that relates to the Hindu version…

Hindu Earth

Hindu mythology has various accounts of World Tortoises, besides a World Serpent (**Shesha**), Kurmaraja and World-Elephants.

The most widespread name given to the tortoise is **Kurma** or Kurmaraja. ("raja" means king). The Shatapatha Brahmana identifies the earth as its lower shell, the atmosphere as its body and the vault of heaven as its upper shell. The concept of World-

(credit: ulc.org)

Tortoise and World-Elephant was conflated in popular or rhetorical references to Hindu mythology. An alleged tortoise ***Chukwa*** supports **Mahapadm**a, the elephants supporting holy (home of the gods) **Mount Meru.**

Sure enough, the Chinese, Hindus, and Japanese are geographically close enough that a common myth is understandable, and yet, the Amerindian Iroquois are not close to China/India, so how come they share the similar myth? Is it possible, as <u>Anunnaki Legacy</u> suggests, that the Iroquois and the Hindu have a similar origin (Lemuria?). It is already known that the Zuñi and Japanese share a common ancestry as their language and pottery suggests, [152] and Japanese Jomon pottery motif was discovered in Chile years ago, indicating that the Japanese did sail to Chile.

Amerindian Earth

Notably the **Navajo Cosmos** looks like this:

We are back to flat Earth under a dome again, mountains on both sides of the land, and the Sun is the orb, top left in the diagram.

Some cultures just had to have Earth supported by <u>something</u> and they chose a turtle (sea-based) and/or elephants (large land animal).

Navajo Cosmos
based on March 1990
National Geographic
"Ancient Skywatchers"

Art revised for web by
http://www.edwardtbabinski.us/

And that resembles the Celtic version…

Celtic Earth

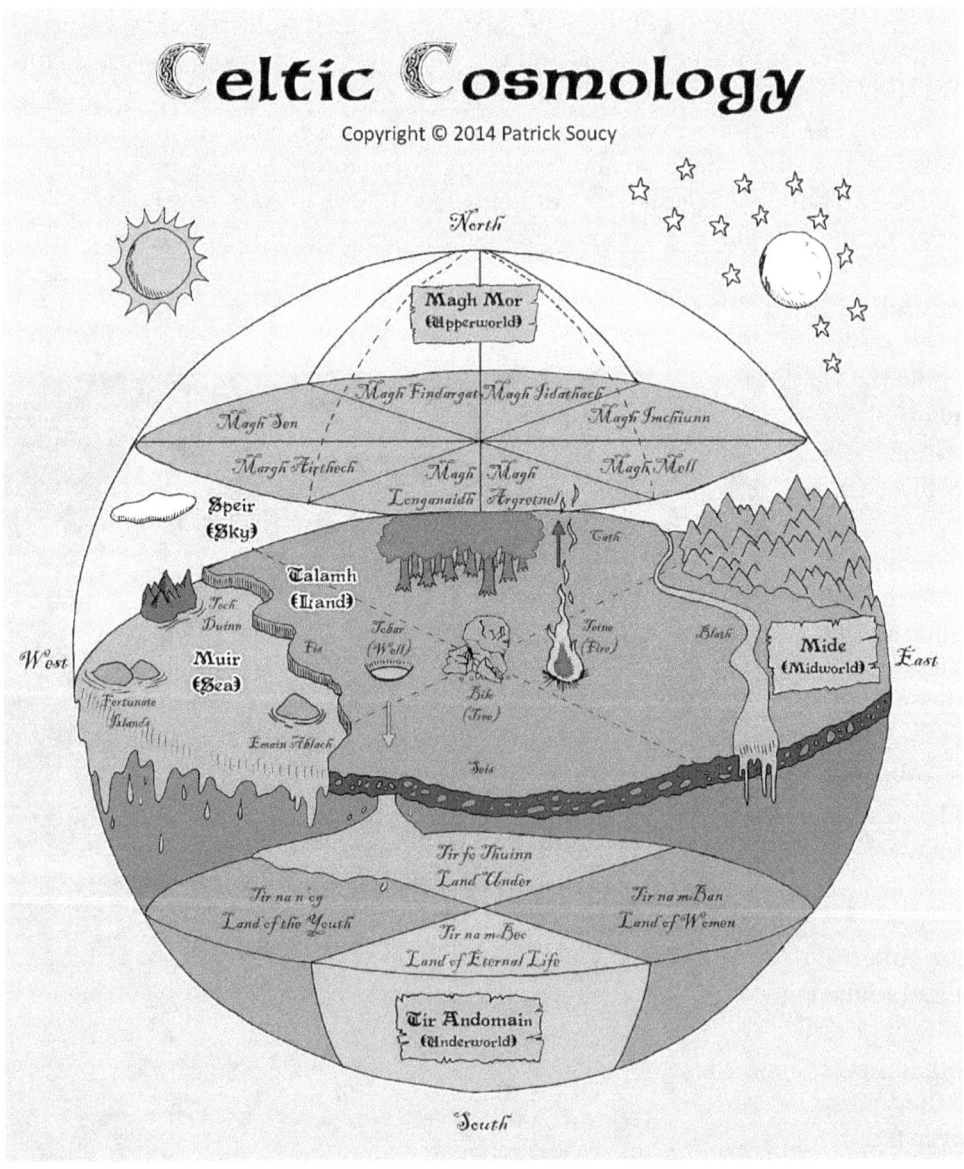

This shows the Earth <u>realm</u> as a sphere because by the time the Celts documented their cosmogony, it was being considered that Earth might also be a sphere (see the Greeks). But there is still an Upper World, and an Underworld and a curious river (right center) flowing off the **flat land**… Except that the drawing encases all in a sphere, **the land is flat** and the Sun and Moon are close to the Earth.

And since the Celts in large manner were a product of Germanic tribes from Europe and Scandinavian Vikings, it makes sense that the Celts drew some of their cosmogony from the Norse peoples:

Norse Earth

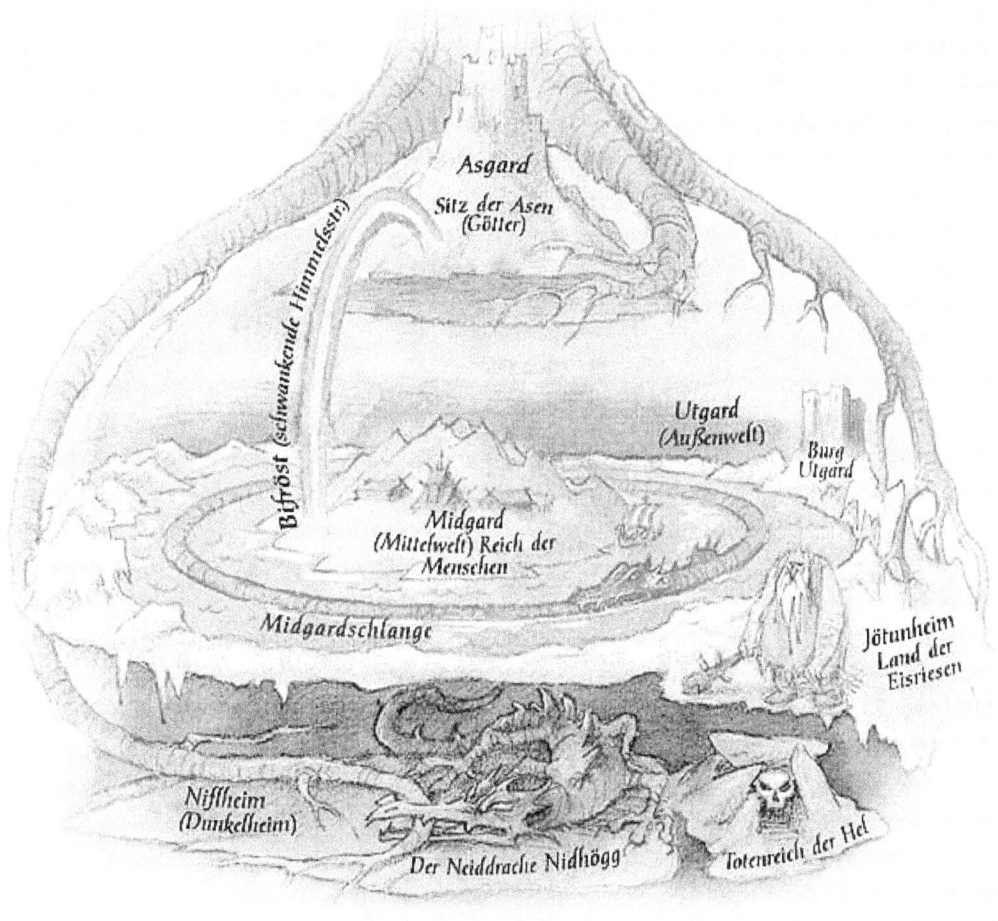

In this case, note that Midgard + ocean, the land of the humans, is surrounded by an **Ice Wall** – which is what one would encounter if one sails from anywhere on the Earth (flat or round) and goes due South! Note that Hel (English: Hell) is bottom right under the ground, and Asgard, the <u>Arctic</u> realm of the gods, is above Midgard. Instead of elephants or a turtle, the Norse said there was a nasty dragon guarding the Lower World, and a nasty serpent keeping one from going too far out to sea. Why?

The traditional Norse representation of Earth (as used in <u>Anunnaki Legacy</u>) is the following:

Again, note the Tree of Life, the Flat Earth inhabited by humans, the **Ice Wall** (aka Antarctica) and the concept of Above and Below Realms. You have to ask yourself, How did the Norse know there was an Ice Wall far south on Earth – unless they sailed as far south as they could go? Obviously the sea serpent didn't deter them.

And then we come to the Mexican/Mayan concept which follows. This diagram does not include **Xibalba**, the Mayan Underworld: home of the Mayan Death gods. But it is interesting that the Maya picked up the concept of Earth, Air, Fire, and Water as was the similar teaching in China.

Pachacuti is an Incan word, so the following diagram is a generic concept applicable to Latin and South American native mythology.

Mexican/Mayan Earth

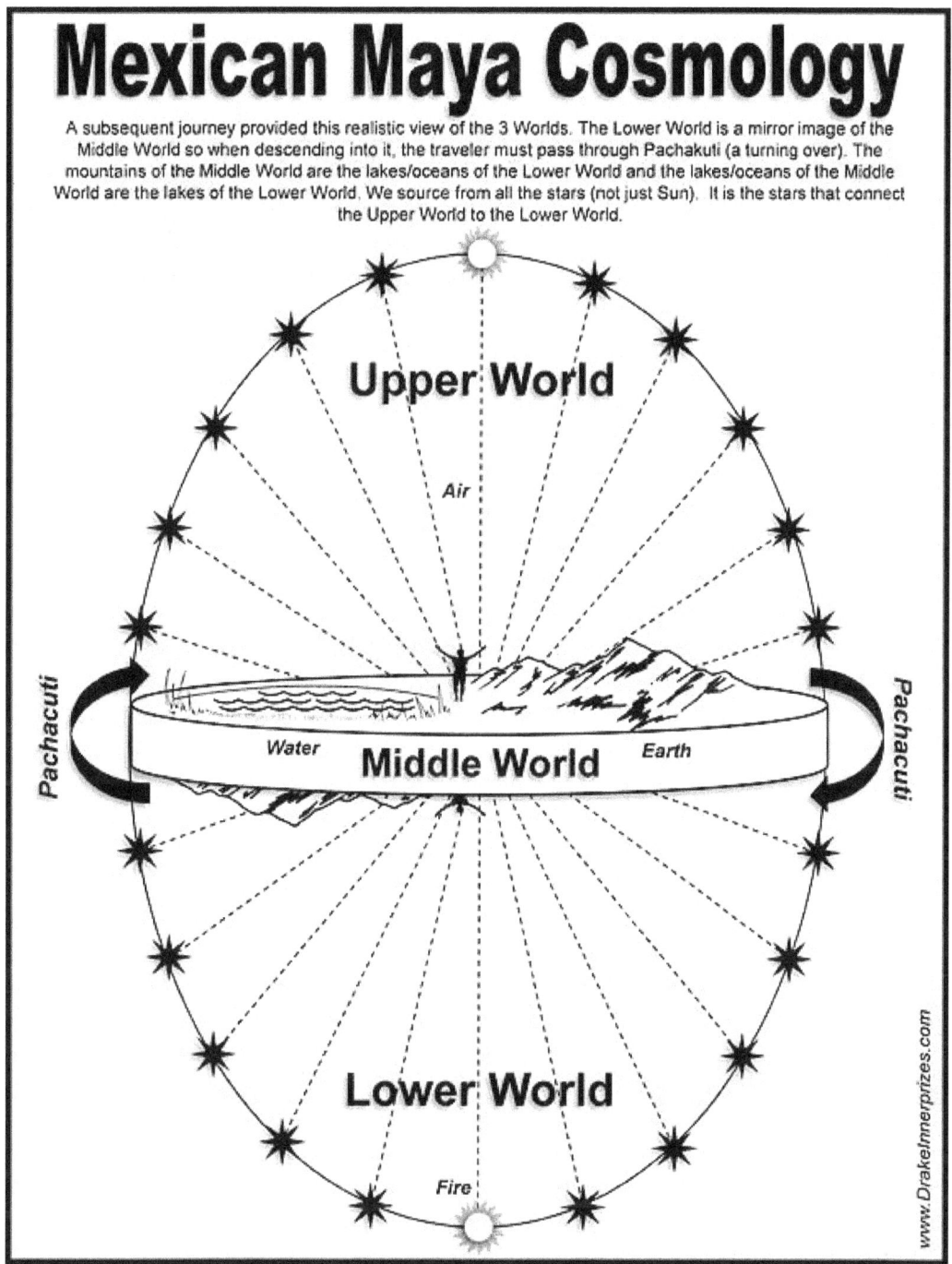

Note the term "Pachacuti" in the middle of the diagram. That relates to the Incan concept of movement of the soul between realms…

Incan Earth

Earth souls pass into and out of the Earth Living Realm... 'Pachacuti' which may suggest a curious connection with reincarnation as souls seek "completion"...

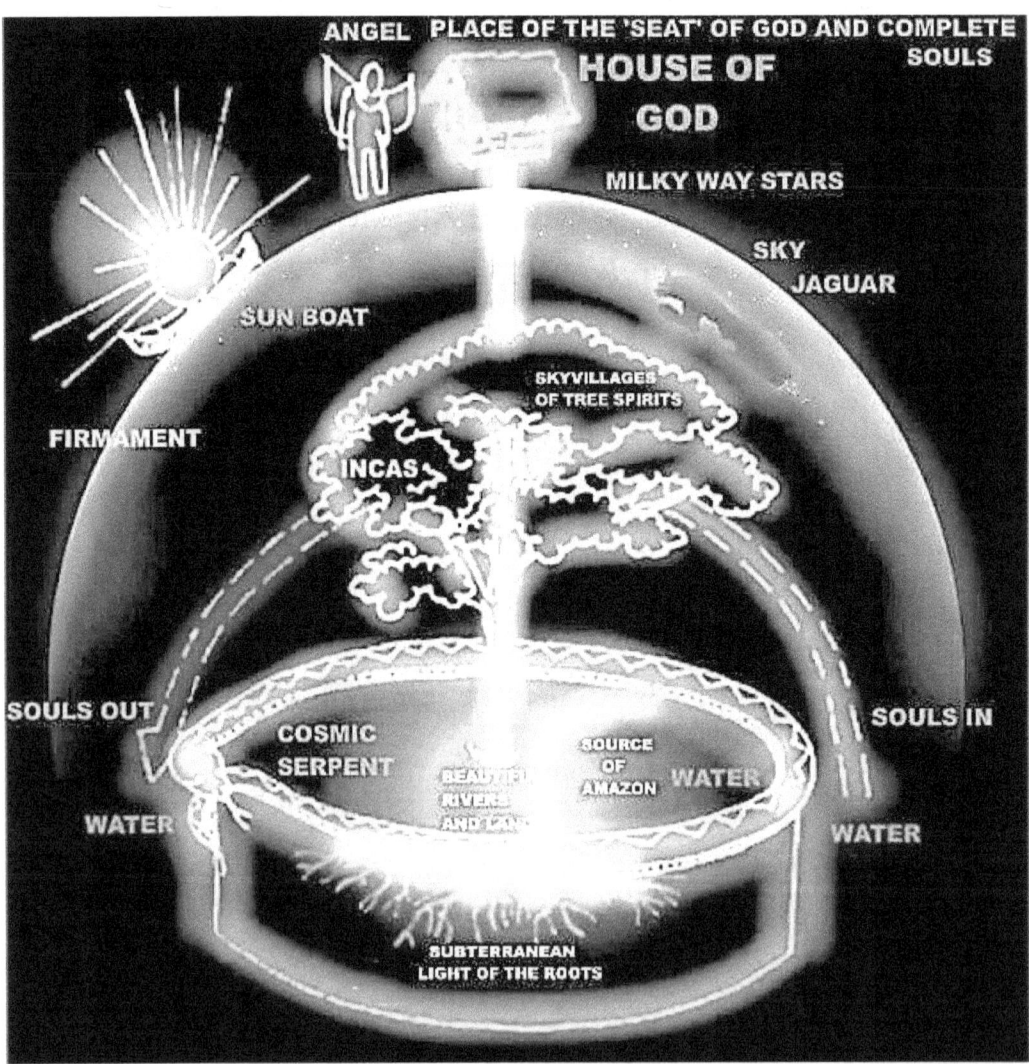

And again, there is a Firmament, a Serpent (like the Norse) guarding the edge of the world, and the Sun is just outside the Firmament, with the **stars in the Firmament**... and there is an Above and a Below, plus a Tree of Life.

The Incan Cosmos has also been represented in a more familiar way:

Inca Cosmos
based on March 1990
National Geographic
"Ancient Skywatchers"
Art revised for web by
http://www.edwardtbabinski.us/

Does this resemble the Amerindian version, shown earlier? Note that the Sun and Moon are <u>within the Firmament</u>. There is also the curious **Ice Wall** so many cultures have pictured... they all seem to know that Antarctica existed but called it a wall containing the Earth.

By the way, with the great **seafaring peoples** of Scandinavia, Japan, and Phoenicia why did they not try to sail around the Ice Wall south of Africa and South America? Or did some of them, and we don't know about it – it stretches for quite a distance!

So what does the Ice Wall look like?

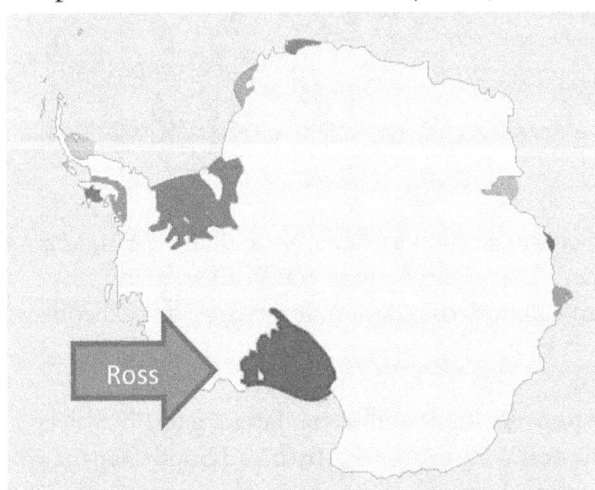

Ross Ice Shelf
(credit: National Geographic.)

Except for the shaded Ice Shelves (below) where boats and men can access the land, most of Antarctica's periphery is an Ice Wall. The above is a real picture and shows that there is a 150' tall wall of ice in most parts of Antarctica. There are also inlets and beaches near the **Ross Ice Shelf** (closer to Australia) where boats can land and men can load/unload supplies and equipment, and the United States has an installation there called **Little America**.

(credit: By Diti Torterat, CC BY 2.0 fr,
httpscommons.wikimedia.orgwindex.phpcurid=5025098)

So how can there be an Ice Wall encompassing the Flat Earth perimeter if explorers can land and go inland? Because there is a 300-mile snow & ice land region <u>before</u>

one encounters **the base of the Firmament** and can go no farther (see Enoch at chapter's end).

Last but not least we come to the Greek/Roman version...

Greek/Roman Earth

The outer edge of the diagram suggests the encircling land is called "Ringland" and "Terra Circumnre." Not an Ice Wall. Note also the label "Atlantis" in the ocean West of the British Isles... just where Plato said it was. In addition, the artist pictured the Moon, the Sun, Mars and Venus, running clockwise from the upper left.

Isn't it curious that the Mediterranean area is well-drawn but Africa and Asia are a guess? So if they have the Boundary Wall, what made them think the Earth was round? And where is Australia?

And for comparison, as we are about to see what the Flat Earth people say today, the suggested Earth layout is as follows…..

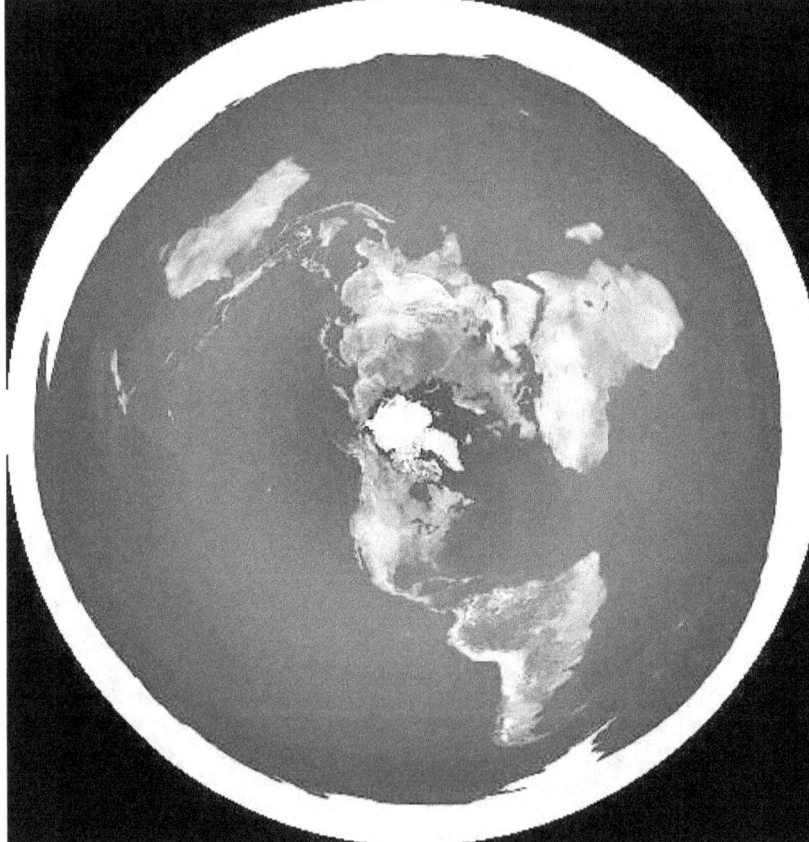

Official Flat Earth picture.

Note the **Ice Wall** again (aka Antarctica) surrounding the land and sea.

Ross Shelf is at 9 o'clock.

By the way, given the above picture, even a Flat Earth can be circumnavigated. As the diagram (right) shows, **circumnavigating the world does not prove it is a sphere**…

This is your 1st jolt with Reality.

(credit: Eric Dubay, The Flat Earth Conspiracy.)

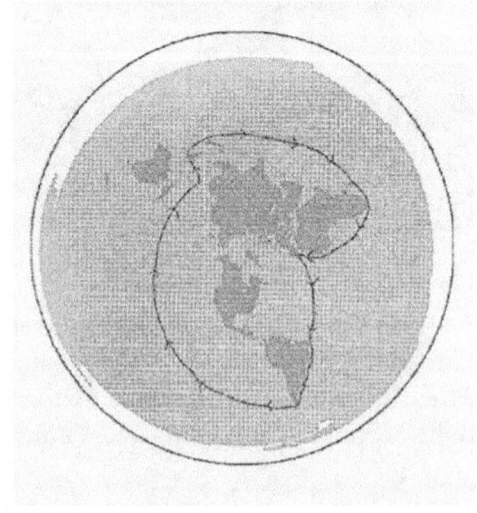

All of that to show that the concept of a Flat Earth was known around the world. Was this something humans just imagined because to them the horizon looked flat? Or as other books have intimated, there were Others here, called the Anunnaki, the Shining Ones, the Watchers, or just the Skygods… and the gods communicated it to Man? And how did they get past the Firmament – or were they **inserts**? [Glossary]

Certainly if the Earth is Flat or even a VR Sphere, then we did not go to the Moon, there are no ET visits, and Man is in a controlled environment… namely a School and the gods who run this place want Man to grow up, learn Patience, Compassion, Humility and Respect for self and other lifeforms. That could be called **The Game**. The Flat Earth certainly qualifies for that latter purpose and description.

And it is a suggestion of this chapter that, after looking at the following evidence for Earth being anything other than what we have believed it to be (i.e., a rock rotating at 1000 mph, whizzing around the Sun at 67,000 mph from 93,000,000 miles away), the **Earth might have originally been flat** until it was necessary to redo the Simulation when Man began to fly and wanted a higher perspective on his realm. Certainly if and when Man would attempt to venture into space, and to the Moon, the Flat Earth Simulation could have been reworked into a Sphere (as Virtual Earth Graduate suggests) so as to not give **The Game** away.

But, what if what we have been told is wrong? What if the Sun is not more than 7000 miles away? And what if it is the same size as the Moon, which is why the Moon exactly eclipses the Sun? What if the Earth is not rotating and does not circle the Sun? Hang on to your hat, we are about to go down the rabbit hole and consider some things which will leave you honestly wondering about Earth, and its original form thousands of years ago. Of course, a Flat Earth may not be proven, but going thru the following **Evidence for a Flat Earth** will also apply some evidence to what has been said earlier about the VR Earth in a controlled Simulation.

At the very least, consider it all as "brain candy."

Evidence for a Flat Earth

> Again, I know the risk for even suggesting that there might be something to the evidence for a Flat Earth (FE)… and yet, when I started out to prove it wrong in this book what a shock I got. Some arguments for the FE are substantial, and I could not disprove them… because they can also apply to the VR Sphere. Keep an open mind for the time being…

Having said that, to emphasize: since I could not disprove the logic or science behind everything the FE people were saying (and some things <u>were</u> nutty), I became a qualified believer: I still look at some pro-FE arguments with considerable interest because IF they are correct, and the Earth really is flat, then Man has to shape up, stop war, stop lying, stop polluting and be responsible!   Let's see how.

Preliminary Scientific Data

As we will be coming back to the numbers again and again, it is useful to present the alleged facts about Earth, as everyone knows them and mainstream Science has given us regarding Earth:

Distance from the Sun	93,000,000 miles
Earth's circumference	24,901 miles
Earth's diameter	7,926 miles
Earth day	24 hours (23 hr 56 m)
Earth speed around Sun	67,108 mph
Earth axial tilt	23°
Earth rotational speed	1038 mph (24,901÷24)
Oakland, CA – Norfolk, VA	3,000 mi (2989 mi)
US timezones	3

Argument #1: The Earth does not rotate.

Science says that the Earth is rotating and in 24 hours will complete one revolution, such that at 10 am in the morning the Sun is in the same position in the sky on Tuesday that it was the day before, on Monday – in the same month, same week. Also it takes the Sun 1 hour to move 1000 miles, and each timezone is 1 hour different from the one to its side… thus NYC is 3 hours ahead of Los Angeles because the United States is about 3000 miles wide.

The math: if the circumference of the Earth is about 25,000 miles and it takes the Sun 24 hours to return to its same position, then the Sun effectively 'traverses' the Earth at about 1000 miles per hour. Said in terms of today's Science: the Earth is said to be rotating on its axis at (about) **1000 mph**.

Again: If the Sun moves across the US from NYC to LA in 3 hours, and the USA is about 3000 miles wide, then the apparent speed of the Sun is dependent on the Earth's rotation which has to be 1038 miles per hour. This is important.
Note that the speed of sound is 768 mph… Do you hear anything?

Now I am going to ask you the first question: **Does it feel like the Earth is rotating at 1000 mph?** NO – and Science says that is because we are standing on it, moving with it. Ok, then what about the clouds? Or chimney smoke? They are not attached and the Earth should be whizzing by, moving away from the clouds! Gravity does not keep the clouds 'attached.' And how does a leaf lazily fall to the ground if the Earth is rotating at 1000 mph? BTW: **1000 mph counter-clockwise (left to right)**.

Still don't get it?

Let's look at a **helicopter.** If the Earth is moving at 1000 mph, all a helicopter in Las Vegas has to do to 'fly' to Los Angeles (275 mi away), is lift up and remain stationary for about 17 minutes (no horizontal motion on its part) and when it comes back down, it will land in Los Angeles. Yet that doesn't happen… because the Earth is not rotating.

Still not convinced?

Consider a plane flying from Dallas to Cleveland, a distance of 926 miles, South to North. If the Earth is moving and the jet flies at 300 mph, in a straight line by the compass (with its navigational compass set to the North Pole), the plane, not correcting for the alleged rotation of the Earth, should land in Chicago. Why?

> This is called the **Coriolis Effect** – a spinning body exerts a force at right angles to its direction of motion.

The Coriolis Effect does exist, but if the Earth were rotating, the Effect would be manifested in airplane trajectories as well as missile trajectories, and no such thing happens <u>at the global level</u>. Gunners and pilots do not have to adjust for the rotation of the Earth. Consider that if such a global Force or Effect existed, it would be very difficult to land an airplane on a runway facing North-South – it would be moving below the plane!

Further, if a plane flies East to West, and the Earth is rotating, the flight should be much <u>longer</u> than if it flew from West to East (because we are told that Earth rotates counterclockwise: Sun is up in the East before it 'rises' in the West). So the target city is moving away from the plane. This does not happen, the Jetstream notwithstanding. And a plane flying from Chicago to Dallas will <u>and does</u> take the same amount of time as flying from Dallas to Chicago… the pilot does not adjust for any drift or rotation of the Earth.

Coriolis Test

The Coriolis Effect was tested by scientists with a **cannon** in England. The cannon was fixed vertically to the ground, the muzzle pointing a true 90° from the Earth. The cannon was fired and the ball ascended for 14 seconds, and then took another 14 seconds to fall back to Earth. Due to a slight wind, the cannon ball fell back to Earth <u>within 2 feet of the canon</u>. The Coriolis Effect (mathematically calculated) should have put the ball about 5 miles to the West, assuming at this latitude (in England) and that while the equator is spinning at 1000 mph, England should be moving about 600 mph [less spin toward the poles]. This did not happen. [153]

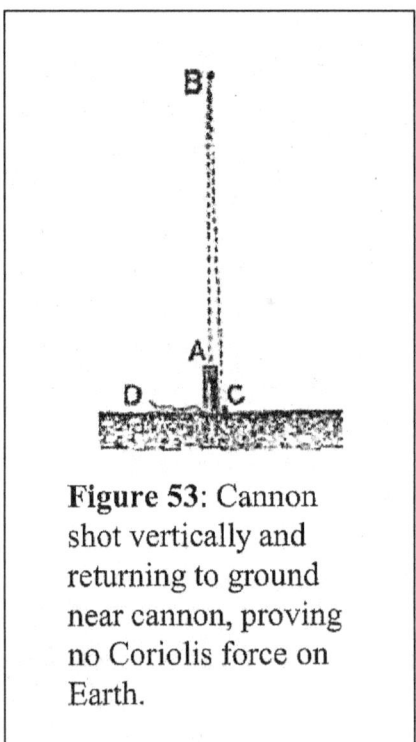

Figure 53: Cannon shot vertically and returning to ground near cannon, proving no Coriolis force on Earth.

If the Earth is rotating at 1000 mph, there should be a Coriolis Effect on airplanes, cannon balls, missiles and space shuttle launches. Therefore, the Flat Earth <u>and the VR SPhere</u> are **not rotating**.

Argument #2: The horizon shows a curve to the Earth

> By The Way: The greatest argument FOR the curve is when the Earth causes an arced shadow (a crescent) on the Moon ... but IF the Earth were flat and IF the Moon is a self-illumined orb that is slightly curved itself... how would that happen? FE still has a <u>temporary</u> problem.

Let's look at this further...

Crescent Moon

For the Moon to show a crescent (a partial eclipse by the Earth between the Moon and the Sun), **Heliocentrism** says that this is due to the Moon – Earth –Sun being in a partial 160-180° alignment. And what we see is the following:

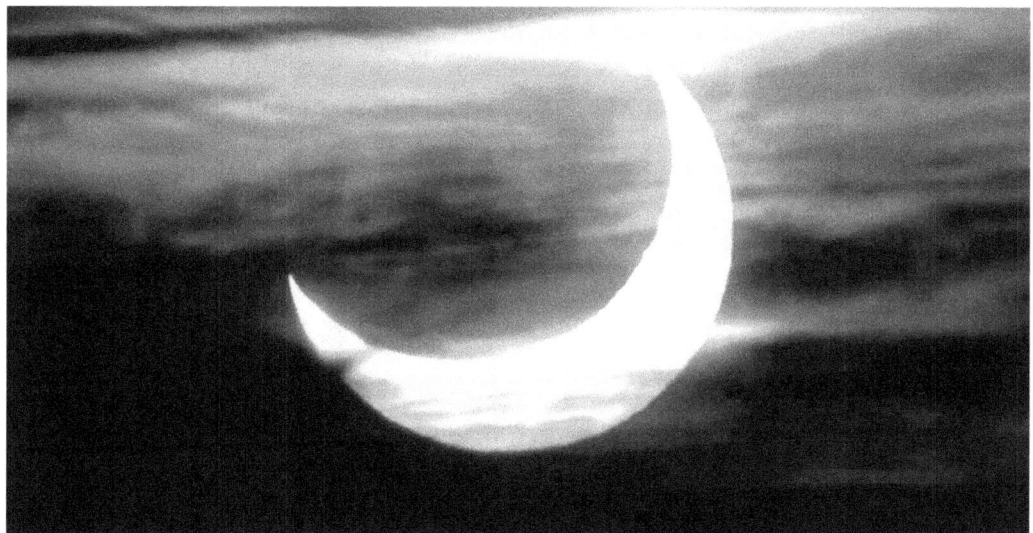

What will blow you away and cause you to rethink <u>what you think you know</u> is the following:[154]

> Eclipses happen regularly with precision in 18 year cycles, so regardless of geocentric or heliocentric, flat or globe earth cosmologies, eclipses can be regularly calculated [predicted] independent of such factors.

And then there is this FACT:

> According to the globular theory, a [full] lunar eclipse occurs when the sun, earth and moon are in a direct line; but it is on record that since about the 15th century **over 50 eclipses have occurred while both sun and moon have been visible above the horizon**. – F.H. Cook, *The Terrestrial Plane*. [emphasis added]

Thus … "it follows that it cannot be the shadow of the Earth that eclipses the Moon, and that the [Heliocentric] theory is a blunder." -- Wm Carpenter, *One Hundred Proofs the Earth is not a Globe*.

And lunar eclipses continue to happen when <u>both</u> the Sun and the Moon are visible in the sky, above the horizon… so how does the FE Theory explain this?

That the eclipsor of the Moon is a shadow at all is an assumption – no proof whatever is offered. That the Moon receives her light from the Sun and that therefore her surface is darkened by the Earth intercepting the Sun's light, <u>is not proved</u>…. The contrary has clearly been proved, that the Moon is not eclipsed by a shadow; that **she is self-luminous** and not merely a reflector of solar light, and therefore could not possibly be obscured or eclipsed by a shadow from any object whatever; and **that the Earth is devoid of motion** [not rotating]….Hence to call that an argument for the Earth's rotundity, where every necessary proposition is only assumed, and in relation to which direct and practical evidence to the contrary is abundant, is to stultify the judgment and every other reasoning faculty.
– Dr. Samuel Rowbotham, *Zetetic Astronomy, Earth Not a Globe!* [emphasis added]

So how do they explain a Lunar Eclipse? It is unlikely… unless there is something else flying around with the Sun and the Moon.

Dark Orbs

To handle the eclipse enigma, FE researchers have discovered something they call the Shadow Object, and is elsewhere called Dark Orbs. The idea is that the Orb is largely invisible as it is made from whatever the Firmament is made of (currently theorized Dark Matter, and functions as opaque but invisible) and these Orbs have a set route that regularly causes eclipses of the Moon, and because the Sun and Moon circle in adjacent orbits, an occasional eclipse of the Sun by the Moon is possible.

Left: four Orbs caught on camera with the Sun.

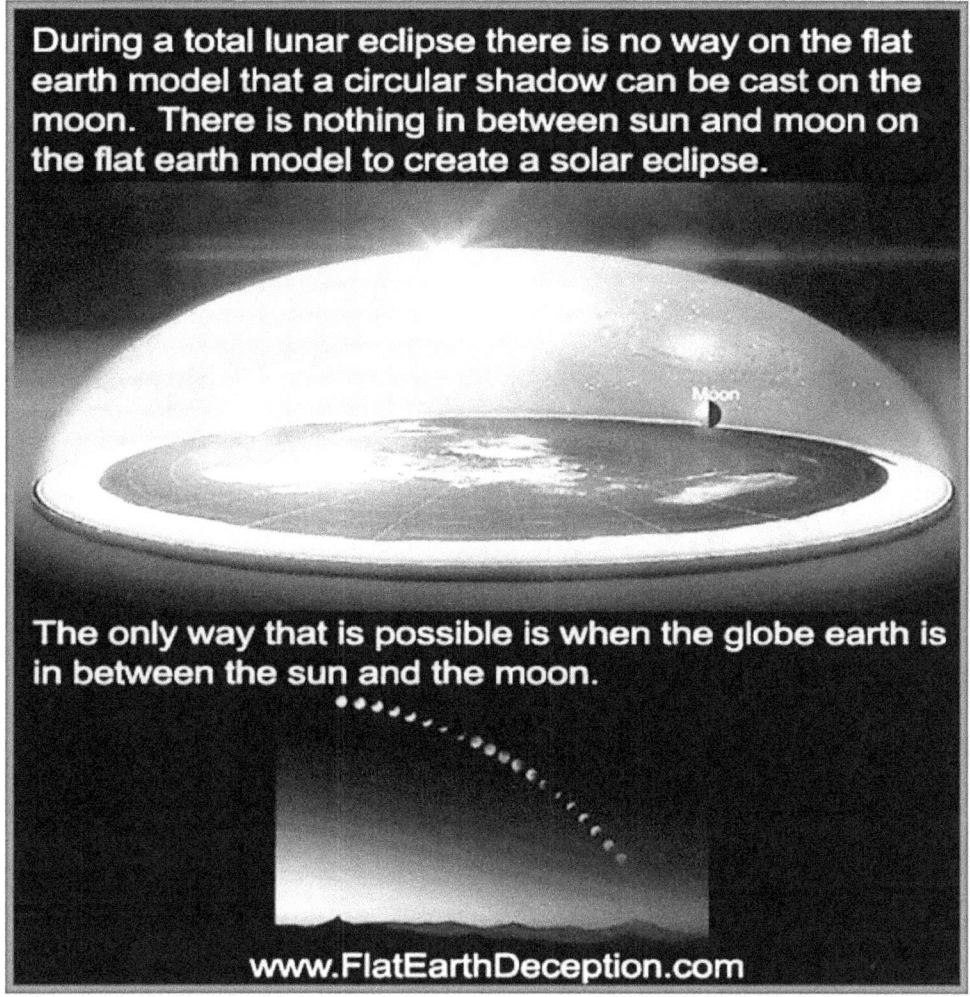

During a total lunar eclipse there is no way on the flat earth model that a circular shadow can be cast on the moon. There is nothing in between sun and moon on the flat earth model to create a solar eclipse.

Moon

The only way that is possible is when the globe earth is in between the sun and the moon.

www.FlatEarthDeception.com

This diagram supports the VR Sphere model, not the FE model

Such are the arguments for a Flat Earth in both Dubay's and Hendrie's Books (see Bibliography)… too numerous to detail here. They say that the Moon may not be what we have assumed it to be.

> **Do we really know what this place is on which we live, or are we just buying into whatever others have told us?**

Earth Alleged Curvature

Ok back to the math again:

> If the Earth's circumference is 25,000 miles, and there are 360° in a circle, then each 1° = 69 miles. Thus, if we see a horizon that covers 500-600 miles, we should see an 8° curve to the horizon, but we don't.

Examine the picture below. It easily shows about 100+ miles off the Eastern Coast… is there a curve? Maybe not for short distances (less than 500 miles).

(Examples are from Eric Dubay video: **see endnote #156**)

It will later be seen that even on a spherical Earth, short horizons will appear to be flat because the Earth is so big.

Or again:

Where is the curve? To begin to see it, due to the Math above, the Sphere horizon must exceed 500 miles. If the Earth doesn't rotate, and there is no curve to the horizon.... Could the Earth really be flat? Or is something else happening?

Ok, you say, show me Earth from space. Here it is: (more pix follow...)

(credit: Hendrie, p. 69, and p. 78 next pg.)

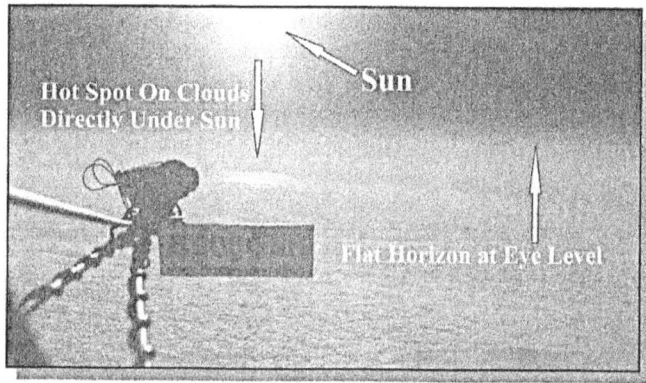

And again, we have a picture from space...

Figure 51: Flat horizon at eye level from an altitude of 80,000 feet. Note the hot spot on the clouds directly under the sun, which indicates that the sun is much smaller than the earth and close overhead.

Here is a better view:

This is 13 miles above the Earth. The issue is not the horizon, but the **hotspot** – meaning the Sun is close, directly above the Earth. .

Camera & Lens Set-Up
Used When Taking High Altitude
Aerial Photography

Curved Horizon Caused by Fish Eye Lens

Perfectly Flat Horizon After Correction for Fish Eye Lens Distortion

Sun Hot Spot
Indicating the Sun is
Close Overhead and →
Smaller Than the Earth

Corrected Image for Lens Distortion

Text: Sun hotspot indicating the Sun is close overhead and smaller than the Earth.

Pictures of the Earth's curvature are done with **convex** ("fish eye") lenses, and thus a slight curve is evident.

In addition, note the **"hot spot"** caused by the Sun which has to obviously be closer than 93,000,000 mi as there would be **no hot spot** if it were farther away. Same is true of the picture on the preceding page.

We will examine this a bit more in the Section on NASA pictures.

Remember, it takes 5-6° arcs (about 500+ miles) to begin to see a curve to the horizon – if it is curved. The above 'correction' may have removed any real curvature…

And again, a **hotspot** (concentrated light) where there should be none:

(**Credit: TheWorldWeLiveIn:** https://youtu.be/vknnZoBrAP8)

Airplane & Curvature

There is another argument subordinate to this #2, and it is that a plane that takes off on a long flight, say from NYC to Paris, assuming a curvature of the Earth, and a steady flight path, will fly off into space.

This is true but not possible, nor is it a good argument for a Flat Earth, as the **gyroscope and altimeter** in today's navigational system in today's planes will automatically make the adjustment to keep the plane flying at the same altitude.

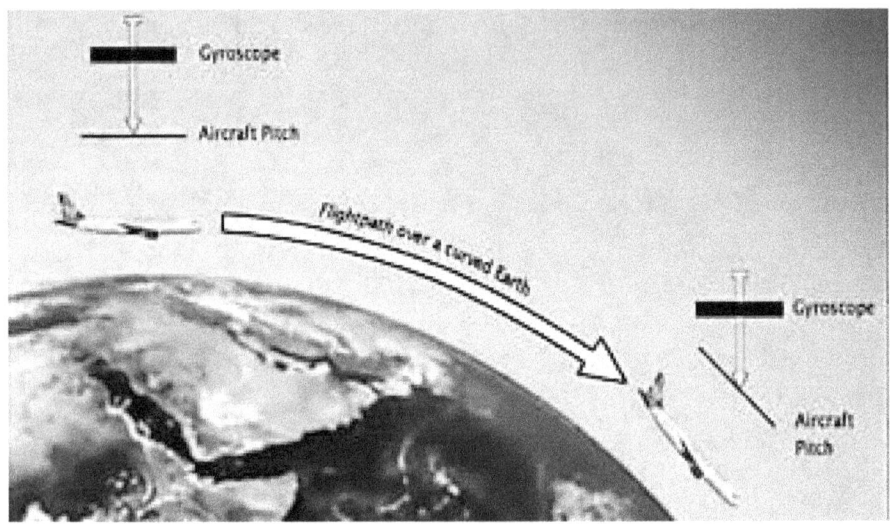

However, as we'll see in a coming section, where more curvature is examined, convex lenses are not always the 'culprit.'

Even as a Flat Earth or a VR Sphere, keep in mind that **the ISS** is moving around the Earth (just as Magellan circumnavigated the world), and the astronauts cannot tell if the Earth is rotating. They are flying above it, counterclockwise (against its alleged rotation). We are about to punch a hole in the rotation disinformation….

Argument #3: A rotating Earth, tilted 23° on its axis which wobbles ('precession'), and is moving 67,000 mph around the Sun, and the solar system is moving thru the galaxy, Polaris the Pole Star should move about. Science <u>and</u> FE Theory says it doesn't.

This one is pretty much intuitive… Polaris is said to be 433 lightyears away (one LY = 6 trillion miles!) How can the stars and constellations stay in the same position relative to each other, AND the Pole Star never move from directly above the North Pole if the Earth is rotating, wobbling and moving thru the galaxy? Polaris is perfectly in the same position <u>all the time</u>, else this time-lapse picture below could not be taken:

And by the way, if Polaris is a gazillion miles away, how can we see it with the naked eye?

(credit: Bing images: universetoday.com)

Polaris does not move, has never moved, and that means that Polaris is directly (permanently) overhead and <u>not far away</u>. If Earth "wobbles" on its axis (i.e., 'precession') how is this possible? Wakey-wakey!

Note: the FE Theory says the **constellations do not change shape** as they would if the Earth was moving, wobbling around the Sun which is also rotating (with our solar system) about the Galactic Core, which suggests only one possible arrangement... (using the Southern Cross as an example):

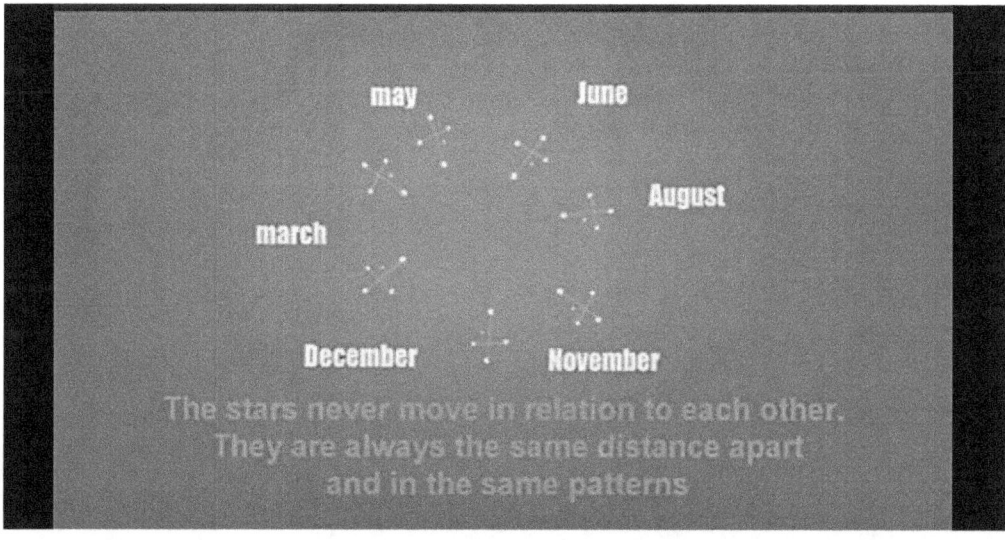

thus the bottom line for FE people says (below) that Polaris is <u>always</u> <u>directly</u> overhead:

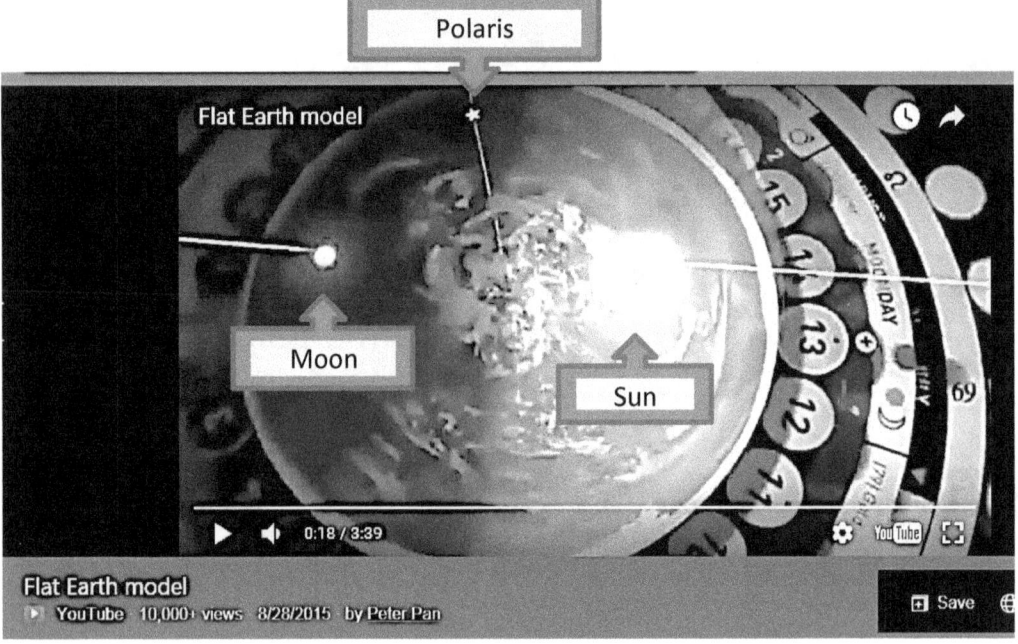

It is said that the FE Theory fails in the constellations in the Southern Hemisphere…. Pictured last page was the **Southern Cross** – visible only from the Southern Hemisphere. But if Earth is flat under a dome, how is it possible that someone in Texas cannot see the Southern Cross like someone in Brazil?

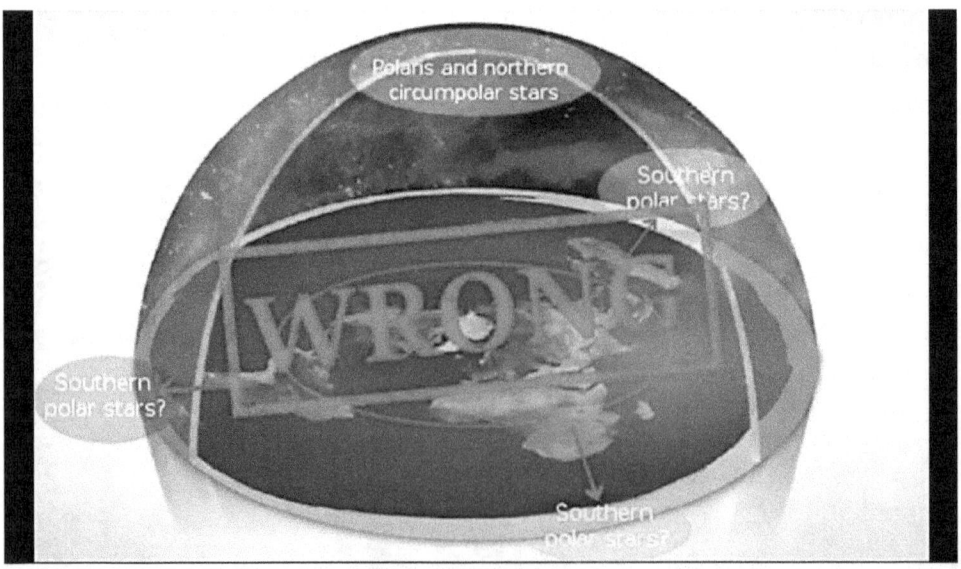

(credit: flatearthdeception.com)

Where would the southern polar stars be located on the Dome so that only those in the Southern Hemisphere (i.e., the outer edge of the FE plane) could see them – and so they could not be seen by those in the North?

Simple: the Flat Earth from rim to rim is about 10,000 miles across and the Dome can display stars on its southern half that cannot be seen from the people viewing the sky from the northern half. You cannot see 1,000 miles, let alone 10,000 miles and the **vanishing point** when observing things at an unobstructed distance is about 67 miles (due to haze, dust, clouds…).

So far there is nothing to disprove the **VR Sphere** – the axis tilt is not necessary to explain the seasons – the Sun ranges up and down in latitude from the Tropic of Cancer down to the Tropic of Capricorn.
Because the VR Sphere is a Simulation, the Sun is in the **Cspace** and rotates around the Earth – as the Church once used to say. The Moon does likewise — only closer to Earth such that a total eclipse is possible. This has to be the case to explain the "hotspots" shown throughout this chapter (…more to come). And the constellations are part of the Outer Wall, called the **Konstruct** earlier in Chapter 1.

But that does not explain the FE scenario. Gravity and Curvature are the two main items to keep an eye on in the following pages…

Argument #4: The Moon causes the tides

If the Moon causes the tides on large bodies of water, why are there no tides on large lakes? Consider the Great Lakes, the Black Sea, and even Loch Ness – which has a connection with the ocean. So they tell us that the oceans have a tidal action due to (1) the rotation of the Earth (i.e., Coriolis Effect), and (2) the pull of the Moon's gravitation.

However, it was discovered in the early 70's that the Moon's gravitational field is weaker than the Earth's gravitational field, and only extends 23,900 miles from the Moon, or 1/10 of the 240,000 mile distance. Where the two fields meet is called the **Neutral Point** and Earth's stronger gravity negates that of the Moon – so **the Moon's gravity doesn't even reach the Earth**. [155] How do the tides work, then?

So that leaves the rotation of the Earth, which is now in doubt… so could tides be due to something else? According to the FE Theory, the land sits on the Waters of the Deep (as the Bible says), and tectonics coupled with the slight motion of the oceans of the Deep could account for the tidal action…. Hmmmmm.

Here is where that comes from…

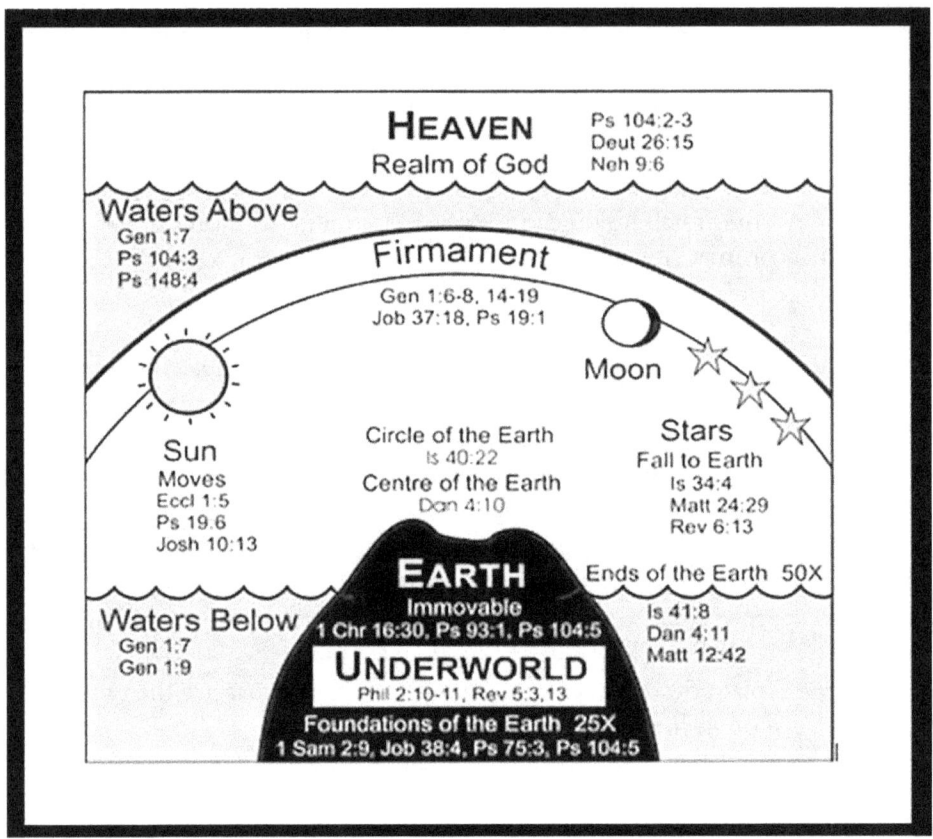

Inasmuch as the Waters Above the Firmament do not today exist, it is fair to say that the gods might have changed the scenario – and instead of a Flat Earth, it may now be a Sphere, but <u>without rotation</u> (more on this in the NASA section). And as a Simulation, it would also have the Sun and Moon closer as pictured earlier.

River Curiosity

In fact, something related and even more interesting is to consider the flow of the mighty **Amazon River** – the River is <u>huge</u> and we're told that it all starts as a trickle in the mountains in eastern Peru, Bolivia and Ecuador…. Oh really? The River's flow is such that snow melt and small streams down the western mountains <u>cannot</u> account for the total flow… the River is 4300 miles long, and up to 62 miles wide in some places, and the rate of flow is 7.3 million cubic ft/sec. Per <u>second</u>! That is **55 million gallons/second**. And some of the source is kept in **151 dams** in 6 of the main tributaries. The incredible flow's <u>source</u> gets harder to believe every minute.

The same question applies to the Yangtze, Nile and Mississippi Rivers… What is their real source? Suggested: the **underground water** on which the landmasses float… their weight could have the effect of forcing water up thru natural wells and springs, as well as seeping up thru porous sedimentary rocks, to the surface. In fact, flowing <u>under the length</u> of the Amazon River is an **aquifer** called *Hamza* which is saline and follows the Amazon to the Atlantic Ocean…. sustaining its salinity. So this is a twin-river system flowing at different levels of the earth's crust in Brazil.

> The watchword for this book:
> Things are not always what they seem.

Argument #5: The Sun is not 93 million miles from Earth

It has always seemed like a weird coincidence that the Sun is so huge and the Moon is so small and when the Moon passes in front of the Sun, it just happens to do a perfect total eclipse of the Sun… and there is a 400x factor difference in size! How coincidental is that? It almost looks planned….

So, what if the Sun and Moon are the same size (which would facilitate perfect total eclipses) and are both a lot closer to the Firmament. If that were true, then we should be able to get a clue from closely observing a sunset… and we do.

The Sun shining on the ocean is a clue, and we have never noticed it before…
(the following examples are credited to Eric Dubay,)[156]

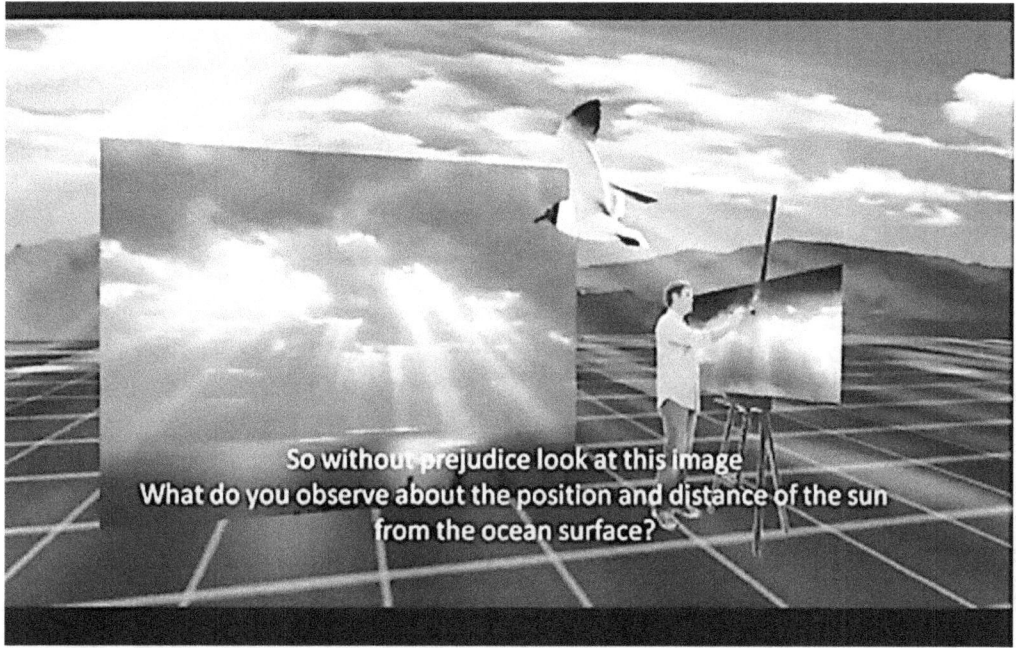

The way the Sun's rays play out at different angles (30° - 45°) suggests that **the Sun is not far from the clouds.** This is made clearer in the example which follows:

The location of the lightsource (Sun) has to be close to the apex of the triangle.

Ask any professional photographer.

If the Sun were really 93,000,000 miles away from Earth, the light would be more evenly distributed, clouds or no clouds… as in the following example:

If the sun was 150 million km away the suns rays would be all be the same angle from your perspective like this.

And again, we have another irregularity in the following picture:

The FE Theory suggests that the **ocean is flat** since a curve in the Earth would curve the ocean and <u>not</u> transmit a **'hotspot'** as shown above. Any curve in the ocean would absorb or block the straight flow of the light. <u>Unless</u> the Sun is actually closer than 93 million miles! So where is the Sun?

This is a perfectly normal glare – the Sun is <u>above</u> the horizon. There does seem to be a problem with a "hotspot" when the Sun is <u>on</u> the horizon…. The glare line cannot curve with the water…. Answer?

Water Level

Be aware that water always seeks its own <u>level</u>. In other words, water on the ocean is not curved or convex – This is a biggie!

Below is <u>not</u> what we see out on the ocean: (Not the hotspot), the resting **water doesn't curve**.

What we see is this:

(credit: TheWorldWeLiveIn)

And this one:

(credit: TheWorldWeLiveIn)

And as an extra added piece of information, the FE debunkers claim that the Sun stays the same size as it rises and sets – that is false. In the double picture below, you can see that the Sun is getting smaller as it sets – which means it is moving away from you – over a Flat Earth.

Rewind: Perspective

The FE debunkers say that due to the curve of the Earth is why you can't see a building, ship or an island out over the water at distance. Wrong –if you have a good enough telephoto lens , the following two pictures are possible and are real:

This is the original view without any optical assist.

Can you see the island 50 miles away?

(credit above and below: The WorldWeLiveIn)

Now let's zoom in on the same view:

Now the far distant island <u>and beach</u> appears and is not hidden by any curvature.

Atmospheric haze and the vanishing point is why you can't see it in the above photo.

This very example is shown on: YouTube -- *Best Kept Secret Since Flat Earth 2016.* (https://youtu.be/s0AOj01yzII for confirmation) about point 12:01.

Rewind: Sunsets

This is another time lapse from the same spot with a 5-minute size comparison...
(https://www.youtube.com/watch?v=UWJuE-d8JTU)

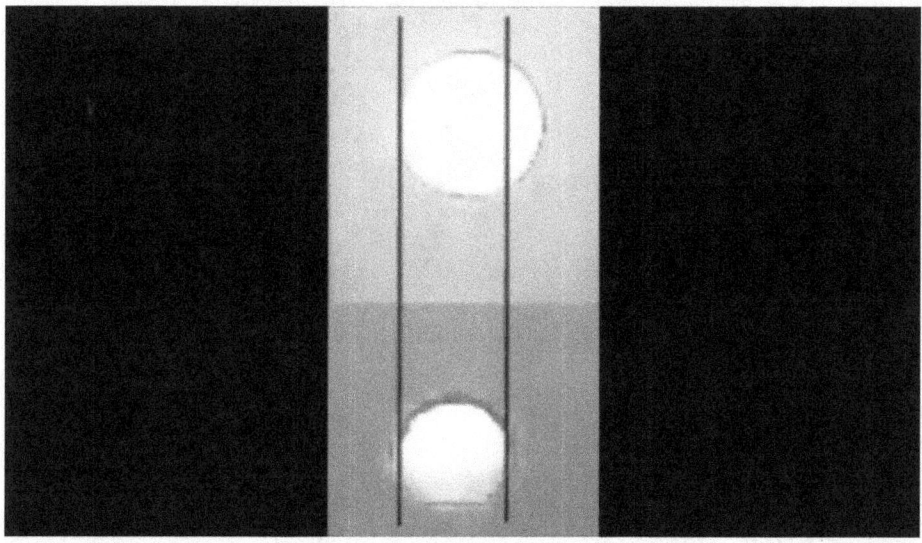

Five Minutes of Shrinking Sun Footage. The Death
of Heliocentrism. Flat Earth

 Zeteticism DotCom

And this: The Sun setting at a constant distance from the Earth (93,000,000 miles)
where Earth is a globe, should not change the size of the Sun – except in the case
where the Sun at the horizon is seen thru a heavy atmosphere, smog, etc. and then it
would appear to be larger – not smaller as the following 4 pictures below show:

Picture sequence 1:

Note the size of the
Sun – same as when
it was directly
overhead....

@ 4:27

Picture sequence 2:

Note the Sun is smaller…

And the **time stamp** Has been left on the bottom for your reference…

@ 4:29

Picture sequence 3:

… getting smaller…

@ 4:32

Picture sequence 4:

… and finally at sunset….

Compare with first Picture sequence 1….

@ 4:34

Needless to say, the Sun would change size if we are on a Flat Earth and the Sun moves away from the observer – <u>as it also does on the VR Sphere</u>. And couple that with the hotspots we have seen, and it says the Sun is also close to <u>both</u> versions of the Earth – not 93,000,000 miles away! It appears to be <u>above</u> the Earth plane.

Rewind: Hotspot

I agree with the Flat Earth people… Look at the following. How is this possible unless the Sun is close to the Earth?

(credit: geoshifter.com)

The 'spotlight' effect is not possible if the Sun's intensity is 93,000,000 miles away! This again is evidence for the Sun being much closer to Earth and supports both the FE Theory <u>and</u> the VR Sphere idea – and don't forget the similar *Gegenschein* which reflects off something above the Earth. They both have a 'concentrated' glare that would not be possible with the Sun's light diffused over millions of miles.

Argument #6: Heliocentrism doesn't work

Heliocentrism is the teaching that says the Earth rotates and also circles the Sun. As the Earth rotates, it has a 23° axis tilt and a corresponding wobble on the axis (precession), and the Earth is speeding around the Sun at 67,000 mph.

So if you go toward the Arctic Circle, at 1400 miles south (in Norway) and set your camera up to track the **Midnight Sun**, you get the following time-lapse picture where the photographer moved his camera over 360° once <u>every hour</u> to capture the Sun in the middle of the picture:

> BTW, there is **no Midnight Sun in the Antarctic** – why not? If Earth is a globe, both poles should reflect the Sun equally, 6 months apart: as the Sun does Winter in the Northern Hemisphere, we get one Midnight Sun effect, then when it's Winter in the Southern Hemisphere, it <u>should</u> have a corresponding Midnight Sun there, too… It is a globe with (allegedly) similar poles, right?
>
> Wrong. It is due to the path of the Sun which centers more over the northern latitudes, and does not get below the Tropic of Capricorn. (See Spiral Path diagram following Midnight Sun below.)
>
> There is an **Aurora Borealis** – in the Arctic and one over Australia – … which changes simultaneously with changes in the northern auroral zone. [157]   This argues for the VR Sphere layout.

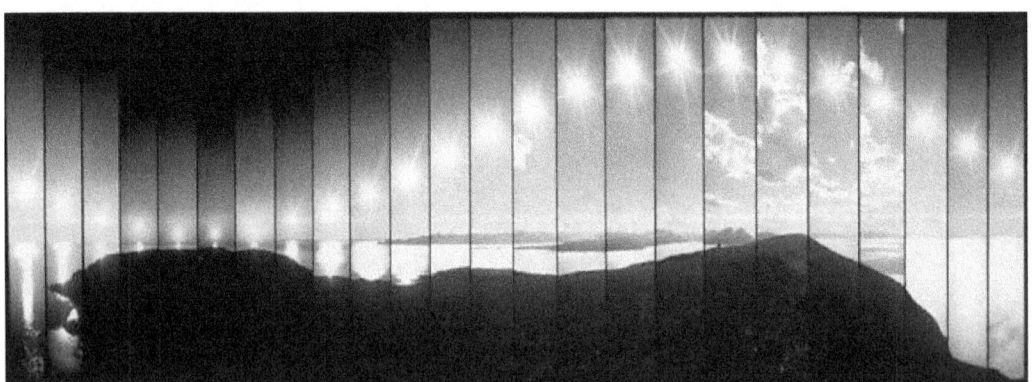

Midnight Sun
(credit: Big Images: gettyimages.co.uk)

Very Important: this essentially <u>proves</u> that the Sun is rotating <u>above</u> the Earth, in a circular path, over the Northern latitudes.

If the Earth rotated, alternately facing the Sun, and then away, you'd get a zig-zag picture. Here the photographer moved his camera <u>once each hour</u> to follow the Sun as it moved over Loppa at **70° north latitude**. … it never went below the horizon, meaning the Sun could not have been on the same horizontal plane as Earth (i.e., not at right angles to the equator) – the **Sun was clearly above** and shining while moving thru the Northern hemisphere!

"Do they never reflect on the earth how it was made flat"

Quran: chapter 88:17-20

So Heliocentrism is wrong. Copernicus and Galileo were wrong. Ironically the Church was right!

Even more interesting is that **the Koran** (left) speaks of a Flat Earth.[158]

So again, this book's alternate point of view, based on the information that was given me in 2008, is that the Earth may be a VR Sphere – but it may have been initially created flat. There were few people and Man was not flying yet, so why would the gods create a bigger Earth than they had to, to provide a School?

So, if the Sun does not shine on a globe, but a Flat Earth, what does Earth look like?

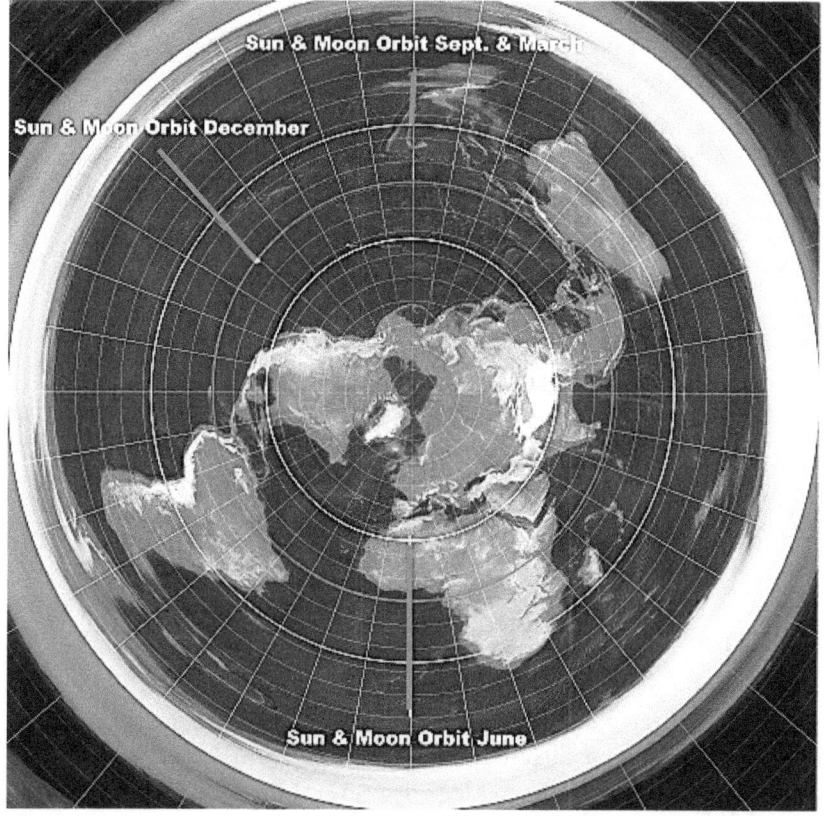

Sun & Moon Orbit Sept. & March

Sun & Moon Orbit December

Sun & Moon Orbit June

Note that the Winter orbit is the **outer ring**, Tropic of Capricorn. And the Summer orbit is the **inner ring**, Tropic of Cancer.

The Sun moves back and forth, Between the 2 rings, and up & down – creating a **spiral path**.

Suspected spiral path of the Sun (based on the 360° time lapse picture earlier):

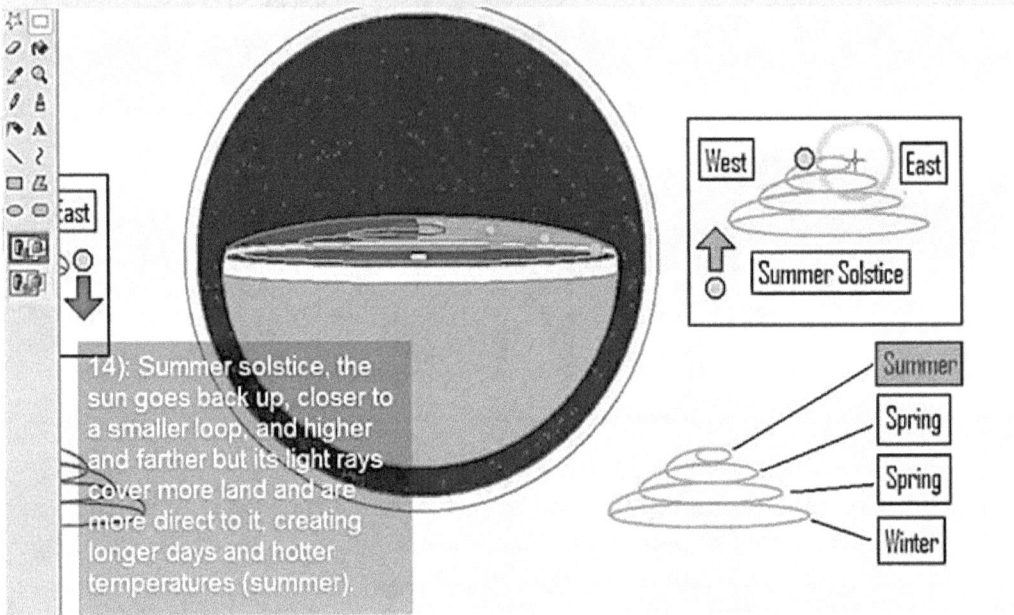

14): Summer solstice, the sun goes back up, closer to a smaller loop, and higher and farther but its light rays cover more land and are more direct to it, creating longer days and hotter temperatures (summer).

(credit: TheWorldWeLiveIn: Flat Earth? Top Ten Evidences)

Sphere Revisited

The smart reader will counter with an objection that we have pictures of Earth from space, and those show Earth to be a Sphere. And we do, but NASA is still playing games…. Let's examine that and we'll later address the question of <u>why</u> they would do that and it is worth your very soul to see it, and could be the most important thing you will ever learn while on Earth.

> I repeat: if you knew what this place really was, you would not waste any time in doing what it takes to get out of here. **Earth is not your home.** This is not melodrama, BS or a trick. Earth is not your home. **It is a School,** and you don't stay forever , <u>or live</u>, in a school.
> (The answers were given in Ch. 15-16 of <u>Virtual Earth Graduate</u>.)

NASA Exposed

NASA Flight Path Insight

Whenever people watched the Space Shuttles (any orbital flights!) make Earth orbits on TV, on the large NASA Houston wall screen, there was a curious "sine wave" layout to the path followed, and that never made sense – until the Flat Earth layout showed it to be **really a circle**: This is the 2nd Biggie!

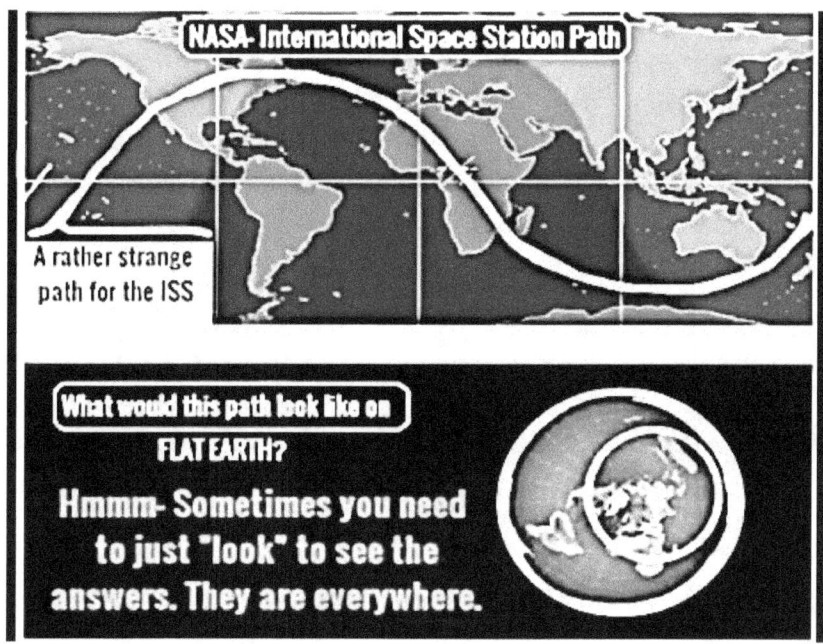

(credit: **Odiupicku: Flat Earth – Guess Why We Can't See Antarctica…**)

NASA Mission Control – see center wall tracking screen…

(credit: **NASA** http://spaceflight.nasa.gov/gallery/images/shuttle/sts-115/html/jsc2006e40472.html **Public Domain**)

To replay that, because it is very significant, here is the Mission Control Screen by itself –

Note that the middle S-curve crosses America, Africa and goes under Australia.

Now compare that with the FE map below:

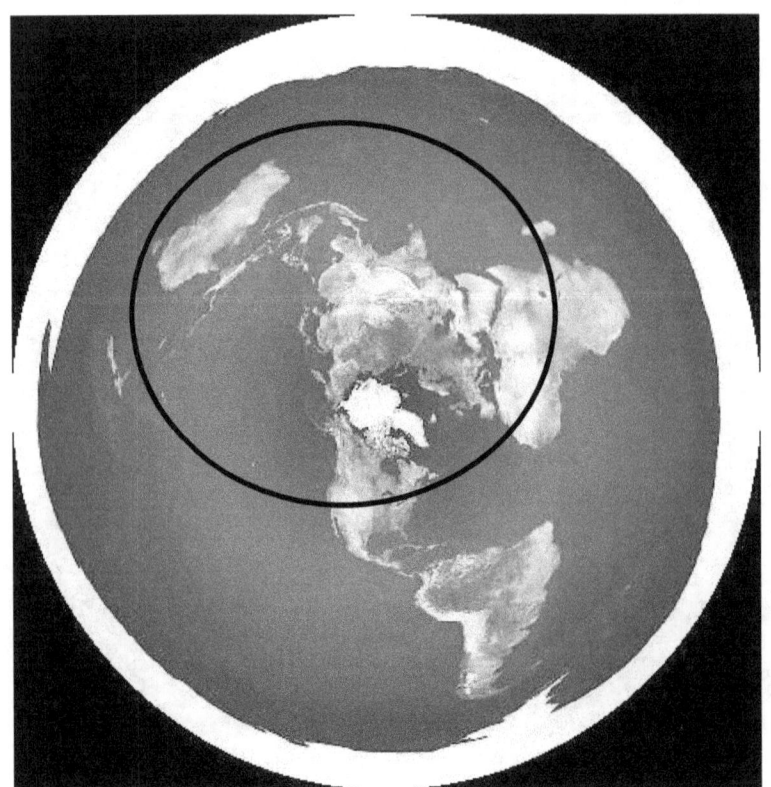

The orbital "S" paths describe a circle.... As one would expect!

The orbital path is the same for the ISS, a satellite or a Space Shuttle.

NASA knows!

Photoshopped® Photos

The book <u>Virtual Earth Graduate</u>, in its Appendix A, goes into some detail exposing the photographic errors of Earth from/on the Moon, and just a quick reminder will be offered here.

> **Once again, the math**:
> Earth and the Moon are about 240,000 miles apart, according to Science, and the Earth is about 8,000 miles in diameter, and the Moon is about 2184 miles. So that means the Earth is almost **4x** (3.6x) bigger than the Moon.

So any picture of Earth from the Moon would look **3-4 times bigger than the Moon** looks to us from Earth. Got it?

That means no telephoto lens, just a picture, with a Hasselblad camera, taking a picture as if it were a box camera –i.e., no special apertures or focal length, magnification, etc… So does the picture (left) of Earth from the Moon look right?

The picture is Apollo 17: AS17-134-20384
Why is the Earth so small in NASA pictures?

To take this picture (above), the other astronaut had to be leaning to his right… (because it is shooting UP and the ground is at a 45° angle). And that is still wrong. Apollo 17's landing site was in the upper right <u>central</u> part of the Moon, and the Earth would have been <u>directly overhead</u>, **not off over the horizon**. Remember the Moon does not rotate – the same side always faces Earth.

Same thing for the left one: Apollo AS-137-20910.

And even more interesting is a picture of Earth from the Apollo 11 Command Module as it circled the Moon (below)....

Look carefully at the Earth – it may be the same picture of Earth in the first two pictures ... cloud patterns are a give-away.

AS11-44-6550

This appears to be made with a telephoto lens, but the Earth is still not shown its real size.

Remember, if an astronaut is standing on the Moon, and the Moon always faces the Earth, which is 3-4x larger than the Moon, even if the astronauts are on the edge of the Moon, the Earth will always be **directly overhead** – not off to the side. When seen from the Moon, the Earth will be about the size of a 3' diameter beachball held at arm's length. Similar to this:

The picture (left) shows closer relative sizes ... so why is the Earth so small in the NASA pictures above?

Imagine standing on the Moon and looking straight up – at the Earth. It is bigger.

So coming around the edge of the Moon, <u>at about the equator of the Moon</u>, and seeing Earth, this should be the view – with the Earth's continents laying **horizontally** when the astronaut in the Command Module shoots the Moon's <u>equatorial</u> horizon:

(credit: TheWorldWeLiveIn)

There is an error in the above diagram – the Command Module did not fly over the top of the Moon, it flew around the **Moon's equator** and putting the Moon's horizon at the bottom of the picture (insert) , means that the Earth's North Pole would be in the right half of the picture. This error is easy to do – NASA did it, too.

> The value of the preceding diagram is that it shows the error in size (in the inset) versus Earth's real size. Again NASA photoshopped the 'Earthrise' picture.

Picture Technology

What is NASA doing to get its pictures? Ever heard of **CGI**? Computer created graphics. That means NASA would be manipulating or doing **composite assembly** of "Earth" in **Photoshop ®** (or something similar) from an assortment of sources. Examples of compositing the pictures follow…

The <u>globes are the same size</u> but North America (top half of each sphere) is not the same size… How big is America?

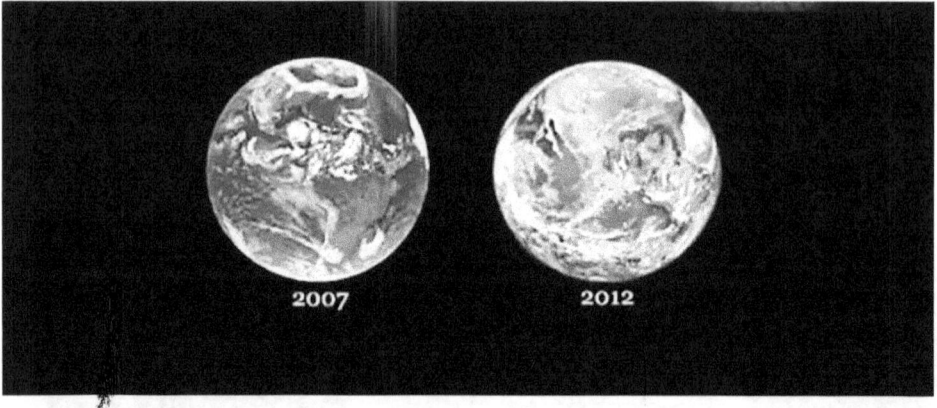

Look again (outlined for convenience):

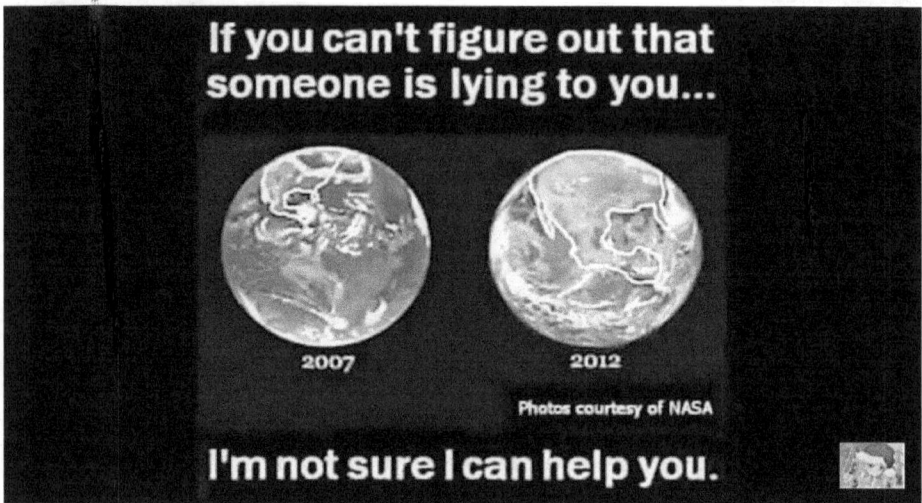

In addition, when the TV runs the NASA animation of the **Blue Marble** (Earth) spinning on its axis, allegedly taken from ½ way to the Moon, **the clouds don't move**. This animation is at: (https://youtu.be/s0AOj01yzII   about 5:02).
(Source: **Best Kept Secret Since Flat Earth 2016**
† TheWorldWeLiveIn a 24:03 video)

Embedded Errata

Now, note the two following: (**credit: Edw. Hendrie, pp 110-128**)

Figure 72: Photograph of earth allegedly taken from a camera aboard the NASA Deep Space Climate Observatory, from a distance of one million miles.

So some people at NASA have a sense of humor. However, how about a composite picture with **duplicate clouds**? Look closely....

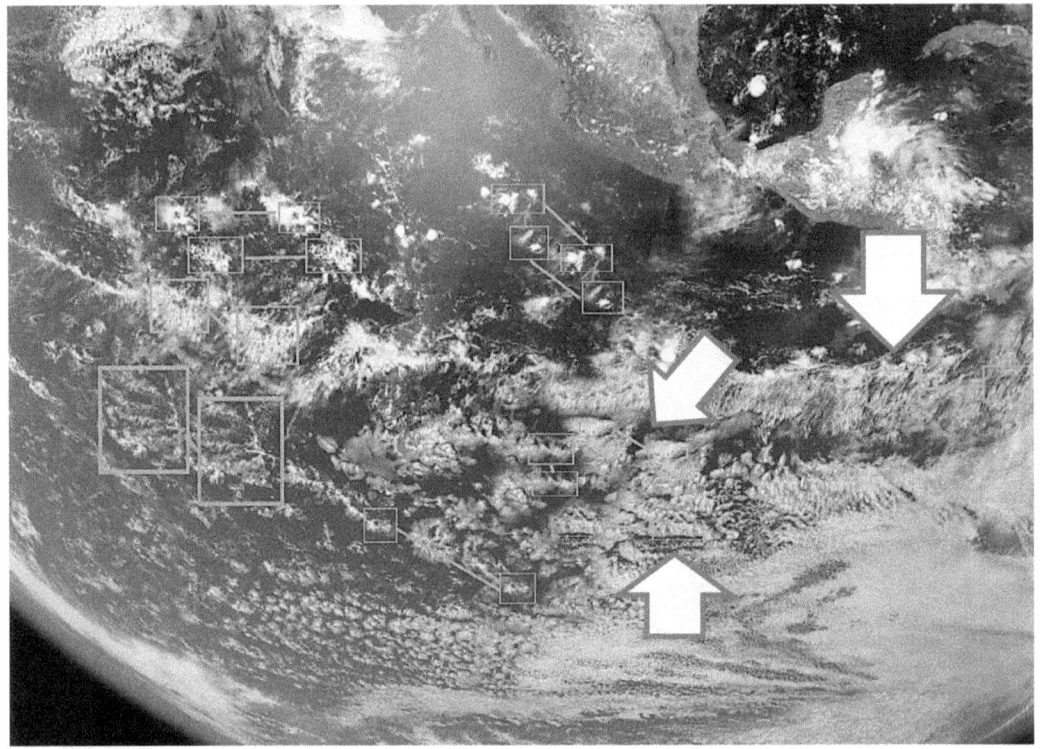

So if they manipulate pictures, because we couldn't really go to the Moon (and we can't even if Earth is a VR Sphere) – or even if it is a "luminary" as the Bible and FE people say (viz., <u>it has its own light</u>) – why would that be done? See the Chapter Summary for a plausible answer.

Thus it appears that NASA may have reached its [temporary?] end of the road. We have not been back to the Moon in 44 years, the Space Shuttles no longer fly, but while this book's initial premise of a 3D VR Sphere is looking more credible – so is the possibility of a Flat Earth – both keeping the non-rotating Earth concept with Sun & Moon circling.

And one last look at pictures and their sources...

Hubble vs SOFIA

There is also controversy surrounding the Hubble Telescope – if we are an FE scenario or a VR Sphere, there is a **barrier** surrounding Earth (Firmament or otherwise) , off which the *Gegenschein* manifests...
So how would Hubble get pictures of our Galaxy and far distant space?

Hubble Space Telescope (HST)

We are told, by Wikipedia and other sources, that the Hubble Space Telescope was launched in 1990, and after a needed repair in 1993 to the focusing mirror, it remains operable, giving fantastically clear and beautiful pictures of our universe. It is said to operate in Low Earth Orbit (LEO) ... perhaps 200-300 miles above Earth where the ISS is said to be (at 250 miles). And whether Earth is flat or a VR Sphere, it is still possible to orbit the ISS, HST and satellites in the LEO range of 100-1200 miles above Earth – well below any Firmament or barrier.

> It remains to be seen what quality of pictures can be made thru a Firmament if Earth is an FE scenario, or whether HST can pierce thru the (pretty much invisible) barrier surrounding the VR Sphere. Perhaps the stars and galaxies are pictured IN the Firmament?

And that issue brings us to SOFIA

Stratospheric Observatory for Infrared Astronomy (SOFIA)

SOFIA is a joint project between the German Aerospace Agency and NASA –using a large Boeing 747 with a telescope on board that peeks out thru a hatch in the aft section of the plane. It began service in 2010... but the question has lately been, why SOFIA when he have Hubble?

FLAT EARTH | SOFIA GREATER THAN HUBBLE??? | CROW777 & Robert Bassano (1/2)

It appears that the SOFIA schema was initiated before Hubble, in 1977 and performs regular optical astronomy at high altitudes as well as infrared astronomy. SOFIA is a 3-meter telescope whereas the HST is a 2.4 meter telescope. HST also does spectrographic analysis and has a Faint Object camera. In 2005 HST also received infrared capabilities in an update, and it was pushed into a higher orbit by a Space Shuttle.

> The conspiracists say that some HST pictures are really taken by SOFIA, and that argument can be followed at this location: UnsilentMajorityTV at https://www.youtube.com/watch?v=2grKOerf5hE
> and at https://www.youtube.com/watch?v=If5eISZIbS0

The point being that not only is there an ISS, but a high-flying SOFIA craft (in LEO which starts at 99 miles altitude), and Hubble, and satellites. LEO works within the parameters of the FE or the VR Sphere, so it is interesting that LEO is possible in either schema.

Moving forward, were there any other <u>notable</u> people who also noticed that something wasn't right with the Heliocentric Theory?

Notable Observers, part I

Ptolemy made it appear that the Sun and Stars revolved around a central point. He ingeniously showed that **the Earth must be at the center of the celestial schema**. He proves that unless this were the case, the stars would not display the consistent uniformity with regard to **the constellations which do not change shape**. "The Earth lies at the center of the celestial sphere **[Geocentrism]**. If the Earth were to be endowed with movement [rotation on axis, and/or rotation about the Sun] it would not lie always at this [center] point, it must therefore shift to some other part of the sphere. The movement of the stars, however, preclude this and therefore **the Earth must be as devoid of any movement** … as it is of rotation." [159] [emphasis added]

Copernicus did not produce any newly discovered fact to prove Ptolemy wrong, neither did he offer any proof that he was right, but worked out his **Heliocentric System** as a means of explaining day and night, and the seasons. [160] **Galileo, Bruno, Kepler,** and **Newton** just bought into Copernicus' theory as it sounded scientific.

Isaac Newton went a step further, and said, "allow us, **without proof, which is impossible**, the existence of two universal forces – centrifugal and centripetal, or attraction and repulsion, and we will construct a theory which shall explain all the leading phenomena and mysteries of Nature." [161] Sheesh – How scientific was that?!

Further, **Gravity**, as postulated by Newton, may not exist. Newton used the example of the apple falling from the tree to demonstrate a principle that even today's physicists cannot find, nor prove. **Gravity** remains elusive because it is not a force. FE says it probably does not exist. (I know that sounds weird, bear with me... I am about to show you what is behind their thinking, and it is a tough nut to crack.)

> The problem with the gravitational theory is that ....the gravitational attraction to the earth of all persons and objects **remains the same at all places on earth.** That means the gravitational force at the North Pole is the same as the gravitational force at the equator. That poses a very real problem if the earth is spinning as alleged....
>
> **Centrifugal force** is perfectly <u>balanced at the equator</u> by the force of gravity... and yet decreases every mile toward the North Pole [where it would be zero] .... Thus as one approaches the North Pole, the force of gravity [less CF] would crush a person, which of course does not happen, so the spinning earth and the mystical force of gravity are thus proven to be preposterous fictions. [162] [emphasis added]

If the apple, when ripe, naturally falls to the Earth because it is <u>heavier than air</u> (not due to Gravity), why doesn't smoke from a chimney also get dragged to the ground? <u>Because</u> the smoke is lighter than air and rises so that gravitational force is not even a factor in whether something falls or not.

Copernicus argued that the tilt in the Earth axis, which had to be there since it rotated, accounted for the **Seasons**… tilting first one way and then another ("precession"), but he could not account for [nor locate] the shifting of the waters and oceans (tsunamis) that must accompany such a change in tilt and thus change the center of gravity twice a year! However, it would seem more likely that any tilt affecting seasons would be more overridden by a large distance of the Earth from the Sun.

> From the earliest times it has been believed and said that the heavens were not an empty space, but a solid surface. The **Chaldeans and Egyptians** regarded the sky as the massive cover of the world; and in India and Persia it was thought to be a metallic lid, flat or convex…[163]

This will show up again when we get to Charles Fort and Enoch.

Sky Dome

In **African lore**, in another book by this author called <u>Anunnaki Legacy,</u> there are many stories and legends of their ancestors being able to build towers to get to the Sky where their gods lived. (Hint: **Tower of Babel**?) According to many African legends, the gods grew annoyed with the humans constantly trying to get to them, to petition them for favors, and according to legend the gods <u>moved</u> the Sky (**Firmament**?) higher up and the humans finally gave up.

So while we think the Bible's story about the **Tower of Babel** was a nice allegory – maybe there was something to it? Maybe when the Earth was younger, and the humans spoke often about the gods coming to them and teaching them, just maybe the Firmament was much closer to the land. (More when we get to Charles Fort…)

Rewind: Gravity & Centrifugal Forces

Ok, let's look at this issue one more time. What you think you know about Gravity is going to be shaken.

Let's consider that the Earth is a globe, spinning, and circling the Sun, as Science says. There is a problem with this scenario and it involves a not-so-obvious problem with the interplay of Centrifugal Force and Gravity. (Most laymen do not know or think about the following.)

The **Flat Earth Theory says there is no Gravity**. However, if the Earth is a round ball, then what keeps things like the Oceans from falling off (centrifugal force) a round Earth? So I have to admit, that part says there is something called Gravity or

Centripetal Force ... and I have often wondered what holds the oceans to the Earth -- but consider this: in places <u>the ocean is several miles deep and</u> that amount of water (i.e., **mass**) weighs a heck of a lot... more than the force that causes humans to 'stick' to the Earth... the deeper part of the ocean weighs millions of tons. Try easily lifting a 5 gal. jug of water... now multiply that by a mile deep water's weight.... what is keeping that water 'stuck' to the Earth??

Oh, you say Gravity is offset by the Centrifugal Force, or maybe Centripetal Force gets involved? Wrong.

If the force of Gravity (or Centripetal Force) is strong enough to hold a two-mile deep ocean to the Earth, how can <u>we</u> even walk on the Earth? This has never made any sense.

So I initially bought the FE idea that Newton was wrong and there is no Gravity....and then there is the Centrifugal Force (if the Earth is spinning 1000 mph) which tries to fling us off the planet. Supposedly (according to Science) the Force of Gravity and the **Centrifugal Force (CF)** equally balance each other -- <u>at the equator</u> where CF would be the strongest.... and it is lighter near the poles... so is the Force of Gravity variable all over the Earth, or is it a constant -- as Science says?? See, this makes no sense... unless the Earth is Flat and things stay put because **they are heavier than air**, and there is no ocean on the bottom half of the Earth globe (to fall off).... See, if CF is less at the 30th latitude than at the Equator, <u>AND if Gravity is constant</u>, people should find it harder to walk in Canada than they do at the Equator because while CF is lessening in Canada, Gravity stays the same.

This makes no sense ... and that is why people blindly believe what they are told...they give up. It takes a lot of horsepower between the ears to begin to reason this out.... (my IQ btw is 160 and I still have trouble). To me, this Gravity - CF issue is more evidence <u>for</u> the Flat Earth (FE)... Gravity (if it exists) would be constant on a FE, and there is no danger of water falling off the planet, and since Earth is not spinning, there is no CF to consider...

> And don't forget, Newton just a few pages back, said we are <u>assuming</u> the existence of Gravity and "it is impossible to prove." Are today's scientists vainly trying to prove the existence of Newton's assumption?

However, something resembling CF does exist -- get on a child's **Roundabout** at the playground, get on and start it spinning, then throw a beachball --not only do you get the **Coriolis Effect** (the ball curves away --not in a straight line) -- and unless you hang on, there is a tendency for the CF to throw you off the Roundabout.... there is no Centripetal Force trying to keep you on, any more than there is on the spinning Earth.

Note the FE science says <u>Coriolis and CF do not apply to the Earth globe as a whole</u>... Winds and water move due to temperature differences, convection currents, and perhaps the Simulation gods are doing something we do not know about to move wind and water. If the Sun and Moon move above the VR Sphere as is said of the Flat Earth, and if they also exert some unseen force, that could account for wind/water movement... not the Coriolis Effect.

Rewind: Curvature of the Earth

The following discussion was part of a 3-episode documentary on **Netflix**, dated 2012 but aired Dec 2016, which was called <u>Orbit: Earth's Extraordinary Journey</u> and was in total support of the Earth being round, spinning, and circling the Sun. And yet the following 2 items were observed:

> A. To show the Earth's upper atmosphere, at the edge of space, a huge balloon was filled with helium and sent up into the stratosphere – with 4 cameras on board recording the trip. The balloon burst as it left the atmosphere, and the gondola carrying the cameras fell back to earth.

When the videos were replayed, something very interesting appeared. As the balloon gently bounced and the cameras angled up and down, the balloon took many pix of the Earth's horizon or curvature. Freak out time! When the camera swung UP the curvature was **convex**, when the camera swung DOWN, the **curvature was <u>concave</u>**! And in the middle when the camera was still, the Earth's horizon was FLAT!!! Nothing was said about this in the movie.

It is obvious that the publicized convex curvature of the Earth can be made to look that way by using a convex (or "fisheye") lens. But the **concave view** of the curvature was a giveaway of what type of lens they were using! It revealed the way NASA also shows us the convex curve to the Earth's horizon...

B. Later, when the narrator, a science woman, was standing on a plateau in Ecuador (at the Equator), the camera did a pan of the horizon and it was ALL PERFECTLY FLAT for as far as one could see! And there were thousands of miles to the length of the horizon, and it was pretty far from her to the horizon to start with -- and BTW, no one said anything about the flat horizon.

In both cases I wondered if they were showing us the Truth, but not saying anything about it?? I risk putting this in the book because it supports the FE Theory and <u>not</u> the 3D Construct VR Sphere!

Here are the high altitude balloon pictures:

> **The key point in the three following pictures is that the curve or no-curve of the Earth's horizon depended on the bouncing and tilting of the camera in the high-altitude balloon!!**

> **When it bounced up, the horizon was convex.**
> **When it bounced down, the horizon was concave.**
> **When the balloon was stationary, the horizon was flat.**

Convex:

and then …

Concave:

… and then…

Flat:

If you are a real thinking person, at this point you're going to wonder about two issues: Gravity and the Curvature of the Earth….

Want more?

Three allegedly non-Photoshop pictures follow…..

If we remember the **69-mile Rule** calculated earlier in the chapter, each 1° of the Earth's circumference is 69.4 miles (69.4 x 360 = 24,984 and the Earth's circumference is approximately 25,000 mi) thus it would be hard to see any curvature before one sees 5-6 ° distance in a picture… and certainly the top two pictures <u>below</u> show more than 69 miles.

Over 20 Miles High, Horizon Still 100% Flat

Mount Everest

Interesting "proof." So it is not inappropriate to suspect that the pictures have been 'corrected.' And there is such **software....**

That is why it cannot be proven the Earth is flat using pictures.

And the three balloon pictures shown on the last two pages are further reason to distrust pictures.

(credit: http://whale.to/a/earth_curvature.html

Having said that, there are pictures from space, around Earth, such as the following from a tourist who went into low earth orbit – **Guy Laliberte**:

Mexico from Space

(Credit: https://alizul2.blogspot.com/2012/04/magnificent-images-of-earth-from-space.html?m=1 *ASSOULINE*)

Guy was a passenger on the Soyuz TMA-16 rocket, about 2012, similar to what **Richard Garriott** did in 2008. **This is evidence for the VR Sphere** – unless a convex lens camera was used to take the picture…. And as was said pages ago: it is hard to use photographs to prove curvature/noncurvature thanks to convex lenses and Photoshop® …. and now Go-Pro®.

So it is interesting that two modern day explorers paid to get a ride on the Soyuz Earth-to-ISS rockets… and they both say that the Earth is a sphere. Can they be sure? They were only 250 miles up above a 10,000 mile wide Earth, moving at 17,000 mph… kind of fast for such a shallow height… What if they were orbiting the FE in a circle – would they know that? Magellan didn't.

Are there any supporters of the FE?

Notable Observers, part II

As was shown earlier at the beginning of this chapter – the ancients knew the Earth to be flat. And concurrent with that is the 1931 statement of **Dr. Auguste Piccard**:

This is fascinating and serious...

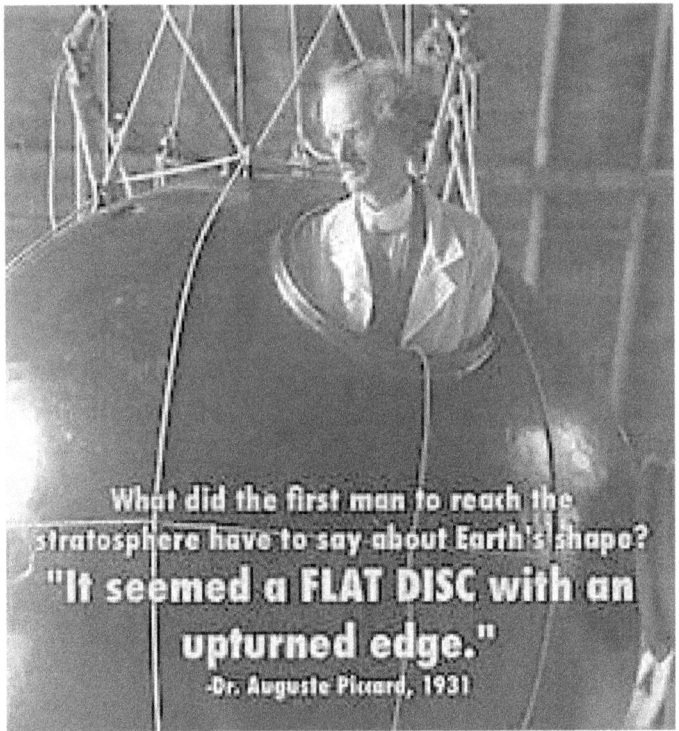

Oh boy, here we go again.

Now, concurrent with this book's original proposition that the Earth may now be a Sphere (as we have orbiting satellites, the International Space Station, and some pictures of Earth from NASA and stratospheric rockets), and considering that the gods who run this place might have had to modify the Simulation to permit Man a larger "field of play," it begins to look as if the **Earth could have been modified from the original flat state** to a current spherical one.

Somewhere between 1931 and today? That is not credible – yet Dr. Piccard supports the Flat Earth and we now find it suspect. Unfortunately there are no pictures from Dr. Piccard's balloon trip. What the heck did he see? What did his accompanying scientist, Paul Kipfer, see? They were up only 9-10 miles – maybe he saw the Alps at a distance ("upturned edge")... or the Urals? No, he took off from Augsburg (southern) Germany and came down on a glacier near Ober-Gurgl, Austria.

However, this is still inconclusive, and many things pertaining to the Flat Earth also apply to the VR Sphere... so far... with the exception of the curvature of the Earth. And Gravity. So we still need to dig a bit further...

Just a brief reminder…

Fortean *Gegenschein*

Charles Fort will remind us of the Chaldeans and the sky vault. He was a researcher into all sorts of unusual occurrences around Earth – fish falling from the sky, rocks hitting people with no one around to throw them, people crossing a field in full view of others and just disappearing, and anything that was not a normal, logical happening. He then wrote books about his findings and one time ventured the opinion that he thought we humans on Earth were someone's property.

So even if the Earth is flat, or round, <u>but contained by a 'shell'</u>, then there are no ETs, and SETI is a waste of time.

Gegenschein & The Dome

Gegenschein

But if this Earth is round and there is a dome or energy barrier containing it all, such as Fort called the shell reflecting the *Gegenschein*,

then could the picture look something like this (left)?

Monroe OOBE

Both Robert Monroe and Charles Fort were examined in Chapter 4 because they experienced the **barrier** that surrounds this Earth … Fort noted it because of the

Gegenschein and Monroe noted it as he hit the barrier several times in his Out Of Body explorations.

Many times while attempting to leave Earth and visit other parts of the Universe, he would hit a barrier such that he could not get out of nor leave the Earth Realm.

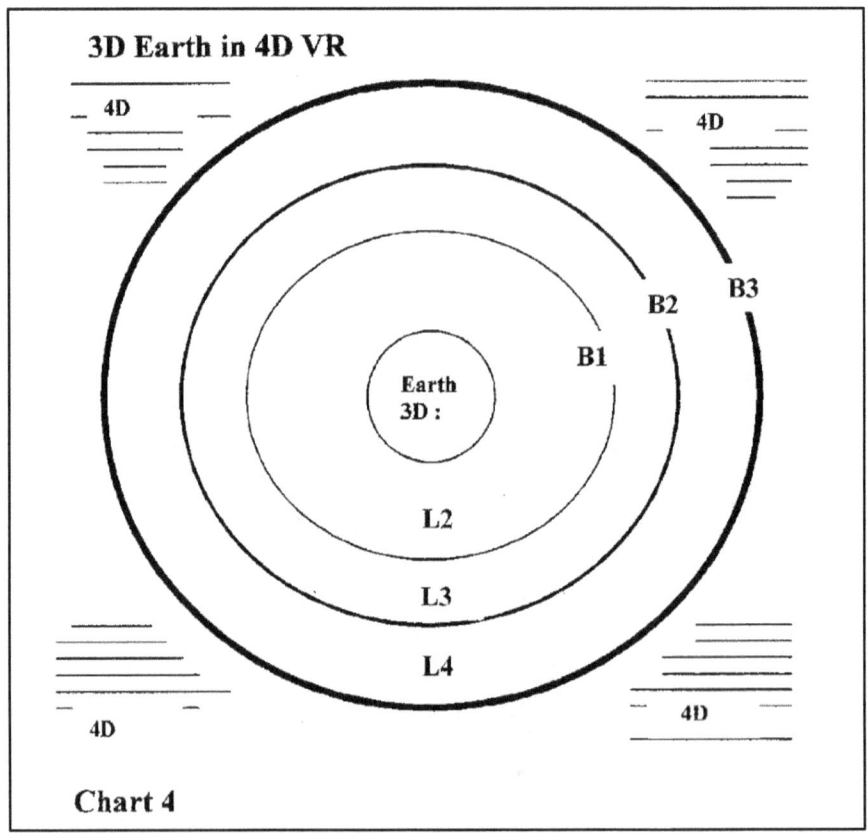

The chart above is a general, high-level view of the 3D Construct in the 4D realm that we are calling the VR Sphere. The outermost barrier B3 is the one that keeps the VR Sphere in **quarantine** and protects it from external 4D+ entities.

The barrier would also qualify for the Firmament in an FE scenario.

Recap: VR Sphere vs FE Scenario

Both: again, **no rotation** and the Sun and Moon for VR Sphere are very close to the Inner Shell but just outside it (see Chapter 1). FE Theory says they are <u>inside</u> the Firmament.
VR Sphere differs from the FE concept in that the shape of the Earth is spherical, not flat.

Both: the constellations are part of the Outer Shell, and do not change shape, and Polaris does not move.

Scientists who study Earth and use the CERN collider are really studying the IMAX Theatre (i.e., the Earth School) and not the real Universe.

Thus the main difference between the VR Sphere and the FE scenario is **curvature**. Concomitant with that is the way the Sun looks at Sunrise/Sunset – on the FE the Sun appears to change size, coming at you and going away, whereas the Sun would not produce hotspots on the VR Sphere, nor change size, being farther away.

And a biggie that we are coming to: **Gravity** is a different issue for both scenarios. The VR Sphere needs it, and the FE doesn't.

Meanwhile, let's look at some new data from NASA regarding the Firmament or barrier just discussed…

Invisible Shield

Science News in 2014 reported on an invisible shield surrounding the Earth that was discovered some **7200 miles above the Earth**. Sounds like a good distance for the dome or Firmament to be located. The shield protects Earth from high-energy "killer" electrons that can destroy satellites, threaten astronauts, and degrade space systems. It sits between the two layers of the **Van Allen Belts**. It has been described as "impenetrable" and "a glass wall in space" that has "an extremely sharp boundary" and is "extremely puzzling." [164]

NASA calls it a **plasmasphere**:

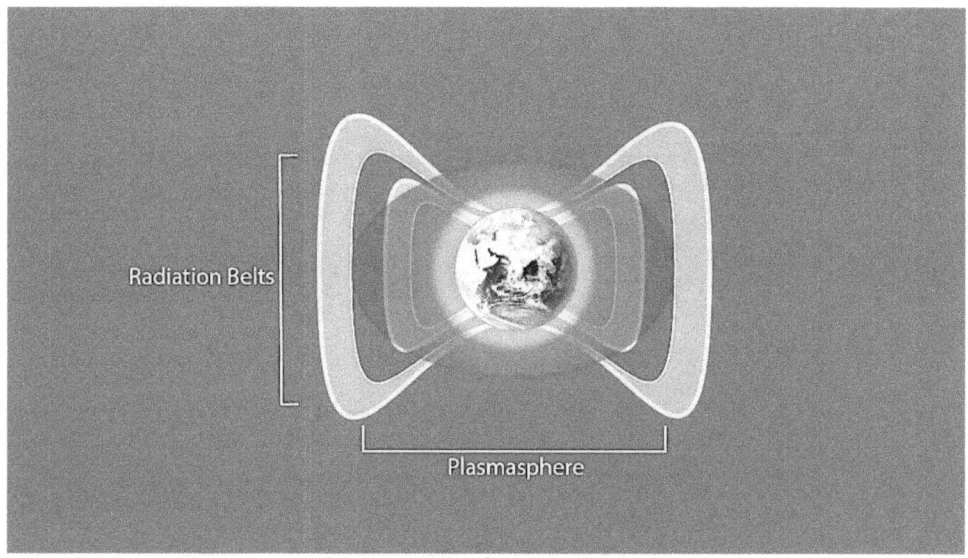

(credit: NASA/Goddard)

This anomaly was briefly mentioned in Chapter 13 in the last version 8 of this book. Whereas the Van Allen Belts of **high radiation** are said to be harmful enough to kill astronauts flying thru them, the new shield may prohibit <u>any</u> passage to outer space.

Pictures of the Horizon

So, next we should ask if there are any credible pictures of Earth from airplanes cruising at 35,000' or credible pictures from space. Given that Photoshop® is so easy to get and use, it is wise to remain tongue-in-cheek when viewing the following examples. Even some of the NASA pictures shown in an earlier section of this chapter are suspect, since it is obvious NASA was enhancing and manipulating them.

By the same token, the Flat Earth people seek to show pictures where the curvature of the Earth is absent, and they also have modified at least the one double-image picture (earlier section) where the curvature of the Earth was "corrected"….

Of course the pictures from space are courtesy NASA and they use **convex lenses** which impart a slight curvature to the Earth's horizon, so we'll use **another high altitude balloon** instead:

No curvature:

And again…from a plane: it sure looks flat.

**Remember the 69 mile rule – each 1 ° arc of the Earth's 360°
circumference represents 69 miles and to see any curvature,
we have to be looking at 5-6 ° minimum, or 400-500 miles.**

Thus the Flat Earth people submit the following – a **4-way picture** from different
parts of the globe: (arrows mark breaks between pix)

And because The Elite (aka **Powers That Be**) might have a hand in hiding the truth about Earth, having made sure that Science says Man evolved from the Ape and Earth is a spinning ball...

...and the possibility that we are being kept in the dark is what leads us to this chapter's conclusion... **and a very important point** – whether it is a Flat Earth or a 3D VR Sphere Construct.

Showstopper

I was saving this tidbit until this point, and the readers who quit the chapter early will never see this curious fact:

> Consider that whether the Earth is a sphere or is a flat plane, even **the flat plane is also round (round border)** – hopefully you have noticed that in many of the foregoing Flat Earth diagrams. So if a picture of Earth is taken <u>from space</u>, somewhere above the sphere or the rounded FE plane, **the curvature should still show up** – and therefore seeing a curvature to the edge of the Earth does not prove Earth is a globe. It would have a round edge or curve whether it be a true globe or a flat, spherical plane.

Back in the section *Notable Observers, part II*, **Dr. Auguste Piccard** was quoted as saying this about his view of Earth from nine miles up:

"It seemed a flat disk with an upturned edge."

For the benefit of the cynics, here is the whole passage from Wikipedia:[165]

> On 27 May 1931, **Auguste Piccard** and Paul Kipfer took off from
> Augsburg, Germany, [in an aluminum gondola with oxygen tanks]
> and reached a record altitude of 15,781 m (51,775 ft [or **9.8 mi**.]).
> (FAI Record File Number 10634)
> During this flight, Piccard was able to gather substantial data on the
> upper atmosphere, as well as measure cosmic rays.
> An article in Popular Science in August 1931 described their journey:
>
> "The story of their adventure surpasses fiction. During the ascent, the
> aluminum ball began to leak. They plugged it desperately with vaseline
> and cotton waste, stopping the leak. In the first half hour, the balloon
> shot upward nine miles. Through portholes, the observers saw the earth
> through copper-colored, then bluish, haze. **It seemed a flat disk with
> an upturned edge.** At the <u>ten mile level</u> the sky appeared a deep,
> dark blue. With observations complete, the observers tried to descend,
> but couldn't. While their oxygen tanks emptied, they floated aimlessly
> over Germany, Austria, and Italy. Cool evening air contracted the
> balloon's gas and brought them down on a glacier near Ober-Gurgl,
> Austria, with one hour's supply of oxygen [in the gondola] to spare."

So instead of an **Ice Wall**, perhaps there could be some sort of edge (or energy
barrier) keeping the waters in place...? He saw that in **1931**! So maybe the gods did
not change the Sim a couple hundred years ago, and maybe Earth is still not a VR
Sphere? It may still be flat and just the Firmament was raised to a higher altitude...
maybe 7200 miles? Sheesh, if you're a god in charge of the Simulation, you can do
with it what you want!

I know what you're thinking, and I cannot fault you... we have all been **conditioned
to think the Earth is a big ball** whizzing around the Sun, and throughout the last
500 years, we have all laughed at our ancestors who thought the Earth was flat. This
has been promoted by the Media as well... the minute someone offers an antiquated
idea (such as "a woman's place is only in the home") – we accuse the speaker of Flat
Earth thinking... How could they be so out-of-step with modern thinking?

And to tell the truth at this point I am now not totally convinced of either point of
view: Big Earth Ball or Flat Earth – both have serious merits. And yet, for reasons
offered earlier in this book, I do believe that **the Earth is a very sophisticated
Simulation**... surpassing the technology of the *Star Trek* Holodeck. The sole reason
for inserting this FE chapter here is to make a point: We do not really know what
Earth is, nor do we know <u>why</u> we are here on Earth, and <u>if it is flat</u>, the Powers That

Be are against us knowing what it really is, and for the very plausible reason that follows…

Summary Point

Earth is a School for those who seek to grow and a **Prison** for those who are wayward and dysfunctional. Overall, the PTB are OPs (see Glossary) and are the NPCs who guide the Father of Light's Earth Drama -- it is all catalyst for our growth... we are to handle it and rise above the BS on the planet. And the PTB do an excellent job of being and providing the catalyst with all their games, deception, lies, and obstruction of the Truth… but there is a Rule, if you will, that the gods insist on: somewhere, sometime before the Powers That Be (PTB) do their nasties, they have to warn people... and thus we have the Georgia Guidestones as the latest "warning" and of course, given the dense level of consciousness on this planet, most people pooh-pooh everything and do whatever they want... later they will realize they are held accountable for what they did. (Karma only applies to the Earth School -- nowhere else in the Multiverse which is outside the 3D Construct. This was examined in Transformation of Man).

It 'kills' me to see the physicists trying to use a powerful CERN Collider to discover more facts about the IMAX theatre in which we live!!   They will not learn about/discover gravity IF it doesn't exist and is based on Newton's false science (see earlier quote from him)...   All they are doing is discovering how this IMAX theatre works! and NOT the Multiverse -- For example, there was no Big Bang, and as was pointed out in an Anomaly page, there is **noise coming from the edge of the Universe** (see end of Chapter 8) – Of course, there would be equipment that sustains this Simulation!

There was no Big Bang because the Multiverse has always existed and Man cannot fathom that... but the Earth was created. And the benevolent gods have run many simulations here with different levels of mankind -- smart, dumb, very advanced, caveman..... When they don't work out, like Neanderthal, they are removed and replaced with a better version, like Cro-Magnon.

Now I think you can see that whereas the OPs (PTB) who think they run Earth, have no souls (see Virtual Earth Graduate, Ch. 5) and thus have no connection to anything higher than themselves (unlike souls who have a Higher Self and conscience), the PTB think religion and spirituality are silly, see no reason for it, and their latest "invention" was the godless Soviet Union. How great was that? (And now they are over here in the USA... promoting unisex, homosexuality [and HIV which we forget about], tattoos, drugs, localized wars, riots and no God......) Thus **their goal is to keep Man from realizing where he is AND what he is** -- because once that happens, any man or woman will bust a gut doing what it takes to get out of here! Earth is not our home, and the PTB will do anything to keep us here -- but they have to keep you stupid to do that. And what possible reason, besides a love of being a Lord over the Sheeple, could there be? Is there a larger pragmatic reason to deny the truth about what Earth really is?

Project BlueBeam

There was a rumor 10-12 years ago about a plot to fake an Alien Invasion of Earth which would result in the NWO being created. The story went that the UFOs were being built on Earth and flown by humans, and once there were enough of them to land in many countries at once, the UFOs would descend and out steps a human-looking being (because they are Earthlings!), and they just happen to speak English (because they have studied our planet for many decades), and now they come forth to help us resolve our world conflicts, establish peace and their answer is the NWO.

If this happens in a time of terrible turmoil and terrorism on Earth, the populace will be much more willing to accept the "Aliens" (aka Visitors) and surrender all freedoms to their control. Of course, the Visitors would want to be able to track humans and easily identify them, so an RFID chip would be placed on humans – replacing the chips already in their cars, cellphone and creditcards (because they can be counterfeited).

So as the Elite or PTB, you'd want to deny that UFOs exist , or at least let the public think that the ones they see are ET craft.

And along with that, you'd want the Sheeple to believe Earth is an actual planet, spinning around the Sun and vulnerable to ETs and their craft visiting it. If you told the people that Earth is a contained Sphere or Flat Earth and that there is no one else out there (in what we see as other stars and planets – which are really just projections, as Charles Fort said) , the ruse could not be pulled off.

How do you do it? Not only with scores of actual UFOs, but you also need advanced holographics that can project realistic UFO images in the sky to look like many more in the skies (even though these don't land).

This could be done from satellites, cellphone towers, remote wilderness locations and even ships at sea. In fact the possibilities are limited only to what technology is available....

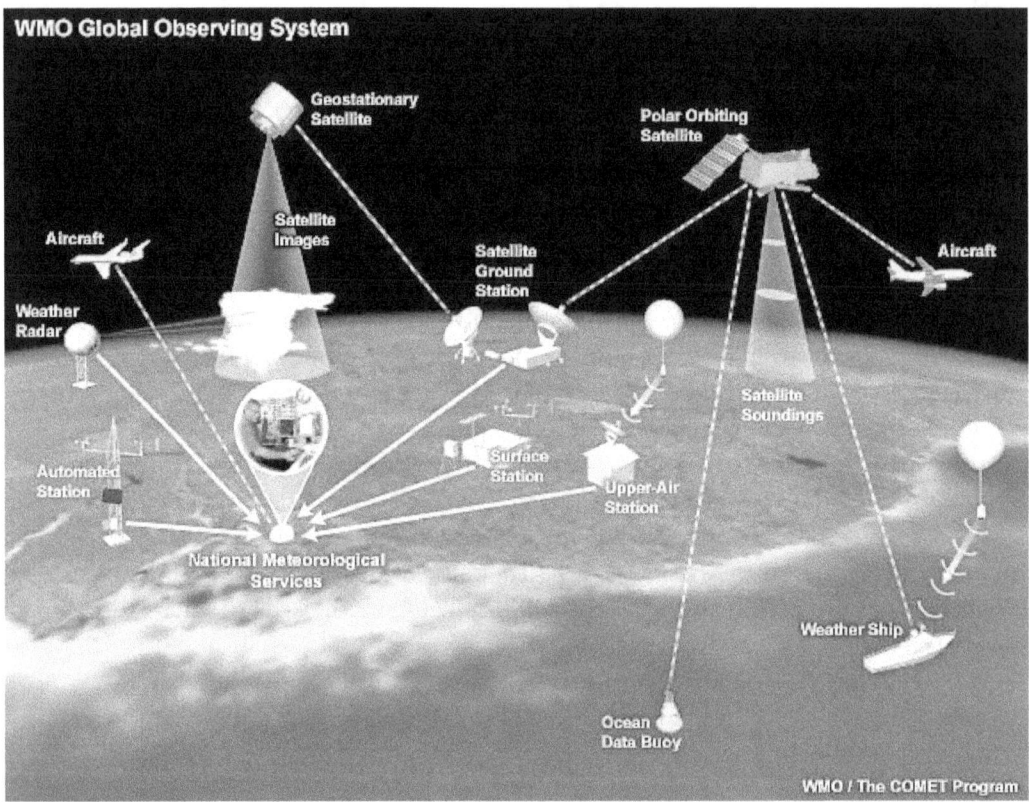

But **the key** to all that is (1) chaos on Earth: wars, terrorism, drought, catastrophes, disease and (2) the people must think that there are ETs and that Earth is just a planet in a solar system in a large <u>inhabited</u> galaxy. The Media must get people to laugh at the idea of a Flat Earth and even UFOs – until they appear in the skies – then the surprise and panic will cause the humans to surrender to the Visitors. The Sheeple must not think that Earth is really a Construct with a Firmament over a Flat Earth or an Energy barrier around a VR Sphere.

It works even better if the people have been told that humans evolved from Apes and the Visitors now tell Man there is no God, that they have never found one in all their travels in the Universe, and that humans are not special and need to surrender their childish ideas (and freedom) and get in step with the United Galactic Federation – or whatever they call it. This could be the **Soviet Union** on steroids – now with the technology to make it work.

Good thing it was just a rumored conspiracy, eh?

The following picture says it pretty well:

> If you are atheistic and part of the ruling Elite, you'd not want your Sheeple to believe in God, or the specialness of Earth and the souls… the Sheeple would want to do what it takes to get out! Graduate from Earth School!

> Don't let the Sheeple see that Earth was created, that the human body was <u>designed</u>, and that Man did not evolve from the Apes.

(credit: Eric Dubay – see endnote #156)

> Keep them dumbed down, make public education a poor excuse for real learning; teach the kids that wearing the right clothes and having the latest app on their cellphone is more important than learning to read and write well … that way you can control them!

> That is all that is going on.

If the deception is being sustained by those who seek to keep Man dumbed down so he is <u>easier to manipulate</u>, then the above diagram spells out the agenda that those who would call themselves **Lords** have over those humans considered Slaves/**serfs**.

If you can be convinced that (1) there is no God, (2) we occupy a rock spinning thru space with no purpose, and (3) Man evolved from the Apes, and (4) that we are free to do whatever we want, with nobody watching…. no special purpose or special design… and (5) Science can explain everything…then the PTB have succeeded.

They are largely atheists and want you to be one, too. If you submit, my condolences – you will never get out of here. For you, Earth will be a **prison.**

And Earth is not our home. It is a special creation, like Man, for the purpose of growing souls… A **School** in which souls learn Patience, Compassion, Humility, and Respect – for self and the Creation. And when a soul learns what s/he is, why s/he is here and what is expected, no more time will be wasted acquiring Knowledge and Compassion – the two required attributes to **graduate** from this School.

If you knew what Earth really was, you'd realize that **souls are special and have a divine potential**, and then seek to develop it. And you'd automatically do what it takes to **graduate from the School**. Mankind needs to wake up.

OK, enough soapbox.

Conclusions

Now what remains is to address the major issue:

> **So is the Earth Flat or Round? And if it is other than a real rock circling the Sun, you now have an idea why it would be played down by atheistic scientists and politicians.**

There is no outstanding reason why our Earth can't be a Flat Earth or a VR Sphere… they are both contained, both a School, and isolated from the real 4D Realm in which it has its being/location as a 3D Construct. And yet, it **is** one or the other, not both.

At this point, you are free to consider the Earth one or the other, whichever feels more comfortable to you. Decide which you like, and move on to the next chapter. If you read on, you will be presented with an inescapable conclusion that you may not like…. especially if you are an atheist.

And while it looks like the VR Sphere is losing some ground, it was the best interpretation I had – given the 3 bits of data back in 2008 (see Introduction again). I was not specifically told FE or VR Sphere, and went with the best interpretation I had. It looks now like I was wrong.

The information (a coming section) pounds another nail in the VR Sphere scenario when we get to Enoch.

One Damning Clue

Oh yes, there was <u>one</u> clue that we cannot ignore, and it <u>does</u> say which way the decision goes. Like it or not.

The True Earth

Remember this?

Arctic 360 ° Photo

This was presented earlier as a time-lapse photography near the Arctic Circle – each vertical sliver taken an hour after the former. One day's rotation of the Sun around the Earth. The Midnight Sun <u>never went below the horizon</u> meaning it had to be **above the Earth** and the Sun could not have been at right angles to the Earth's equator **– in the same plane.**

The following diagram is what we are told is the Sun – Earth relationship –– the Sun at 90° to the Earth's equator – and **the following layout cannot make the above photographic sequence:**

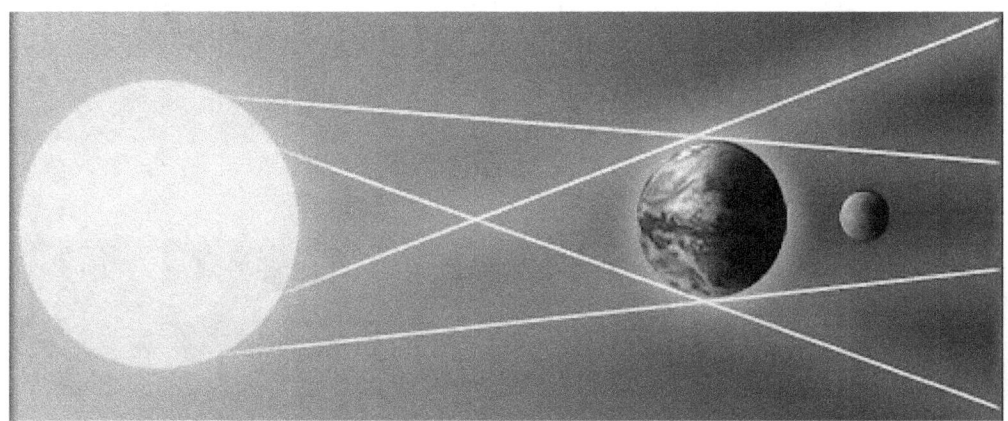

Earth – Sun Layout

If the Arctic 360 ° Photo (top of page) had been on a Sphere with a 23° axis tilt, it would have yielded a **WWWW** or zigzag image on the film as the Sun moved in and then out as it circled low (sunset) then high (sunrise).

Also the 360 ° picture does not fit with the VR Sphere alignment… <u>unless</u> the Sun (and Moon) are just outside the inner barrier, close as they are with the FE. But if in the VR Sphere diagram the Sun is still in the same plane 90° to the Earth and does not move up and down covering the distance only from the Tropic of Cancer to the Tropic of Capricorn, like FE to effect the seasons, then the VR Sphere is not the answer.

The above 360 ° picture can only be achieved with the Sun above the Earth plane.

Corollary Evidence 1

There is another 'proof' when taken with the time-lapse photo above: Because the Sun is actually closer to Earth than we have been told, it often leaves a **"hotspot"** on the ocean – which could not be seen if the Earth were curved (even a VR Sphere) because the water would be curved (convex) and the sunlight would not be a **straight line from the horizon** to a person standing on the beach.

Again, the Earth would have to be flat, because standing water (i.e., the ocean) is not convex. It is flat. And the Sun has to be close to create a hotspot.

And again:

The Sun is shooting a straight line from the horizon to the beach. No curve. It cannot be 93 million miles away to do this… the light would be more diffused.

Corollary Evidence 2

Has anyone seriously considered: If space is a vacuum and Earth has an atmosphere,

(credit: aplanetruth.info)

then what keeps the atmosphere from rushing off into the vacuum? Science says "Nature abhors a vacuum" and we can't use the Gravity argument here since Gravity is not strong enough to pull the clouds to the ground, nor does it pull smoke to the ground, let alone pull on the much lighter air molecules!

How about a more simple explanation?

The Earth is flat, Gravity is not a factor (just Mass or density), and the Firmament keeps the Earth's atmosphere from rushing off into the vacuum of space.

Reality Check

(credit: TheWorldWeLiveIn: https://youtu.be/s0AOj01yzII)

It is very hard to change one's beliefs even in the face of indisputable Truth – and this is called **Cognitive Dissonance** -- which happens because the neural pathways become "locked in concrete" over the years. Dr. Caroline Leaf explained this in The Science in Metaphysics (by this same author) and shows how the pathways and neurons actually grow bigger and stronger the more the belief is activated. Trying to change, delete or "move" a big neural pathway to another memory location can be very traumatic, to put it bluntly. And sometimes it results in a nervous breakdown if the force for change is strong enough. Thus p. 290 suggested you consider both the VR Sphere and FE Scenario and, even if you see that the Flat Earth has serious merit, but are not comfortable with it (Cognitive Dissonance), you are invited to hang out with the one you feel comfortable with – but remember the key point: **Earth is a School** (regardless of its structure).

The point in examining Earth's structure is simply this 3-way analysis:

> The **3D Earth** could, <u>at its very simplest</u>, just be a 3D Earth with an energy quarantine around it: a rotating sphere that circles the Sun in a regular solar system with other round planets.
> This is the one most people think is Earth.

> A more sophisticated Earth could be a Simulation as a **VR Sphere**, with a protective Energy Barrier (which reflects the Gegenschein). This Earth does not rotate and is fixed somewhere in 4D space and the stars are part of the outer Konstruct shell whereas the Sun and Moon are very close to the transparent inner shell around the 3D Construct but they move around Earth, just outside the inner shell – i.e., in the Cspace next to the inner shell. (Chapter 1)
> This is the one that this book (and VEG) initially presented as it is the result of a very sophisticated Simulation run by benevolent Higher Beings on behalf of God who wants to see souls develop in His School.

> The **Flat Earth** is similar to the VR Sphere, except that the Earth is flat, there is no rotation, it does not circle the Sun, and there is no Gravity. It also appears to be a Simulation for reasons given earlier in the book. In addition, there is an Energy Barrier called a Firmament around the Earth which is contained by a gynormous Ice Wall – to which the largely transparent Firmament connects, sealing humans inside the Dome on which the stars and constellations revolve. The Sun and the Moon are inside the Dome and circle the stationary Earth, creating seasons, as well as Day and Night.

If Someone went to all the trouble to make Earth a VR Sphere or a Flat Earth, it means we are protected and watched over, and expected to learn and grow in this School. The point is that the FE Theory is a much more dynamic revelation than a VR Sphere as it is more puzzling and breaks with the expected norm. They both make us think we are on 3D Earth out in space, but the FE Theory substantiates ancient teachings of past civilizations, and only a powerful God who has a purpose for Man would do a Flat Earth, whereas some powerful ETs could create the VR Sphere – starting with a real planet and/or simulating it, or parts of it. The FE Theory and its likely reality drives home the point <u>in spades</u> that there is a bigger, more powerful Someone who oversees us (i.e., a God who created Earth and Man) and thus has a <u>purpose</u> for us.

> **Remember:** the godless USSR said only Science counted, there was no God, and Man evolved from the Apes; he thus has no soul and we are just floating on a rock somewhere in space with no purpose. If you have a soul, you <u>intuitively know</u> that is false.

Summary Proof Points

Having said that, to clarify the differences in three potential Earths, it is clear that the FE scenario is more important than the VR Sphere. And, yes, the most simple Earth definition is the 3D rock circling the Sun… and yet the simple version has some real problems that are answered by the other two scenarios. Gravity is a problem for all but the FE scenario.

Let's review and summarize the key, solid points pertaining to the Flat Earth Theory.

1. **Houston Mission Control Tracking Board**
    This is a killer support **for** the FE Theory. Orbital flight paths are shown as large "S" curves on the Tracking Screen, but in reality reflect a simple orbital **circle** on the Flat Earth – NASA knows this.

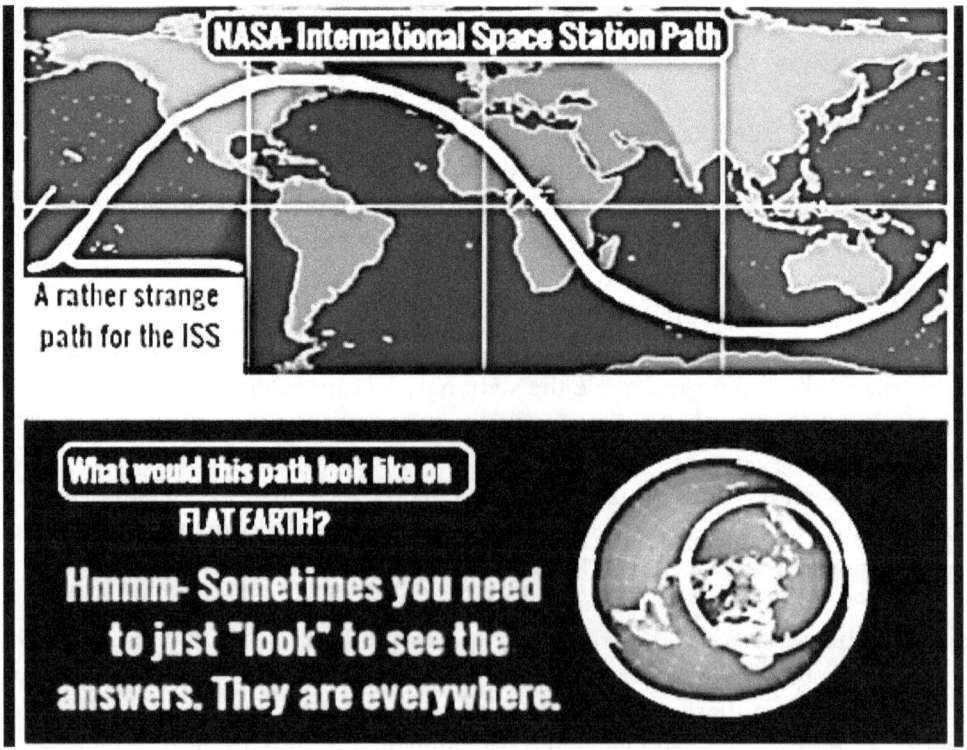

The diagram above is perhaps the most important in this chapter. Note the circular orbit in the lower right – that is what the craft in orbit around the Earth are following… but when you present the standard Earth continent layout that we see in school and elsewhere, on flat 6' x 8' maps of the world, the orbital path must be a large "S". Note too that the "S" path crosses over the same countries in the circle. (Review extended presentation on pp 259-260.)

Did you really think that the Space Shuttle was weaving around the Earth as the "S" path shows? If so, I have some ocean-front property in Montana I want to sell you!

Another in-the-know world agency also knows what the layout is of the Earth:

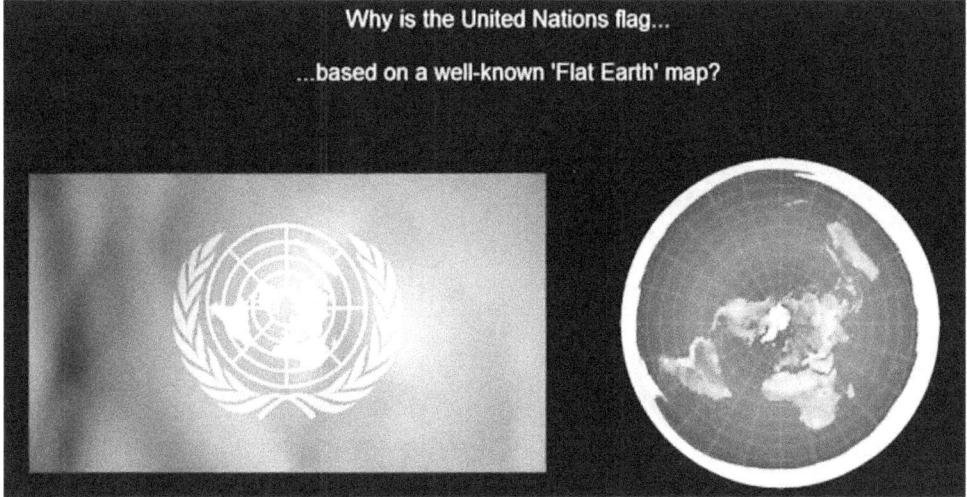

Notice Antarctica is not a continent on either diagram… and looking at the UN flag, above, you have to wonder if they shouldn't have pictured Earth as follows:

And as you will see, that is wrong. It looks like a flight from Capetown, South Africa is a short hop to Australia. The actual distance is much different (see FE map above), and an airplane does not fly as you would think… because it is in the Southern Hemisphere.

Let's look at a shorter route in the Northern Hemisphere….

2. **England to Texas Flight Path**

The picture below says it all -- Why do planes fly thru part of Canada to get to Texas? On a globe, one can draw a straight line/path from England to Texas which does not touch Canada, however the Flat Earth map tells us why:

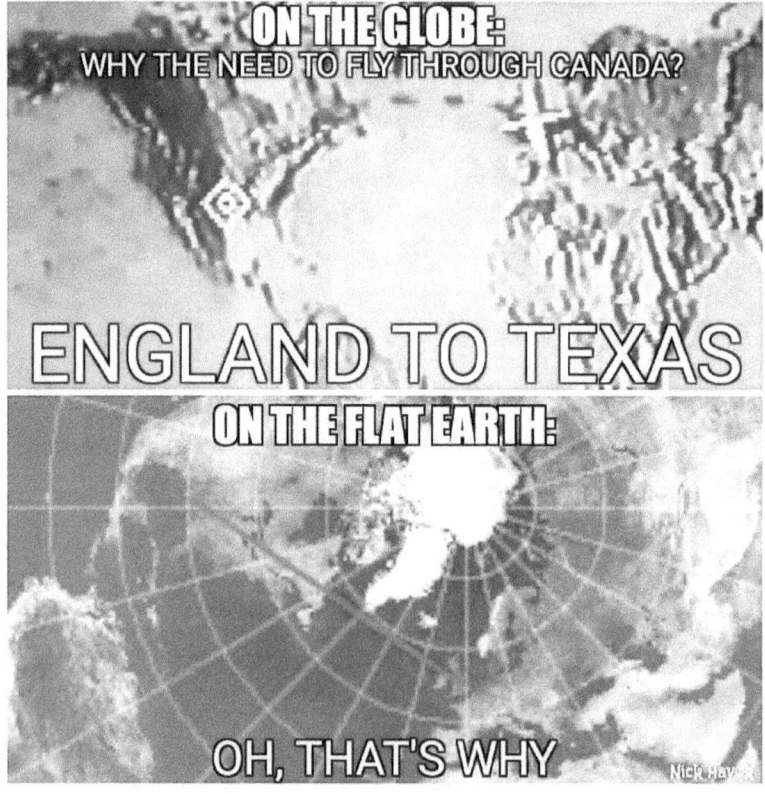

(credit: Odiupicku: Flat Earth – Guess Why We Can't See Antarctica...)

Airlines all want **the shortest, most direct flight path to somewhere**, and curving up or down (as one would on a globe) spends more aviation fuel. But if you know that the Earth is flat, and can plot the most direct route, it will take you thru Northeastern Canada.

And shown on the next page are routes in the Earth's Southern Hemisphere that look Ok on a globe, but in fact do not exist...

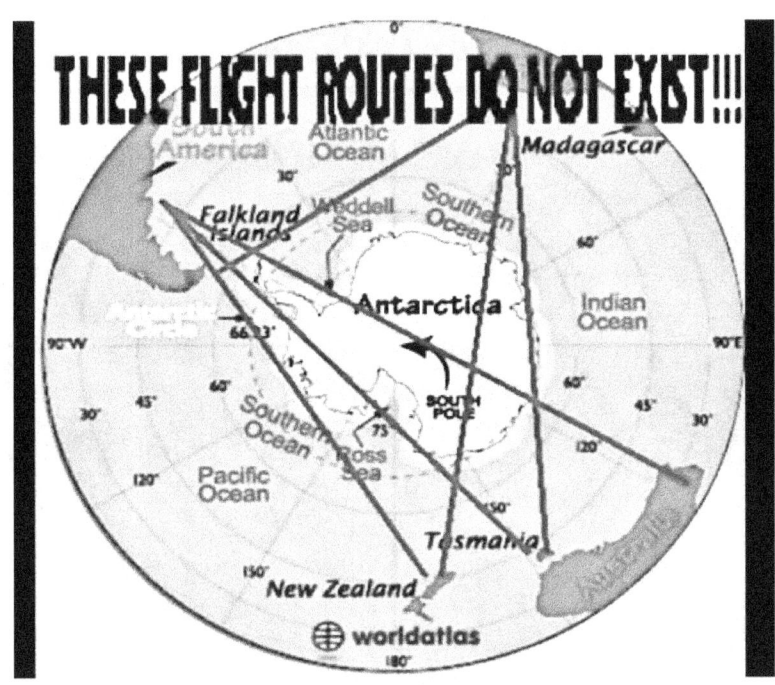

(credit: Odiupicku: Flat Earth – Guess Why We Can't See Antarctica…)

<u>Why</u> do they not exist? They are the shortest routes from Australia to South America, and New Zealand to South Africa – they should be used – even though it means flying over the South Pole…. Unless the South Pole doesn't exist. Look again.

There are <u>no</u> non-stop flights from Australia to South Africa or to Buenos Aires. Unlike the non-stops in the northern hemisphere, **none** exist in the Southern Hemisphere.

Let's say we want to fly from Auckland NZ to Capetown SA, so using a standard world map, that should be as follows (left: double arrow).

Simple eh?

It doesn't work that way-- where they really send us is shown below (because they use an FE map). Non-stop route NOT taken – use fight path on right:

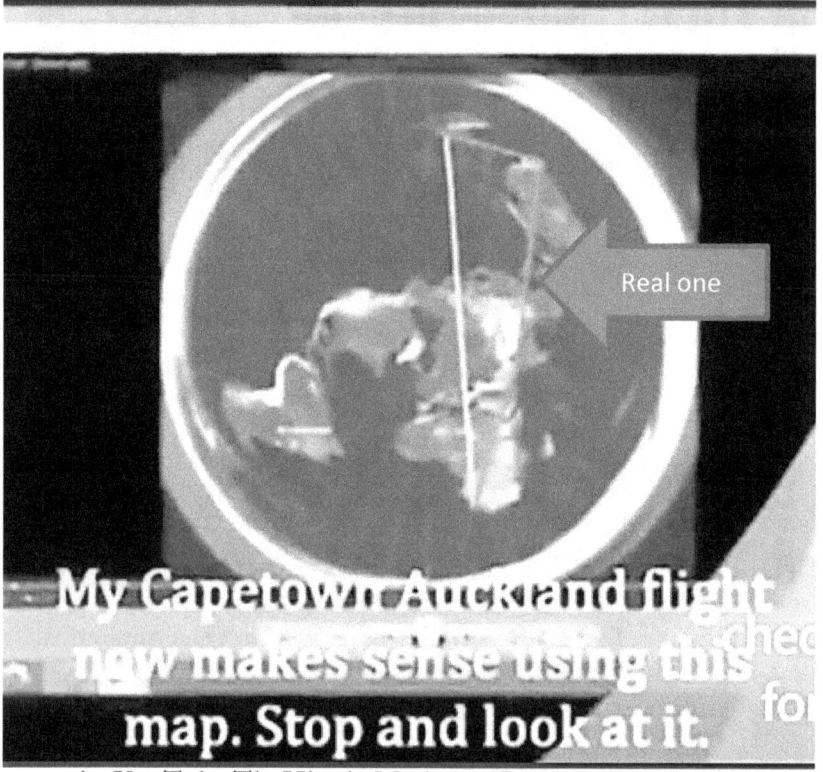

(credit: two pix: YouTube:TheHippie Moderne: Don't Believe in the Flat Earth Theory?)

3. **Hotspots Show the Sun to be Close Overhead**

   This was shown multiple times earlier in this chapter. Hotspots occur when a light source is too close to the oceans, as was shown. If the Sun were 93,000,000 miles away there would be no hotspot because the light is not as concentrated. How can there be a hotspot if the following layout is the light source?

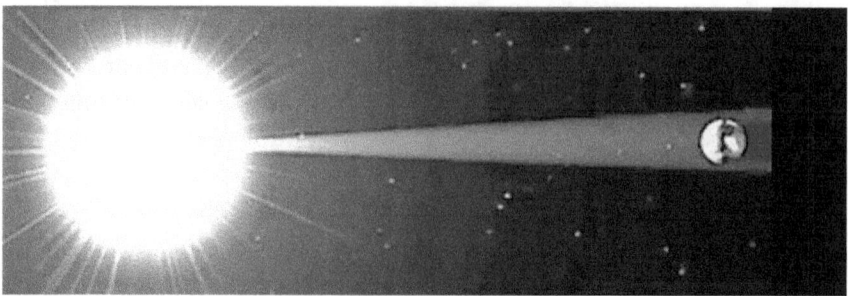

4. **Dr. Auguste Picard Saw the Flat Earth in 1931.**
   This was the German scientist who rode a high-altitude balloon (with a coworker) to 10 miles up and saw the Earth was flat with some sort of border around it. Any pictures he might have taken have disappeared.

In addition, there is a photo from 1946 taken from an Army rocket sent high up into the stratosphere:

(credit: Bing: aplanetruth.info)

Note the Earth has no curvature.

5. **Water Does Not Stick to a Spinning or Curved Sphere**
   Despite Science's best efforts to locate Gravity, it remains elusive… perhaps because it does not exist. But, be that as it may, the issue remains: What holds the million-ton ocean (pick one, most are miles deep) to the rotating Earth – rotating at allegedly 1000 mph? If the Earth rotates, there is Centrifugal Force (CF), and yet the oceans are not flung off the planet… so if Gravity exists, then CF and Gravity balance each other?

Science knows that the CF at the Equator is greater than at the poles. Yet Gravity is the same all over the planet. So Gravity is still 'pulling' with all strength in Northern Canada, 1000 miles south of the pole for example, and because CF is weaker there, would not Gravity exceed the CF and make it hard to walk? "Oh, no" says Science, "there are other factors at work!" Oh really? Just to defend Gravity we have to come up with other forces and arguments… and that suggests it is going *ad absurdum*…

Whatever happened to Occam's Razor?

Occam's Razor: The simplest explanation is often the correct one.

Guess what? The FE scenario is the simplest and need not complicate things because there is no Gravity (and its concomitant corollaries) and things fall to the ground because they are **heavier than air**!

> Having said that, we can also say that if the Earth is a Simulation, Gravity is still not needed, as the programmers would have 'designed' and 'programmed' the oceans to stay where they are put – a very sophisticated holographic/replication of water that moves but is programmed to adhere to the globe.
>
> Is this any more 'involved' than saying Gravity exists but needs CF (and Centripetal Force, and the Casimir Effect... and postulates Loop Quantum Gravity) and whatever else to make it work?

Thus, how does Gravity hold the oceans in place, but permit smoke and birds to fly thru the air? Could it be that birds can glide (not flap their wings) and stay aloft because the coefficients of Lift, Drag and forward momentum temporarily overcome Gravity ? Science likes that one. And even on a Flat Earth, there would still be Lift, Drag and forward momentum so that still leaves the Gravity issue… Does it really exist?

Smoke is particles, and particles have mass and weight – so why aren't they dragged to the ground when they leave the smokestack? "Oh," says Science, "they are too small and don't offer much surface to the force of Gravity" Oh really? We just heard that Science says the Force of Gravity is strong enough to keep million-ton water stuck to the planet…. but it is not strong enough to bring a smoke particle down immediately? Instead they blow around – because they are <u>lighter than air</u> – suggesting Gravity is again not a factor.

A feather has a lager surface area than a smoke particle, and yet it is not dragged to the ground but can waft and float around – eventually hitting the ground, but if Gravity is strong enough to keep the million-ton ocean stuck to the Earth, and a feather has a larger surface area, why is it not dragged down directly to the ground when the bird drops it?

> What Science is doing is coming up with all kinds of modifications and reasons to explain how Gravity works in different scenarios because they have made the <u>assumption</u> that Gravity exists and are now determined to prove it – even if they have to invent corollaries and subtheorems to explain it.

Case in point: in Quantum Physics, the scientists have identified a gazillion forms of **quarks**, one for every action that they observe: Up quarks, Down quarks, Leptons, Muons, Bosons, and now they are searching for *gravitons*… to explain Gravity. … and the latest Higgs Boson which doesn't even fit in the matrix:

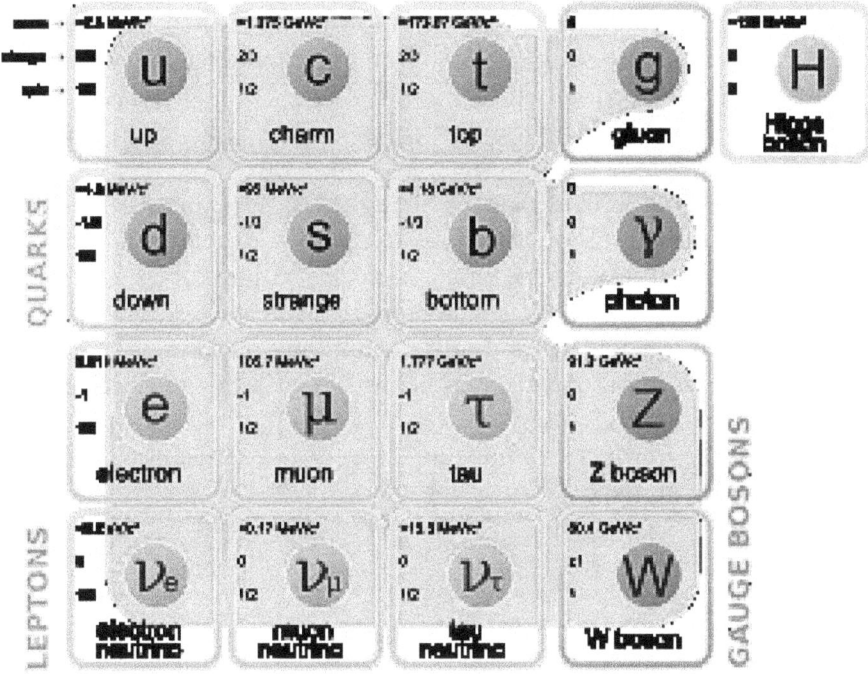

You have got to be kidding. "Charm" quark, "strange" quark…. Are they serious? Could it all be just one particle (quark) with different properties depending on how it is energized (or spinning) – why see how many different particles you can find? And how does that help you?

> I am attacking the Game in Physics ("Develop the Particle Zoo")
> because there is another much more sane Physics version called
> **Subquantum Kinetics** that has only 6 particles in its schema (see
> VEG Ch. 9) and it explains everything that Quantum Physics does, but
> without the expensive Large Hadron Collider, and without 47
> different particles .

The goal seems to be to complicate and quantify when unity is the way of Nature.

6. **The Sun Does Change Size as it Rises and Sets**

   This was demonstrated with a link to a video where the actual process can be observed (due to time-lapse photography) earlier in the chapter. As the Sun comes into view across the Earth plane, it would grow slightly larger (just as an airplane does as it approaches your location) and then begin to get smaller as it moves away.

This is the behavior of a Sun over a Flat Earth – whereas the Sun facing a rotating ball (Earth) would not change in size. Proof? See: 8 minute video proves it:

https://www.youtube.com/watch?v=Nzw1Ug4tOo8   DITRH

or: https://www.youtube.com/watch?v=vHNvUgPRw98

7. **Polaris, the Pole Star, is Always Straight Up, Overhead**

   As was said earlier, this one is a killer to the rotating, weaving and bobbing Earth as it moves around the Sun, and moves with the solar system around the Galaxy. How would a star billions of Light Years away always stay in the exact same position while the Earth has a 23° tilt on the axis and circles the Sun? Look closely at the diagram below – the Earth and its tilt are always in the same position BUT the 4 polar axes do not point to the same location in space as the Earth circles the Sun. How does Polaris move back and forth so that it stays above the North Pole?

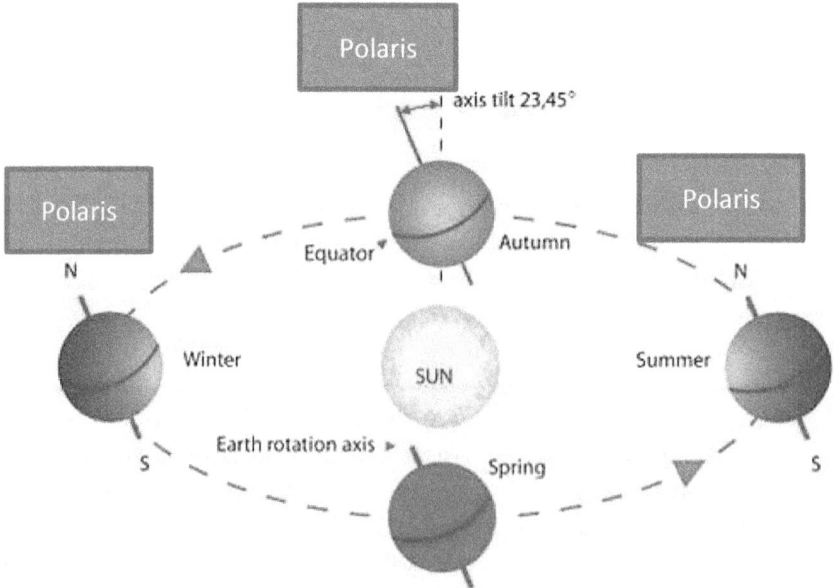

Not only the above angles, but the solar system is moving about the Galaxy as well and Polaris is in the Galaxy… Is Science saying Polaris just happens to keep perfect synchronization with Earth's North Pole in all that gyrating? If Polaris were billions of LY away (as they claim – so that the axes converge at that distance) , you would not be able to see the star with your unaided eye.

8. **Lastly, Enoch Said the Earth Was Flat**
Enoch said he was taken to the "ends of the Earth" – a round planet does not have "ends" – where he saw the Firmament connect to the land. And he said that the Sun and Moon are the same size, and he said he saw the "cornerstone of the Earth" – a flat Earth could have a cornerstone from which all else is measured (and that does not apply to a sphere). He also saw that the stars enter the Earth realm thru portals and run their course in assigned positions.

Enoch's revelations follow.

One last thing of major interest, for those who have read this far…. There is an old book that should have been in the Bible, but because it speaks of Giants or Nephilim **and the Flat Earth**, Constantine removed it from the list of 'acceptable' books for the Bible he was creating (AD 325, Council of Nicea).

Ancient Views

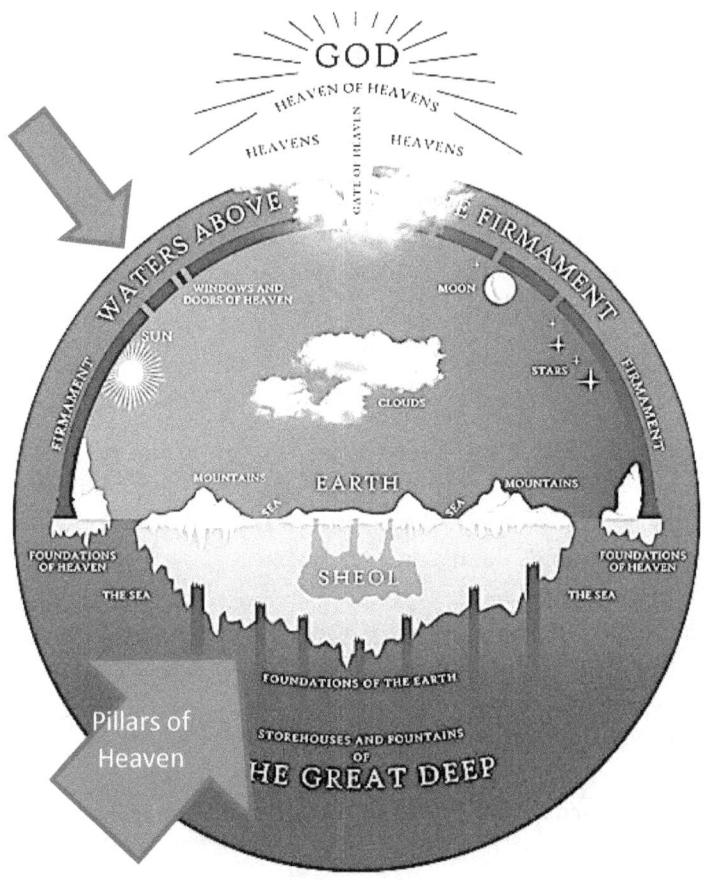

Remember the Bible gave us this information – largely from Genesis:

This diagram will come in handy as we present Enoch's statements about the Earth which follow..

Note the "windows and doors" of the Firmament which Enoch calls "**portals**" (see upper arrow).

Note the Sun, Moon and stars are inside the Firmament.

(see pg. 216 larger)

Well, there was one more biblical person to clue us in – **Enoch** – the father of Methuselah, and the great-grandfather of Noah. He lived 365 years on Earth before God took him up into the region above Earth and showed him what Earth really was (in addition to meeting and judging the Watchers/Nephilim *aka* Giants.

In the above diagram, note the terms "ends of the earth" (lower right), "immovable" (top of black 'rock') and the Firmament with the Sun and Moon <u>inside it</u>.

Book of Enoch

<u>The Book of Enoch</u> was written by a man who is said to have walked very closely with The God. So closely that he knew what God was up to and had knowledge of what was going on in the heavenly (Astral) realm that was affecting Earth. If someone didn't want Man to have a clue about what other beings might also be on the Earth, or who may be afflicting Man, they would suppress this book and keep it out of the Bible. Thank you, Constantine.

While <u>The Book of Enoch</u> is basically a lament over the sinful condition of Mankind, and a call to repent with blessings for those who are righteous, it also expands on Genesis 6 which says that angelic host descended and created giants (Nephilim) and havoc in the Earth.

<u>The Book of Enoch</u> dates to about the 2nd and 1st centuries B.C., and was held in great reverence by many of the early Church Fathers. However, due to efforts of the emerging Christian church, by the 4th century A.D. it was looked on as heretical, and later condemned.[166] So much for truth.

Quotes from Enoch

Enoch is taken up by God above the Earth and shown many things, including the layout of the Earth, from whence blow the winds, from whence flow the rivers, **the ends of the Earth** (Hint: there cannot be an 'end' to a spherical Earth), and how the Sun and Moon operate **"in their orbits."** (Hang on to your hat.)

The following is a series of quotes from <u>The Book of Enoch the Prophet</u> (in the Bibliography) and the parentheses are the page ref's cited:

> And they took and brought me to the place of darkness and to a
> mountain the point of whose summit **reached to heaven**. And I
> saw the places of **the luminaries** [Sun & Moon]… and the **stars**
> **[in the Firmament]**.   (15)

...and to the fire of the west which receives every setting of the Sun... and the place whence all the waters of **the Deep** flow, and the mouths of all the rivers of the earth... (15)

I saw where all the winds are kept...and I saw **the corner-stone** of the earth; I saw the four winds, and **the firmament of the heaven**

I saw how the winds stretch out the vaults of heaven... [and] the **pillars of heaven**... and I saw the winds of heaven which turn and bring the circumference of the Sun and all the stars to their setting [locations].... I saw **at the end of the earth the firmament of the heaven above.** (15)

I saw seven mountains of magnificent stones...and beyond these Mountains is a region, the **end of the earth**: there the heavens were completed. ... I saw a place which has no firmament of the heaven above.... (16)

And I saw other mountains .... And beyond these mountains I saw another mountain {to the east **ends of the earth**}... (24)

I came to the Garden of Righteousness [Eden] ... and from there I went to the **ends of the earth** and saw there great beasts... (25) ...and to the east of those beasts I saw **the ends of the earth whereon the heaven rests** , and the **portals of the heaven** open. And I saw how **the stars of heaven come forth**.... Each individual star by itself ... and their courses and positions. (26)

From there I went toward the north to the **ends of the earth**... And from there I went towards the west to the **ends of the earth**... And from thence I went to the south to the **ends of the earth**.... (26) And I saw there three open **portals** .... **Through each of these small portals pass the stars of heaven** and run their course to the west on the path which is shown to them [in the Firmament]. (27)

I saw all the secrets of the heavens.... And I saw the mansions of the elect and holy ....And I saw **the chambers of the sun and moon** whence they proceed and whither they come again, and their glorious return, and how one is superior to the other, **and their stately orbit...** and how they do not leave their orbit, and they add nor take nothing from the orbit ... The sun goes forth and traverses his path according to the commandment of the Lord... and the moon goes forth... in accordance with **the oath by which they are bound** together [higher entities run the sun and moon and work in agreement with each other] .... I saw the visible path of the moon... by day and by night...the one holding a position opposite to the other before the Lord... (33)

And this is the first law of the **luminaries**: the luminary the sun has its rising in the eastern **portals of heaven** and its setting in the western portals of the heaven... at first there goes forth the great luminary, named the sun, and **his [orbit] is like the circumference of the heaven,** and he is quite filled with illuminating and heating fire.... The great luminary rises, sets and decreases [in brilliance] not, and rests not, but runs day and night, and his light is sevenfold brighter than that of the moon; but **as regards size, they are both equal**. (72-75)

These are the two great luminaries [Sun and Moon]: their [orbit] is like the circumference of the heaven, and **the size of the circumference [orbit] of both is alike.** (81)

And Uriel showed me another law: when light is transferred to the Moon, and on which side it is transferred to her by the sun. (81)

It is important to note that Enoch saw the ends of the earth – from <u>four directions</u> in paragraphs 25-26 – that does not describe a sphere. He also says **the Sun and Moon are the same size** and the Moon reflects light from the Sun – who operate together by agreement.

Perhaps the Amerindians were not idiots when they said the elements of Nature are inhabited or controlled by spirits. Even the Egyptians venerated the Sun as a being (Ra) and if it is empowered by the presence of a higher being [spirit], then we now know what the ancients were doing and why.

In addition, the 4 winds, the stars and Sun and Moon come and go under the Firmament thru the use of **portals.** This suggests that they have special ways of entering/exiting the Dome, but the phraseology and such are so archaic it is hard to tell just what is being described as a portal. The FE Theory shows the Sun and Moon the same size and circling about the North Pole (which is the center of the landmass).

Paragraphs (15) and (25) seem to suggest that the **Firmament rests on the "ends of the earth"** – meaning that the current FE Theory displaying the Dome extending down to the Ice Wall, which is the ends of the earth territory, agrees with Enoch.

So it looks like there is support from the Bible (as well as the Koran mentioned earlier) for the Flat Earth as a real scenario for our Earth. And if the Earth is flat, it is obviously **not hollow**!

Postscript

So if Earth is a **Flat Earth 3D Construct,** does that negate the Anunnaki info (in other books by this or other authors)?

NO. The gods have <u>inserted</u> all that we need to examine, use, and wonder about. The ancient History we think happened, the Great Pyramid, the legend of Atlantis, hammers and jewelry buried deep in coal deposits, the Antikythera Device, Bigfoot….could all have been inserted. Earth would be a pretty boring place if it weren't for things and ideas that keep us amused, entertained, questioning, amazed, and researching… and each to his own area of interest.

The Anunnaki and the Greek, Roman and Norse gods (who were <u>one and the same</u>) were part of the School scenario – just not real ETs. They were <u>inserted</u>.

And since Earth is not a VR Sphere:
NASA orbital paths having proven that one for us,
Southern Hemisphere flight paths follow the FE layout,
Planes cannot fly over Antarctica,
Oceans do not show curved water with flat camera lenses,
and Gravity still being a problem on a VR Sphere,

then there were no ETs running around messing with Man. The Sumerian tablets and scrolls were either left from a former version of Man (pre-Flood) or they were **inserted**, and we may never know the answer to that one.

Consider that a lot of what you have been taught in Science, Religion and Earth History is either backwards or false. There was no Big Bang, no Black Holes, the Universe is not expanding, and Time does not exist but is a convention for the Earth Realm. Avatars and teachers have been sent into the Earth School to guide Man – and **the PTB are just catalyst** against which humans' mental, emotional and spiritual growth is measured. They are not evil, they are a necessary player in The Game – just as a video game has **NPC's (non-playable characters)** who drive the Drama/Game, so too does the School have the PTB and OPs. Your job is to wake up, see what is happening, and rise above it (and not attack it). How you deal with it determines whether you graduate or not. (This is all explained in <u>Virtual Earth Graduate</u>, Ch 15-16 and several other books by the same author.)

Conclusion

Is the round Earth and putative Gravity not like Santa Clause? And why did we not go back to the Moon? And why did we stop the Space Shuttles? Why are the pictures of Earth from the Moon wrong? Why is Polaris always overhead if Earth is weaving and tilts as it circles the Sun? Why is the Plasmasphere at 7200 miles altitude the same height as the apex of the Firmament? Is that what the *Gegenschein* reflects off of? And why does Houston Mission Control show an odd S-shaped curve to the flightpath of any orbiting object?

Is it all because we are on a Flat Earth and cannot get past the Firmament?

Questions Answered

The first important question was: Is Earth Flat or Round?
> That was answered just above (see 5 points).

The **second all-time popular question now:**
> <u>Why</u> do we need to know what Earth really is?

> Because when someone says "Hey, aliens are coming to invade Earth," or "The Earth is hollow…" if you know what the Earth really is, you won't be fooled and run off in fear/panic! Aliens can't get past the Dome, and the Flat Earth is not hollow.

His Grace

You might want to thank the gods who run this place for caring enough to provide and sustain a place (free from interference from 4D ETs by the way) where you can learn and grow. And when you graduate from Earth School, they have a place for you in the Father of Light's Multiverse. (See list of possible positions at end of Chapter 13.)

It is not important to God that you know that the Earth is flat or a VR Sphere…the important key is that Earth is a 3D Construct surrounded by a protective 'shell' of some sort because it is a **<u>School for souls</u>**. However, denying the Flat Earth, or even the VR Sphere if that is your preference, denies the existence of a God that created Earth and Man – and isn't that the goal of the godless PTB? What is your position on this?

Welcome to the Flat Earth

Chapter 12: Simulation and Religion

This is a hard chapter to write, let alone put a title on. Simulation raises a lot of questions about Creation, Genetics, History, the nature of God, and calls into play the Bible's statements about Man and his world. Any of those could have been in the title of this chapter. In addition, one has to wonder where it is all going and is the Endtime portrayed in Revelation in the Christian Bible going to be a reality?

While this book does not have all the magic answers, it can extrapolate some conclusions based on what we have covered, what the scientists have said, and what would have to be true if indeed Man does live in a Simulation.

In no way will this chapter be a put-down or denigration of anyone's religion, but we will have to consider the very nature of Religion as a construct and how it could better serve Man. In the foregoing chapter, it was revealed (in point #4) that Religion was used to control Man – a different kind of Man than we have nowadays – and that it served a purpose which is now becoming outdated. What is needed now is a **Spirituality**, not rote religion, because Western Man is waking up to a greater, more fascinating reality and a more dynamic sense of what he really is... And while the Eastern religions have seen Man as a soul and eternal for centuries, the Chinese have one of the most unenlightened and repressive societies on the planet... probably because they are so populous. That puts a demand on air, food and water, and some sort of regimentation is almost mandatory to deal with the masses.

> Having said that, I have to quickly admit that there are still barbaric, crude, uneducated versions of Man all around the planet. And they still need a hellfire & damnation version of religion because they are so petty and violent that they even use their religion to attack and kill other humans.

Secondly, it is obvious that Christians may have the most problem with this chapter since it is not the purpose to denigrate nor corroborate their Faith. While there is a God, and Man is a precious creation, remember that what passes today for Christianity started centuries ago when Man knew the Earth was flat and when it was 'proven' that the Earth was a sphere, many people still refused to believe it. Everyone knew that Man was the center of the universe, they could all see that even the Sun revolved around the Earth, and if one sailed out too far in the ocean, they'd fall off (not possible with the Ice Wall – so disinformation abounded even 500 years ago).

The Christians had had their thinking done for them by the Church – until the **Gutenberg printing press** was developed about AD 1440. Printing the Bible then meant the public would have to learn to read, but it also led to different versions of the Bible being printed… and again, instead of inter-faith competition ("Our God is better than your God"), it was "Our Bible is better than your Bible!"

> It is obvious that the Anunnaki understood the competitive side of humans, and may have even coded Man genetically for it (!), so that giving them different religions and different languages only served to keep humans at a distance from each other – the idea was for Man to NOT UNITE.
> It has always fascinated me that teens from different high schools fiercely believe that their school is the best – despite the fact that 42,567 high schools across the country can't all be the best! That is the same illogical "me and mine" versus "you and yours" at work again.
> We'll see in the Genetics section below what is being done about that.

Let's examine some major themes in light of a Simulation.

Creation

According to Zechariah Sitchin, humans were created on Earth, in the Anunnaki image, 250,000 years ago. If Man could blast off from Earth for another solar system, in a couple of years, he too, right now, has the knowledge to take the genetics of any hominid he finds on a new planet and mix DNA and create his own worker slaves – to build buildings for him, work the fields (grow food), and even fight his battles for him. That is what the Anunnaki did, and if we can now manipulate genetics, it is just a matter of time before we develop the spacecraft to go and do what our 'ancestors' did.

The new wrinkle is that if we are in a Simulation, we're not going anywhere until we are released – back into 4D where all sentient life in the Universe is. Obviously if we are now in a Simulation, the Anunnaki could have come here while we were still in 4D, and as was revealed earlier, it was about AD 900 that Earth was replicated into the Sphere. Or the Anunnaki were inserted into the Sphere as protagonists in the on-going Greater Drama.

Souls now birth into the Simulation, and die to exit it… procreation is a feature of Earth… as it always was, but now it is protected. When Man grows up, matures, and develops the ability to protect himself, and respect others, he can be released back into 4D – which is not a bed of roses. There is already a warlike race out there that is held in check by equally advanced benevolent races, but they cannot 100% watch over Man... hence another reason for the Sphere which **quarantines** us. (This was related in the docu-novel *The Earth Warrior*, where the Dracs are like space pirates – not evil, they just want what they want and have the power to take it. Man has had no way to stand up to them and defend himself).

> FYI: The human form is a popular one in the 4D Universe, and is found especially this Galaxy. So is the reptilian form – physically superior to the human form, but the Dracs do not have the "divine heritage" of the soul. They fear it and would exterminate Man on Earth if he weren't protected by the Quarantine.

So it wasn't a God in a white suit that created Man on Earth ... Any more than a rabbit lays eggs on Easter, or a bearded fat man visits millions of homes in one night to leave presents. Man loves fairy tales: The Tooth Fairy, the Little People, Cupid shooting 'love' arrows, trolls under bridges, huge, tentacled sea monsters attacking sailing ships, mermaids, unicorns, and demons that make you do bad things.

And if Man as a soul with a divine potential is expected and encouraged to spiritually grow and become more compassionate, patient, humble, gain more Knowledge, respect other people and the Earth, it would stand to reason that he is held accountable for those times when he "misses the mark" – i.e., he makes mistakes, or sins. Thus if the Earth School Control System is defined by Karma and Reincarnation, then vicarious atonement is another fairy tale.

No one can learn your lessons for you. Whereas years ago a man when challenged to a duel could have a surrogate (stand-in) do the duel for him, that was a copout and is not the way the Realm we live in has ever worked. You screw up, <u>you</u> are held accountable. And it is very human to err, so that is another reason for having a protective Simulation in which to learn. And there is a God.

What Religion should be and can be today is addressed in the last section of this chapter. In the meantime, how does Man get better?

Genetics

The Greys are **inserts** and they are rewiring the genetics... to create **Hybrids**. They are **Bio-cybernetic androids**. This is Man's real hope for the future... a sharper, more intuitive, less testosterone version of Man... semi-psychic such that manipulation and deception will be much harder to 'run' on the Hybrid populace.

You don't have to be in a Simulation to abduct and manipulate people's DNA, but it helps with the **insertion** part. Even the gods who run this place see the need to change Man's genetics – from warlike to peaceful… but still enough of the 'fighter' element that he will stand up for what is right. If you err on the side of too peaceful, too little testosterone, you get a human male that is effeminate. If you get the genetic hormonal balance correct, but get the pituitary (master gland) not right, certain hormonal secretions by the pituitary will not be adequate to permit the human to stand up and fight because the adrenals do not pump cortisol and adrenaline into the system, and the human must run away, not strong enough to fight when needed.

The human genetics are complicated, which is why with today's errors in diet, drugs, pollution and stress, we have so many people who are victims of **fragile genetics**, now defective genes, and these people are semi-functioning, some over-reacting, under-thinking, and trying to solve their problems with more drugs… or a gun.

Hence, Big Pharma is having a field day!

The goal in creating Hybrids has been to build a stronger genome (which will be a bigger genome with more self-correcting code and longer-acting telomeres), and insert the genetic forerunners of psychic abilities – clairvoyance, intuition, and the ability to heal self and others. These things were allegedly denied humans when they were first created… the legendary Anunnaki had them, and so did the Hybrids of yesteryear – Sargon, Moses, Marduk and Inanna, Alexander the Great, Noah, Pythagoras and Apollonius of Týana for starters.

> Let's make a note here that while the Anunnaki apparently created Man on Earth as a slave, one has to eventually ask the question: Who created the Anunnaki? Somewhere there was a 1st Creation, somewhere. And that suggests God.

While diluted nowadays, the bloodline from our ancestral Hybrids still exists.

But that isn't important because the improved version is birthing among us – the Indigos, or Homo *noeticus*. These children and young adults are more intuitive, they know when they are being lied to, and they have the basics of clairvoyance, so they know where things are that they are looking for. Some are telepathic and some are healers. They have a connection with their Higher Self.

The gods that run this Simulation have changed the rules – they no longer insert higher spiritual beings as avatars (Think: the movie *Avatar*), they are doing a 5th column approach – fixing the societal scenario from within – so that we avoid a *Soylent Green*, or an Orwellian world. Whether the insertion of better humans will reach a 100th Monkey effect on society <u>before</u> the PTB destroy the gameboard, remains to be seen. It is said that just 6% of the population thinking in a new (proactive) way acts like a trimtab and can make a difference in society.

> My understanding is that the gods have replicated this Simulation minus 80% of the negative, self-serving STS PTB, and if worse comes to worst, a lot of us will be moved over there, to **Sim II**, because the gods really want to see an upbeat scenario play out.

Endtime

Prophesying the Endtimes requires that someone get a look at the Greater Script for this Simulation. Calling this one that we're in, **Sim I,** and the replicated one (above) **Sim II** – the timeline principles given in *Transformation of Man*, Ch 2, apply here as well. In other words, if the **Timeline** that was split/replicated becomes too negative, there is not enough positive coherent energy to hold it together and it will fail to energetically sustain itself… or it may be disintegrated by the gods (Y – Z below).

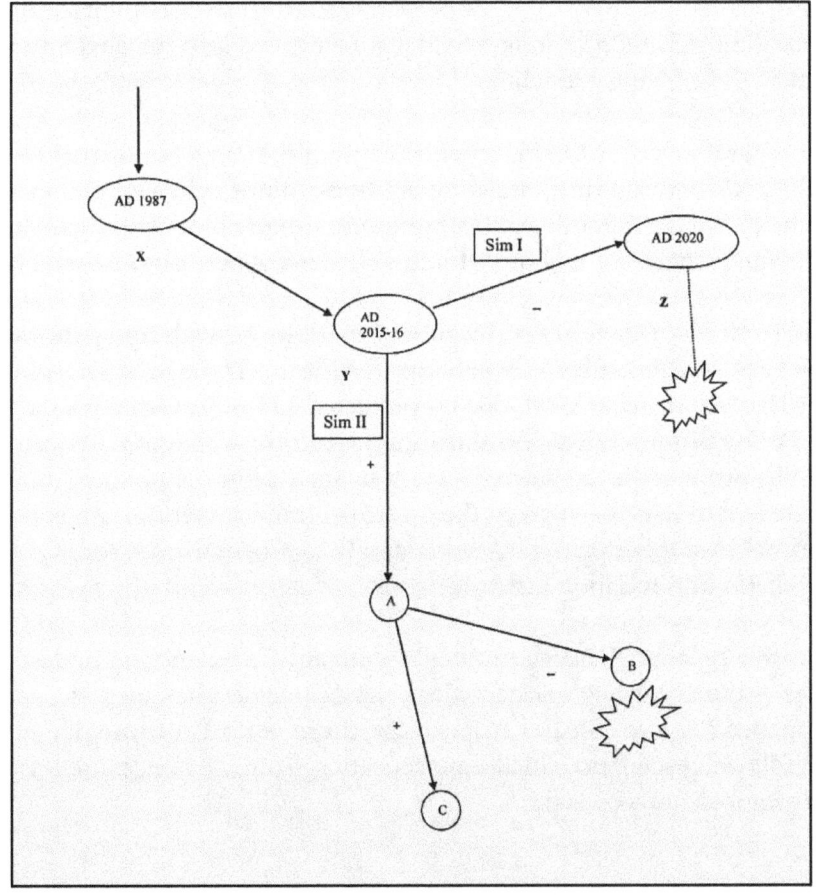

The **Simulation** is the same idea – Sim II is projected to be positive and will carry many souls forward in a proactive scenario – with no Endtime (Y – A – C above). If this Sim I we are in begins WW III and turns negative, that is it, *finito*, for this Simulation.

Diagram above:

Sim I becomes negative and moves from Y to Z, and in a few years, then just disintegrates. Sim II continues to point A and if it stays positive, it progresses to point C. If it too goes negative, it will be disintegrated at point B. Whether a Sim goes positive or negative depends on the overall energy balance of the Sim – if too many souls go STS (negative) then the Sim is cancelled. When a Sim (or Timeline) is too negative, souls do not incarnate there and the lack of sustaining energy causes the Sim to fail.

Thus if we are in a Simulation, the Endtime would be part of the **Greater Script**. And that isn't cast in concrete, either – if the Simulation begins to have problems (glitches, pixellation, things missing, etc.) the gods would make a decision to either reset the non-functioning parts of it, or perhaps restart the whole thing, and whatever was scripted and not done would just be moved into the next Simulation's Script – if they thought it worthwhile.

Keep in mind that if God is a benevolent God, or in a Simulation, if the gods that run the Simulation are **benevolent**, why would they script such a terrible Biblical Endtime, an absolute bloodbath with pain and carnage for millions of humans? That is very out of character.

For the Bible's Book of Revelation Endtime scenario to be real, someone would have to have had a peek at the Greater Script for Earth. (If the prophecy is accurate.) St. Malachy (Irish holy man, AD 1100's) – predicted 111 popes from his day to the final, 111th Pope. (**Supposedly we are on the last Pope, Francis.**) – Was he shown the Script? His prophecy is considered a hoax by many, but the point is: if he got it right, then he had to have been given the information by someone connected with the Simulation Greater Script… This would apply to Nostradamus, too. Or is the Endtime scenario in Revelation just froo-froo?

I heard a preacher harping on the Rapture and Tribulation again last week – and according to one author, the Tribulation has already happened (70 week idea). I suspect that **there is no Endtime** due to the following scenario which I am repeating from VEG because it is very important to consider.

Endtime Scenario Illusion

Religion has been used to promote a fallacious Endtime scenario, whether to give people false hope in a coming Rapture, or to give people fear of a coming Apocalypse, it is big business. And false.

Apocalypse and Rapture

In a very insightful book by **Steve Wohlberg**, <u>End Time Delusions</u>, he describes how the idea of the Rapture came to be, and why Man is being taught that there is a coming Apocalypse. To expose the illusion, Wohlberg examines two main supports for the supposed Endtimes: (1) the coming 7 years of Tribulation and (2) the Apocalypse with its pre-, mid- and post-millenial Rapture concepts which co-exist in confusion because no one can agree on what the Bible really said.

In fact, the Bible does <u>not</u> literally speak of a Rapture. Even the book of Thessalonians just says that believers will be "caught up" in the air to meet Him on "the day of the Lord..." whatever that is. (Thess. 4:17) Of course, this has been promised throughout the years, and it makes for good press. The fact that it is Paul's teaching is suspicious, since he was a ringer and his teaching was different from what Jesus/Apollonius taught. (VEG, Ch 1 & 11.)

The following will not be a biblical exegesis; it will suffice to show that the idea of the Rapture was created by one man in the 1800s, and that there is no "7 weeks" (years) of Tribulation period yet to come. This is important enough to examine closely because the Endtime concept is manmade and man-promoted.

Seven Year Tribulation Theory

Wohlberg informs us that there is no "seven year" concept nor a "seven years of tribulation" anywhere in the Bible.

> Amazingly the entire theory is really based on a rather speculative interpretation of two little words in one single verse....in Daniel 9:27....and the two words are **"one week."** [167]

In Daniel 9:24-27, Gabriel divides the seventy weeks prophecy into 3 periods:

> ".... seven weeks (verse 25), sixty-two weeks (verse 25), and one week (verse 27). 7 + 62 + 1 = 70.... So far we have seen 69 weeks fulfilled. That leaves "one week" left, otherwise known as the famous "70th week of Daniel." [168]

The significance of the 70th week is that there has been a huge **gap** between the end of the 69th week and the 70th week – which was explained away by the creation of a new (non- Biblical) concept: **Dispensationalism**. It was clever but false.

The GAP

After the GAP, allegedly comes the 70th week of Daniel, the so-called Tribulation, and then the Rapture. The GAP is also called **Dispensationalism**. To better explain this, a Scottish Presbyterian minister **Edward Irving** around 1830 began to teach ".... the novel idea of a two-phase return of Christ, the first phase being *a secret rapture before the rise of antichrist.*" It is hotly disputed where he got this idea, but it appears to have come ".... from a young Scottish girl named Margaret MacDonald who first 'saw' it during an ecstatic 'revelation.'" [169]

> If one has read VEG, then it is clear that OPs, Neggs, and Discarnates can and do channel BS.

The Rapture

So Irving developed and taught the idea of a Rapture, and somehow passed it on to **John Nelson Darby** and both of them became champions of a pre-tribulation Rapture and of a coming Antichrist during a period of Tribulation.

> In light of ... careful research, it seems Margaret MacDonald's pre-Antichrist "rapture revelation" is **the real smoking gun** behind Darby's theology. Regardless, the essential pre-tribulationism of Margaret's doctrine soon became a **weapon of mass deception** in the hands of Darby and his dispensationalist followers. [170] [emphasis added]

... Darby later became known as the Father of Dispensationalism, which teaches that God deals with Man in special periods, or ages. (Ironically, the Simulation deals with Man in Eras.) Accordingly, Man is now in the "Church Age" which will conclude with the Rapture. Then Daniel's 70th week (the Tribulation) will supposedly kick in during which the Antichrist will persecute the Jews (I pray not!).

> His [Darby's] most striking innovation was the timing of a concept called the Rapture, drawn from the Apostle Paul's prediction that believers would fly up in the air to meet Christ in heaven. Most theologians understood it as part of the Resurrection **at time's very end**, Darby repositioned it at the Apocalypse's very beginning, a small

shift with large implications….
Darby's scheme became a pillar of the new Fundamentalism. [171]

This is the kind of disinformation that often comes thru the OPs.

Endtime Summary

Wohlberg points out that this is a prophecy about 70 weeks, and in Jewish prophecy a day was symbolic of a year, so 70 weeks would be 490 days, or 490 years. The first 69 weeks would be 483 years, using the same conversion. This 483 years was also the time period from the captivity in Babylon down to the arrival of Christ. So "69 weeks" were fulfilled. Now there is supposedly just one week (7 years left) to account for.

That 70th week (7 years) was accounted for by Jesus: His ministry (3 ½ years) and then the additional ministry after the Resurrection (3 ½ years). [172]

> The entire prophecy of Daniel 9:24-27 covers a period of "seventy weeks." Logic requires that "seventy weeks" refers to **one consecutive block of time**, in other words, to seventy *straight, sequential weeks*. The truth is there is no example in Scripture (or anywhere else!) of a stated time period starting, stopping, and then starting again. *All* **Biblical references to time are consecutive:** 40 days and forty nights, 400 years in Egypt, 70 years of captivity. In Daniel's prophecy, the "seventy weeks" were to begin during the reign of Persia and continue to the time of the Messiah.
>
> Logic also requires that the 70th week follow immediately after the 69th week. If it doesn't, then it cannot properly be called the 70th week!
> **It is illogical to insert a 2,000- year gap between the 69th and the 70th weeks.**
>
> Daniel 9:27 says nothing about a seven-year period of "tribulation," a "rebuilt Jewish temple," or any "antichrist." [173] [emphasis added]
>
> **So the 70th week was part of the other 69 weeks and has already happened.**

The Biblical Endtime is highly unlikely – unless the PTB bring it on as a way to cement their hold on mankind… which is to say that someone may start a bloodbath in the Middle East, but it won't be any scripted Endtime scenario… the Simulation gods are not depraved, quirky or malicious like some humans.

Religion or Spirituality?

The same preacher who spoke last week on the Endtime and the Rapture, also spent time deploring the growing **Apostasy** in America and the world. People are not buying the 5 main tenets of Christianity:

1. The Bible is the <u>literal</u> word of God
2. Jesus was born of a virgin birth
   (and he did have brothers & sisters)
3. Jesus paid for everyone's sins on the cross
4. Jesus physically rose from the dead
5. Jesus is coming back to Rapture his church

Bishop John Shelby Spong disagrees with all 5 in his book, <u>A New Christianity for a New World.</u> And he explains in some depth why, and it is pretty much what those who are disenchanted with Christianity have been saying. The major teachings are fundamentally naïve. Furthermore, Dr. Spong shares that many respectable Christian scholars of today have trouble blindly believing in the above tenets, too.

So Dr. Spong said that **Christianity must update or die**. The time-worn saying, "If it was good enough for my grandfather, then it's good enough for me!" is a vote for ignorance and assumes people don't grow nor do they want to learn more. It also assumes that they not only had all of it correct, but that there wasn't any more to learn, i.e., no more spiritual growth. Both are wrong, and thus Christianity is wrong, as it stands today. We <u>need</u> a new Christianity – one built on truth, Light, Love and brotherhood. Bishop Spong's concern is that if we don't seriously overhaul Christianity and make it relevant for today, it will die. (This was examined in more detail in VEG, Ch 11.)

Bishop John Shelby Spong of the Episcopalian Church has written many books promoting the necessary 'death' of an archaic [too simple] Christianity and he discusses at length what elements the new Christianity should have to truly serve Man, and not just be some lifeless ritual that serves no one spiritually.

> The reformation needed today must, in my opinion, be so total that it will by comparison make the Reformation of the sixteenth century look like a child's tea party… I believe that Christianity cannot continue as the irrelevant religious sideshow to which it has been reduced… [and] **the way Christianity has traditionally been formulated no longer has credibility**. [174]   [emphasis added]

He goes on to say that modern man now perceives reality differently than Man of even 200 years ago – the advances and knowledge in science alone make one question the Bible. The virgin birth is seen today as artificial insemination, and we know that "being born in sin" refers to one's corrupt DNA from one's ancestors, and we also question the existence of a Devil and Hell.

The Darwinian view of life has also affected the Christian world. Besides the standard "No-Creation, All-was-Evolution-from-a-Primordial-Soup" teaching, Darwin also brought to Man the realization that things are still evolving – the Creation is not perfect, nor is it complete. [175]   And now that we know more about the likely **Creation of Man** (by those who run this School), we have begun to look at the world in a different light.

> While Dr. Spong doesn't espouse Gnostic Christianity, he does sustain the **seven main sacraments**, and promotes Spirituality instead of rote Religion, and he believes Man can go within to connect with God.

Gnostic Christianity

A more dynamic version of Christianity was/is **Gnostic Christianity**. Also called the "thinking Man's Christianity." What is required is more of a Gnostic approach – The mystery and spiritual aspect of the world was valued and Man could discover God on a one-to-One basis… which is what we had until the Church threw the baby out with the bathwater in AD 325. Man is going to have to go inward to seek the Kingdom of God, and his connection with the divine. It goes beyond a faith in the unseen and a belief in Something greater than we are which looks out for us, and cares about us. It includes a realization that **we are each part of the One**. And this all assumes that we can discover who and what we really are, what our potential really is, and find a way to actualize it. A gnostic is not afraid to examine his/her world.

The Gnostic Seeker

(source: www.crystallinks.com)

Self-discovery

Gnosis means self-discovery and everyone is encouraged to "light the lamp within you." [176] Such is the meaning of the Gnostic **Book of Thomas**, verse 70:

> Jesus said, "If you bring forth what is within you, what you have [brought forth] will save you. If you do not have that within you, what you do not have within you [and cannot bring forth] will kill you." [177]

Essentially: **Use it or lose it**. Everyone has a connection to the One, and that is what should be 'brought forth.' True Salvation is achieved by bringing forth (applying) what spiritual strength and development one already has; those who do not have enough, will perish – including Jesus' oblique reference to the OPs who do not have an inner connection. When the OPs die, it is truly "dust to dust." The soul has the inner connection. Gather Knowledge and let its Light raise your PFV so that you can get out of here (i.e., save yourself), and become an **Earth Graduate**.

Partial Solution

It is time for a change. Man has been following the traditions set up hundreds (and thousands?) of years ago when Man was more simple, by people who were told what to believe by the gods. They used religion to control Man and he wound up believing what he was told and not what the ultimate truth was... because he wasn't evolved enough for the truth. We must today stop running around believing things because that's what others have believed before us. **We must stop living in the box that was created for us by others... consider the Flat Earth reality.**

When you know that the Anunnaki 'blessed' Man with Religion, we have to ask the following question: Are we going to let the Anunnaki legacy ruin our world, or are we going to stand up, wake up, stop playing with religions that don't work any longer, and determine our own future? The Anunnaki legacy still controls our world... Can we stop it? Can we take our world back?

It is suggested that there need be no riots, no marches, no storming the Bastille... just socialize with like-minded others who seek Truth and Love, Respect and Fair Play. Where? A good place to start is an enlightened, progressive Christian church which is already seeking to follow Bishop Spong's ideals. If that is not available in your area, then seek out one of the following **New Thought churches** – Unity or Centers for Spiritual Living, Christian Science, Eckankar, Ecclesia Gnostica, the Bahá'í Faith or perhaps the Unitarian Church. While the one you contact may not be the one you stay with, once you step on a steppingstone to cross the river, you will see or hear of others what you'll want to check out...

Those who have a scientific bent might be drawn to Religious Science (now called Center for Spiritual Living), wherein the founder Ernest Holmes sought to unite the Truth of Religion with the findings of Science. He died too soon as he would have had a field day with today's Quantum Physics **Observer Effect** concept that agrees with his teachings about how the Multiverse supports one's thinking...

Ministerial Dilemma

With all due respect to all Ministers of Truth everywhere, there is a major problem when one seeks to lead the flock into Truth. Given that there is just so much revealed Truth that forms the Core Truth Teachings (which are universal from one Spiritual Center to the next), after the minister has gone thru the Core precepts, and perhaps some advanced esoteric precepts, such information is limited in a Simulation. Here is the problem that they all eventually face: Unless you are well-connected with your Source (i.e., you are called to minister!) and it continues to feed you ideas that still fall within the Core Teachings, you will have to seek other books and teachings and here is where some New Age churches (and Progressive Christian churches) fall off the tracks.

You may start a teaching as a comparison from a new book, just released, and while it sounds good, it might only "tickle the ears" of your congregation, AND it might even mislead some of them. The worst case, that the author has witnessed twice, is that the Spiritual Center now has a mishmash of theology; some teachings contradict each other. The sharper members of the congregation see that and if they speak privately with the minister and s/he refuses to clarify/drop/clean up the mess, the congregation quits that church and moves on.

Like any science or philosophy, you either keep pushing existing theories and use many examples to demonstrate existing Truth, or you go out on a limb, experiment, include new stuff, claim you are widening your congregation's knowledge... But what happens when they want deeper Truth and a way to heal themselves or manifest better finances? Or, as this author has seen, you have already given them Truth and they cannot apply it (for whatever reason) and they are stuck. They are not growing, the minister cannot demonstrate healing for them, and the congregation begins to stagnate and some wander off.

Obviously having a turnover in the congregation with new members added, keeps things moving and the basic stock of Truths are more easily taught on Sunday.

The basic problem is in a **Simulation, which is limited to 3D Laws and Teachings**: there is NOT an unlimited source of Truth. The basic teachings are finite, and while their interrelations allow for many combinations and a clever minister can keep his congregation thinking for several years, even this aspect is not unlimited.

> I often asked when I studied for the ministry, how many ways can one slice, dice and chop the Bible's message? The basic message was given in Matthew 5-6-7, the Sermon on the Mount. Yes, there is more, but what else is necessary to teach as the basics?

Such would explain why avatars and Divine Teachers took to the road and gave their basic message to a new crowd every time people gathered. And this also works today for Christian pastors and missionaries to drop their best teaching(s) on a new crowd as they are invited to speak across the country.

So, guess what works best in a Simulation?

The Gnostic methodology works best wherein the congregation/students are invited to follow a guided meditation and connect with their inner Self. It may not be labeled a Gnostic Church, but the methodology is Eastern in origin, and is used by those who follow the Tao, meditation and yoga to work even the *kundalini* in the spine up into **higher consciousness**. (TOM

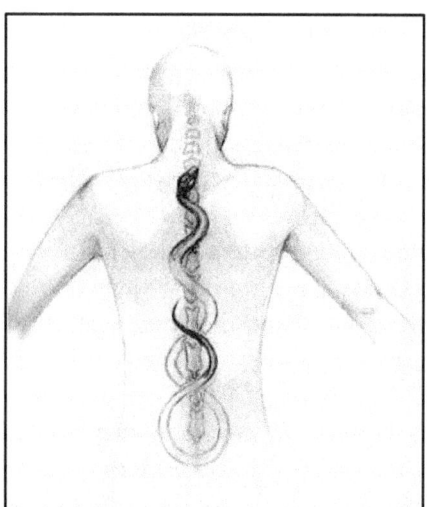

developed this idea.) And to do such a thing MUST have a master who knows the Way and can un-stick the student when higher energies begin to harm rather than enlighten or heal.

Kundalini often is compared to a serpent ascending the spinal column; passing thru the 7 main chakras and reaching the Crown (#7) chakra whereupon enlightenment is granted the aspirant – if he doesn't die first! **(Credit: Bing Images)**

And that is another insight into why the Catholic Church dropped the Gnostic methodology and warned people that it was dangerous – and added that it was "of the devil" just to emphasize that people should stay away from it. And to make sure, the congregations were told that there was no Karma or Reincarnation so that the flock would not accidentally stray back into metaphysical teachings that they probably could not handle.

So for the masses, "straight vanilla" religion worked well for the basic Joes and Janes without much education (we're talking the Middle Ages) and those who weren't interested in going within, and thus a pastor could safely guide the flock into green pastures, and work on getting them to start loving, praying and treating each other well… a handful in itself at times!

And, yet, for those who are older souls, seeking more, then the metaphysical, New Thought church (or Spiritual Center, as some prefer to call it) would be their path to personal spiritual growth. And **the Simulation is also programmed to respond** and help the aspirant to develop their connection with their Higher Self – which is their ticket out of the Earth School.

Synopsis

When the student is ready, the teacher appears.

The Simulation <u>inserts</u> people, objects and creates **synchronicities** for those who are ready. The right book, seminar or person just 'happens' to come into your life when you are seeking something… Or some idea is given to you in your sleep… often that is what happens to scientists and mathematicians.

The Simulation is also programmed to protect you and avoid accidents (as was examined in Chapter 2' Anomalies.) **There are no accidents**. The Simulation is in control – if not, can you imagine God watching a scenario that has a scripted outcome, and something interferes and He winds up saying, "Oh, shoot, I wish I had foreseen that!" I know that is hard to accept, so when a plane with 240 people on it blows up in mid-air, it is the sheer number of people that gets us to say, "No way. That couldn't be an accident!" And yet **human error** causes 'accidents' – it was pre-meditated carelessness!… But note: if you are not to be part of the fatality, you will not be able to get on the plane.

Another personal anomaly that happened in June, 1956:

It was Summer 1956 and I was registered to be on TWA Flight 2 from Los Angeles to New York, and everything interfered with my grandparents getting me to the LAX airport. They missed the flight and we took a room for the night. The next morning, we ate and rushed to the airport, I had not heard the news. My grandmother had seen the morning newspaper as we exited the motel, and she said nothing.

I flew back to LaGuardia, met my folks and it was only then that Mom went hysterical and was so glad I was safe and sound. I still didn't know what was happening. My new stepfather handed me a copy of the NY Times… and I had to sit down.

TWA Flight 2 had crashed in the Grand Canyon the day before; there was a mid-air crash with another plane. No one survived. That was my scheduled flight.

Simulated Conclusion

So Someone is tracking, protecting and guiding us all… whether we're in a Simulation or not. The old saying about death appears to be true, "When it's your time, it's your time!" What we think of as real accidents, or random events that occur 'by themselves,' cannot be allowed to happen, Simulation or no, but events would be easier to control in a Simulation as described in the preceding pages. Willy-nilly accidents cannot be permitted as everything (except souls) is scripted for a desired outcome. Freewill is operable in those cases where you (a soul) have a choice. And, **our very freewill is what we are evaluated on**. The choices we make reflect the amount of knowledge, love or wisdom we have.

However, there would be no point in scripting everything including souls because then there would be no 'tests', no point to being in a School to discover how to live, discovering what is right, what is wrong… And in order to do that, the ideal School would be contained, so elements and forces can be controlled – otherwise, in a willy-nilly environment, tests with a specific outcome might not occur, or they would have no meaning if they could not always be controlled. So the proposed Simulation exists in a 3D Construct, and it is very closely managed (as was hinted at in Chapter 6 with the Hippenskippits).

All that remains is to follow our individual Scripts, do our best to handle or survive the Drama and its tests, handle the Points of Choice (places where one has to make a major decision), and exit the School fairly successfully. (It needn't be exited perfectly – if that were true, no one could exit!)

Now, on the other hand, even if Earth weren't a Simulation, you'd have to do the same thing, anyway – personal growth is what it is all about. Thus, if we were **not** in a Simulation, we would still be **watched over, protected and guided**. That has been the underlying message of most religions. And we are expected to become better people, and to that end we seek Peace, Compassion, Humility and Respect on Earth. That would also be the goal of a new, more spiritual Christianity, as proposed by Dr. Spong.

And one might even suspect that if Man is being 'developed' by the gods, the reason is to better serve in other Realms within the Father's Multiverse. If Man didn't have potential and the Father didn't love humans, none of us would be here – Earth and Man are not a fluke. All has a purpose.

Go for it!

Chapter 13: Simulation and Reality

After a review of the arguments <u>for</u> Simulation and an examination of the anomalies, as recorded by scientists, there is some consideration due the Simulation Theory. In point of fact, we might want to now examine the 11 questions posed on the back cover, and see whether the Simulation Theory is the only answer for these issues.

The Universe Design

According to some physicists, the universe should not exist – If there was a Big Bang (and this book and several Physicists do not support that), then equal amounts of matter and antimatter should have been created, and they would have annihilated each other. But, according to theory, there must have been a greater amount of matter than antimatter, thus permitting the universe to exist as we know it. Really?

According to Paul Davies (<u>The Goldilocks Enigma</u>):

> One of the deepest puzzles of cosmology [is] the origin of matter. Cosmologists want to know exactly how it happened and why that particular amount (10^{50} tons in the observable universe) got made. When matter is made in the lab, by high-energy collisions, the same quantity of antimatter appears, too. [178]

So how did the alleged Big Bang make 10^{50} tons of matter without also making the same amount of antimatter?

> Remember, the scientists are studying the Imax Theatre.

Secondly, the universe as observed is not all that friendly a place – consider that stars explode, Black Holes gobble whatever comes their way, and the universe is filled with deadly gamma rays. **So how did the Earth come to be located in the only place discovered in the universe where life can exist?** Scientists call this the *anthropic principle*. Earth is located in a highly atypical place – even unique compared to many other solar systems and galaxies…. The conditions for life are very restrictive and we just happen to find ourselves located in a special set of circumstances…. Gee, how coincidental.

The *principle of mediocrity* says that there is nothing special about any part of the Universe – what you see is what you get – sameness largely everywhere! And that is not true of Earth's location. [179] Earth has just the right balance of features to support life – as if it were designed to be that way.

Goldilocks Zone

Earth is located in a Goldilocks Zone – not too far or too close from the Sun such that we live in a temperate zone… just right for life. Physicist Brandon Carter theorized that if any of the laws of our existence were just a bit different, we would not exist. It is as if the Earth, and its location, and its ecological balance, chemicals and minerals were all perfect for life to arise here. [180] Like Goldilocks' porridge, it is all just right.   And **that is suspicious**, since that kind of thing isn't found on a regular basis in the world, let alone the solar system, or our galaxy.

As if by design: Science says oxygen to breathe, an electromagnetic shield to protect Earth from the Solar Wind and gamma rays (Mars and Venus do not have one), gravity that isn't too strong nor too weak (but somehow keeps heavy oceans from falling off the globe!), an ozone layer to protect Man from ultraviolet radiation, and a tilt to the Earth's axis to generate seasons, and a Moon to cause tidal action of the oceans. What if it <u>was designed</u>, but the Science facts (just listed) are wrong?

And still think it all just happened by coincidence…?

Lunar Eclipses

Another interesting anomaly is that the Moon is just the right distance from the Earth to perfectly eclipse the Sun. And it is not made of the same material as the Earth, such that it had to either have been captured by Earth's gravity, or someone put it there... and could use it as an observation base – making sure that it didn't rotate (like all the other moons are said to do in the solar system).

According to Jim Marrs (<u>Alien Agenda</u>), the Moon is much older than the Earth, the Moon has at least three layers of rock – with the heaviest in the top layer (contrary to the idea that heavier objects sink), and the Moon's outer shell being the densest is "just what one would expect if it were a spaceship." [181]

The Moon may be hollow as <u>allegedly evidenced</u> by the Apollo astronauts leaving seismic equipment on the Moon, leaving for Earth, and jettisoning their LEM back onto the Moon – the craft hit the Moon "causing a reverberation for 3 hours and 20 minutes, down to a depth of 25 miles… In short, NASA said the Moon rang like a gong!"[182] (Providing we really went to the Moon…)

> This issue reminds one of the ***Truman Show*** where Truman was being watched from a fixed Moon in the sky – which was actually the control room for directing the daily activities of the fake town under the dome in which he lived.

Constants That Change

As was reported in <u>Virtual Earth Graduate</u> (Chapter 8), there are constants that are not constant… such as the Fine Structure constant called ***alpha*** which defines how parts of an atom bind together and helps modulate the decay of radioactive substances.

In addition, the **speed of light**, c, is not constant – it has been found to vary with the medium in which it moves, and there is also light's ***anisotrophy*** in what used to be called the Ether, and nowadays is called Dark Matter. That means that light moving east-west in a medium will have a slightly different speed from its north-south movement in the same medium.

Carbon-14 dating, while not a constant, has also been shown to be unreliable backwards of 5000 years… Isn't that when Noah and The Flood happened? Is there a connection?

These last 3 items were examined in VEG, Chapter 8.

Self-Dual Error-Correcting Code

And then along comes Professor James Gates (Chapter 7) who already told us that he has analyzed superstring behavior and found that their behavior includes self-correcting code – at the quantum level of reality! He further explained that superstring theory equations have hidden computer-like code (like binary 1's and 0's) that is self-correcting like that found in a browser when surfing the Internet.

Says Dr. Gates:

> Since when is Nature concerned with embedding a very particular kind of error-correction code into the fundamental laws of physics? What errors does it need to correct, and how would it know what they are and when they occur? Your guess is as good as mine. [183]

> The unsuspected connection suggests that these codes may be ubiquitous in Nature and could even be embedded in the essence of Reality. [184]

> **Or maybe the gods who designed this 3D Construct embedded self-correcting code as a way to sustain/repair the Simulation?**

And then a well-known author, the late **P.K. Dick**, stepped forward and also supported the Simulation Theory. [185] In 1977, people laughed at him. Today we have supercomputers that do modeling and simulation. They are not laughing now.

Missing Link

Another anomaly is that no one has ever, nor will they, find the bones of a hominid that represents a transitional stage between Homo *erectus* and Neanderthal. There is **no missing link** that proves Evolution is correct about Man evolving from the Apes.

By the way, if that were true, why isn't it still happening?

Whether Man was created by the Anunnaki, in several different stages, by manipulating the DNA of Homo *erectus* and mixing it with theirs, OR whether Man is a construct in a Super Simulation into which he can project his soul for a lifelike experience, the fact remains that we are here, we are experiencing this reality, and it is a School for the education, care and development of Souls.

In fact, it would be easy for the God Hierarchy to create Earth where it is, balance the ecology, physics, chemistry, etc. and insert souls for learning experiences.

Oft said: You are a soul in a body, having a human (Earth) experience.

It does not require humans from the year AD 2200 to create a Simulation, nor are ETs involved.

Rewind: the issue is that Science cannot say God or Soul, and so our reality for them has to be a Simulation.

In point of fact, it is more than that, as will soon be seen....

Anomalies in Coal Deposits

How many weird objects have been found in deep coal deposits... dredged up from ancient geologic levels? Shoes, hammers, and jewelry have all been found and recorded as proof that Man was on the planet much earlier than Anthropologists are willing to admit. Footprints of Man and dinosaur, together, have been found in ancient layers of rock (see VEG, Ch.10).

Keep in mind that there is now an alternate possibility. If this is a Simulation, the objects found in deep strata coal could have been planted there, as part of the scenario. In fact, in a Simulation, many things could be 'plants' – the Great Pyramid of Egypt, Stonehenge, the Wall at Sacsayhuaman (Peru), and the Nazca Lines, for starters... all to make us think that we have a mysterious past.

The Young Earth

As was examined in VEG, Ch. 10, there are 9 geologic anomalies that suggest that **Earth is not 4.5 billion years old**. And if this were a Simulation, the Programmers, or the gods (take your pick), could make Earth look any way they wanted... or they could have reset the Earth in a new Era.

If we assume that this version of the Simulation got going around AD 900 (see VEG, Ch. 10), to populate it with humans (and hence incarnate Souls) it would be necessary to make sure that the humans would find old walls, stone statues, and ancient artifacts in the ground so that they would know that their existence on Earth is <u>not</u> very recent... they more easily could buy into their "history."

> The Maya, Vikings, and the Chinese all have gaps in their history in the AD 800-900 timeframe (VEG, Ch. 10) and that is why that date was chosen for a possible Reset.

To briefly recap, for those who do not have VEG: the Anthropologists say that Man has been on Earth for about 1 million years, BUT there are anomalies...

1. No intermediate **fossil stages** exist for dinosaurs; there are eggs and adults, but no juvenile forms of the same species... let alone transition fossils from one dinosaur type to another;

2. **Population** figures prove Man has not been here for a million years; at 2% growth rate per year, it would take only 1100 years for the population to reach 6 billion (even counting wars and death);

3. Insufficient **bones** for Man and dinosaurs if they have been here for 1 million years;

4. Earth's **magnetic field** has been decaying, and extrapolating backwards, before reaching a super EMF in which nothing could live, we get 10-12,000 years old for the planet;

5. The amount of **Helium** (He) on Earth is odd; 13 million Helium atoms escape into the atmosphere every second... so measuring the amount of He in the Earth's atmosphere, from an alleged start of the planet, yields a very low number and the rocks on

Earth still have too much He if we are even 2 million years old;

6. There should be more **sediment** on the ocean floor than there is if we are an old planet... there isn't;

7. The **salty ocean** should be getting saltier... but after quantifying the ways salt is gained and lost, the scientists calculated that the maximum age of the ocean is not more than 62 million years;

8. **Meteoric dust** on the Moon is far too little; they know that the accumulation rate on the Moon should have generated almost a foot of Moon dust by now and our astronauts found only 1-2 inches;

9. **Continental erosion** is not what it should be if we are 1 million years old; according to calculations, streams and rivers should have reduced continents to sea level in 14 million years...

And so it goes. (The specifics are given in more detail in Ch. 10 of VEG.) We not only do not have a very old planet, but if the Simulation keeps being reset due to glitches, failure of humans to sustain a working and proactive society... it may have been thru several **Wipe & Reboots** (see Glossary) leading us up to this version of reality (it is a 3D Construct in a Simulation).

Anomalies and exceptions to the rule mean something ... If constants are changing and rounding errors (Chapter 7) accumulate and create glitches, then it looks like we have a finite Simulation. If humans screw up (are too violent, petty or stupid), or disease is wiping out the humans, and thus the Simulation doesn't work, it will be stopped and **reset** – no matter who runs it.

Noise from the Edge of the Universe

This was dealt with at the end of Chapter 8 -- and supports the idea that there is "machinery" of some sort creating noise or vibrations as it drives this Simulation.

NASA Orbital Flight Signature

This was addressed in Chapter 11 – showing that the Houston Mission Control wall screen displays a large "S"-type flight path for Shuttles, the ISS, and satellites, and yet this really translates to <u>a circular path around the Flat Earth</u>. This was a major giveaway – backed by commercial airline flight paths that avoid non-stops in the

Southern Hemisphere because the flight paths suggested by a standard wall map of the Earth do not line up with the actual Flat Earth location of cities and countries.

Are You Really Living on the planet You Think You Are?

Most people have no idea what the Earth really is, and it doesn't bother them. They are busy getting and spending, putting the kids thru school, trying to stay healthy, worrying about taxes… and many are trapped in the material world. And as souls living in the **Earth School**, they still have to meet challenges, tests and do the right thing – whether they know what shape the Earth has or not.

And yet, if they really knew, they would be working on gaining Knowledge and practicing Compassion so that they can graduate. But more than that, if they knew the Earth was flat, then (1) they would respect the Creator who built it all and has a purpose for them, and (2) they could not be lied to about the Earth – and believe in "alien invasions," global warming as a hoax to tax people for CO^2 emissions, or stories about a Hollow Earth…. Remember:

Knowledge Protects

Ignorance Enslaves

What If…?

While this book does not have all the answers, it can extrapolate some conclusions based on what we have covered, what the scientists have said, and what would have had to be true if indeed Man does live in a Simulation.

But what if Man does <u>not</u> live in a Simulation? Could the same scientific findings and anomalies be part of a 3D real Flat Earth?

> The chapter discussion assumes only two possible realities for Man: (1) he is in a Super-sophisticated Simulation – run by the gods for his benefit, as a Flat Earth or (2) Man is on a 3D planet in real space – but protected by a Quarantine such that external forces cannot interfere which sounds like the VR Sphere.
> (Terminology and issues: see Glossary and VEG book.)

You're right – there isn't much difference. And what if the gods simulate certain objects and events on the 3D Flat Earth…? They can **insert people, objects and events as needed**, and this is the more proactive version once one realizes how important Man really is and why the gods would go to all that trouble – to build and run a world for his benefit!

If Earth is in a Simulation, and albeit a very sophisticated one, this is a possible concept (for those who prefer the VR Sphere):

(credit: Bing Images/ hoise.com)

Somewhere a huge bank of supercomputers may be creating our Reality…and then, some huge Supercomputer generates a holographic model:

(Credit: Bing Images/ thoughtleraks.com)

…and it may in fact, generate a Flat Earth as Chapter 11 suggested.

And in turn, if the Simulation glitches, we should occasionally experience this:

(credit: scene from The 13th Floor)

On the other hand, if Earth had been a 3D Construct as a VR Sphere in real space, it would have looked something like this:

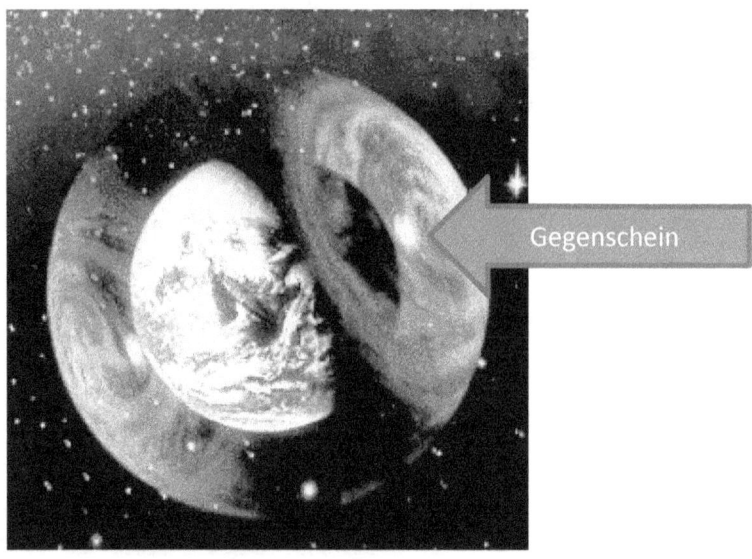

3D Earth in Energy Quarantine
(Credit: Bing Images)

And as a Flat Earth:

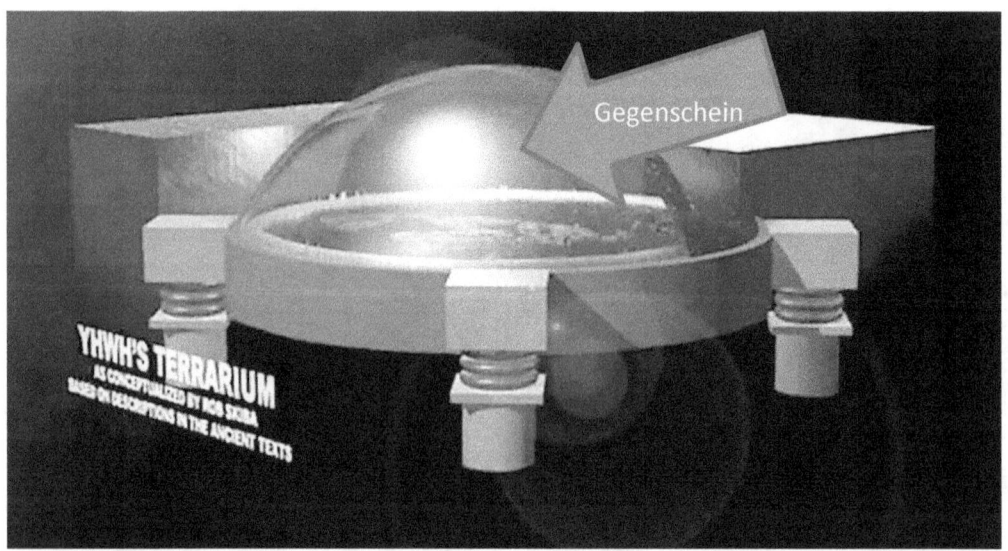

(credit: TheWorldWeLiveIn: https://youtu.be/vknnZoBrAP8)

Rewind: Earth's Quarantine

Having mentioned this a few times, is there any evidence for there being an **energy**

field surrounding the Earth, besides the normal, electromagnetic field that shields us from the Sun's Solar Wind (picture next pages)? Remember, this is alleged to be the *Gegenschein* that Charles Fort discovered 100 years ago… this would have to be the Firmament…

Gegenschein: courtesy NASA http//:apod.nasa.gov/apod/archivepix.html

NASA has also discovered what appears to be a 'force field' over and around the Earth (the 'shield' or **Plasmasphere** mentioned in Chapter 11):

> According to Daniel Baker, director of the Laboratory for Atmospheric and Space Physics, the **electron barrier exists in the Van Allen Belts**…. The Earth's magnetic field holds the Belts in place, but the scientist says that the electrons in those Belts – which travel at nearly the speed of light – are being **blocked by some invisible force** that reminded him of the kind of shields used in television series like Star Trek to stop alien energy weapons… [186]   [emphasis added]

The newly found field at 7200 miles altitude is related to the plasma clouds that comprise the Van Allen Belts. The vicious nature of the Belts has led many to suspect that Man did not go to the Moon as passing thru them is a lethal experience.

As was noted in Chapter 12 of VEG, the highly credible channeled entity RA stated that Earth <u>is</u> in a Quarantine set up by the Solar Council and why. It is interesting that RA did not mention the Flat Earth, and referred to the Firmament as a Quarantine. Are we to not know the Earth is flat?

Van Allen Radiation Belts

For those who wonder what the Belts are alleged to be:

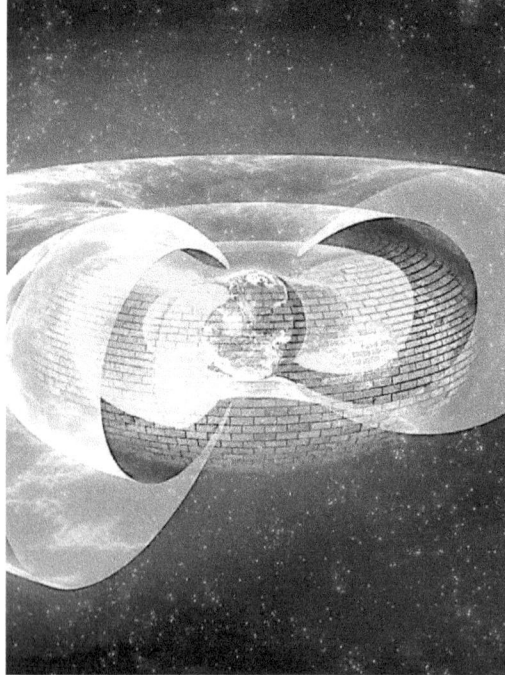

(Credit: Bing Images/ news.islandcrisis.net)

The 'bricks' symbolize the Wall of ionized plasma around Earth... **the Shield**.

There are no less than 2 Belts and a Shield circling Earth. And their function (as designed) is shown.

The above diagram, as provided by NASA (caution!), reflects that the Inner Belt is 620 miles to 3,700 miles from Earth, and the Outer Belt is 8,100 miles to 37,300 miles. The newly discovered Shield is calculated to sit about 7200 miles above Earth.

In addition to that, standard Science there is also the Earth's Electro-Magnetic Field protecting us from the Sun's cosmic rays.

All that is very nice if the Earth were still a VR Sphere... but what about a Flat Earth?

Fortunately and <u>simply</u>, the Firmament takes care of it all (and the following diagram with Solar Wind is interesting but unnecessary).

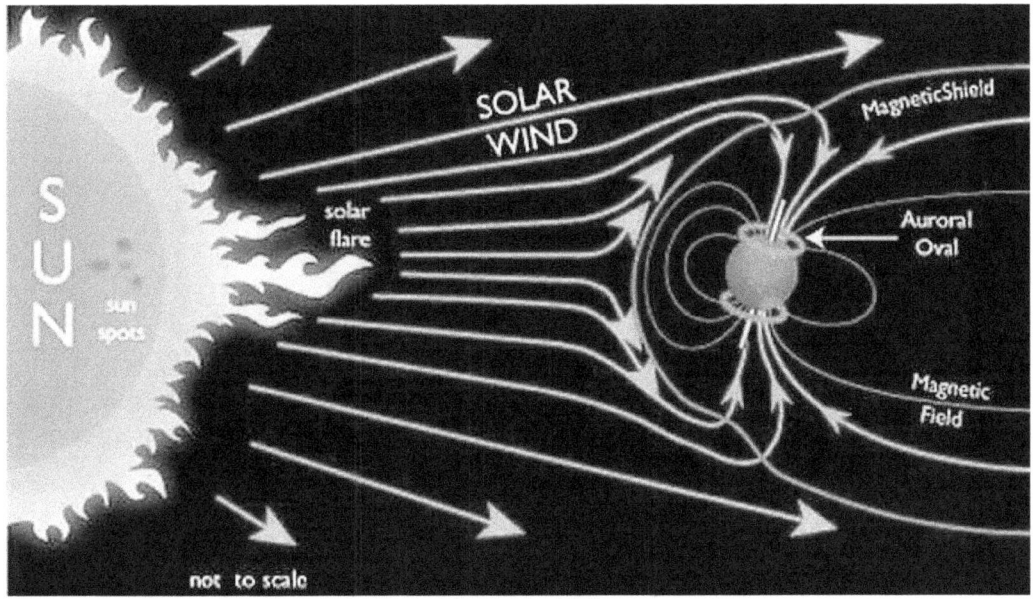

(Credit: Bing Images/consciouslifenews.com)

So if there is a barrier, the Firmament, was SETI such a great idea, or was it a waste of money? The barrier is known to bounce back radio waves...

(Credit: Bing Images/earth shield)

SETI founder, **Dr. Steven Greer**, believes there's a quarantine over the planet because the Earth's inhabitants are a threat to other forms of life outside of our planet, due to our military aggression throughout the world. [187] (Maybe we are too petty & violent?)

Stephen Hawking was worried that announcing our presence to the rest of the universe (by probe, by plaque, or by SETI) could be a dangerous invitation to more advanced, antagonistic races.... Being that we are on a protected 3D Construct, it appears he had nothing to fear. The bottom line is:

Earth is a 3D Construct protected by a Shield.

Rewind

To answer an earlier observation – dealing with the Goldilocks issue – If Earth is not a Simulation, the "just right" nature of our situation suggests that the gods still planted Earth where it is, in what appears to be a **Goldilocks Zone**, for a purpose, and without becoming religious, note that even the Bible says why:

He works all things to His purpose....   Eph. 1:11

The Earth School has a purpose – to grow souls who can then move into bigger and better realms… they are called **Earth Graduates**.

All the great teachers have sought to wake Man up for one reason: to get Man out of here and into the Father of Light's Multiverse where we can **play and serve** at a higher level! That is called the Earth Graduate.

So what do you do when you get out?

Potential Areas for Service

Having said that the Soul who graduates from Earth can fulfill certain basic, initial, functions in the Multiverse, it can be shared that these are some of the areas open to Graduates from the 3D Earth Construct:

Bio-plasmic Quantum Computer Techies – responsible for basic Heavenly Computer support, Scripts and maintenance.

Bio-plasmic Computer Programmer – performs fractal sub-programming (for Recycling) under supervision.

> These last two were demonstrated in Chapter 6. The Heavenly computers develop and manage Scripts, including the Father's overall Greater Script. There are also Science quantum computers that simulate new worlds and species to be designed and built.

Akashic Records Librarian – maintains life records' storage/retrieval.

Gods-in-Training I – responsible to oversee the Drama/Greater
   Script Simulation: Man and feedback of the Control System.
   Many sub-areas here.

Gods-in-Training II – responsible for the Holographic stabilization and
   interface with the Replicator technology. Sub-areas here.

Soul Counselors – responsible for evaluation, guidance and training
   of in-coming souls to the InterLife for further development which
   may include imprinting or vibrational adjustment.
   Many levels here, including Teachers.

This 3D Construct science mirrors that of the 4D realm, <u>partially</u>. So as a Simulation, there are basic things to be learned here that apply back in the 4D realm… (Since the Simulation only partially reflects the real 4D world, **what is learned in Science within the Simulation is like learning the IMAX Theatre** and thinking that ALL the laws and processes are the real world. Simulation Science does not count for the following three types of positions).

Bio Scientists -- these beings experiment with new lifeforms, ways to
   engineer them, transplant them, and ways to improve them.

Astro Scientists -- these beings experiment with Galaxies, Suns, planets,
   comets, etc. to engineer new planets, manipulate orbits, manipulate
   Dark Matter/Energy, and all the while ensure balance/order in the
   material world.

MLD Scientists -- responsible for multilevel universe and dimensional
   interface, including handling Timeline Shifts when necessary.

There are many others, but it is a busy 4D realm over there; no one is sitting around on a cloud playing a harp – unless they're on a coffee break! The reason that the PTB wants to block Souls from progressing is mainly because of this position:

Gods-in-Training III – responsible for overseeing, managing and
   controlling the Beings of Light, the PTB, and humans –to make
   sure that lessons are properly administered (according to Scripts)
   and it amounts to controlling what the PTB can 'get away with.'
   This is as close as the InterLife comes to having a "police force."

Simulated Conclusion

Thus it doesn't really matter whether we believe in the Flat Earth or an easier-to-believe VR Sphere (which still has problems) – we are watched over under either scenario and the goal is to wake up, and graduate with Love and Light (Knowledge).

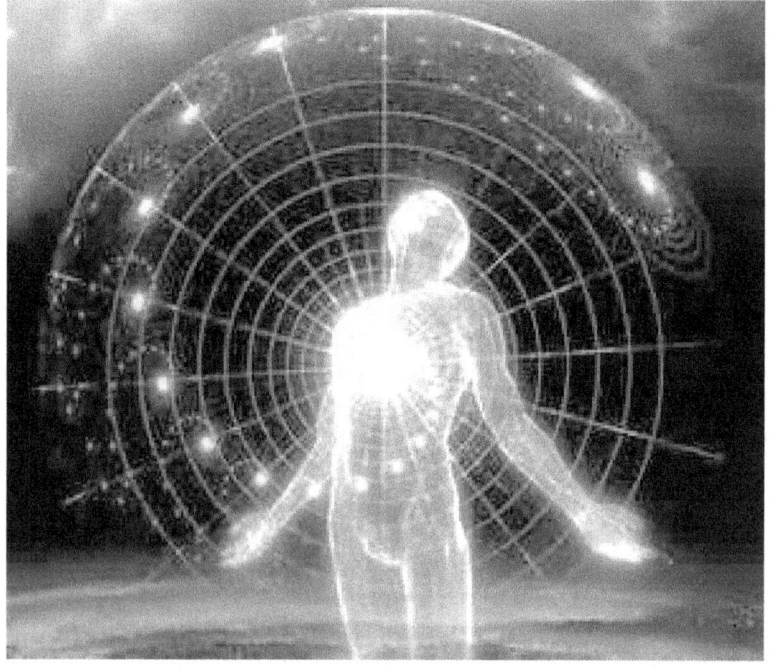

Earth Graduate
(Credit: Bing Images/consciousness)

Namaste!

For those who have been trying to piece it all together and still want a look behind the Curtain…. Try this on for size…

Simulation Addendum

While searching for criticisms of **Nick Bostrom's Simulation Argument**, and considering **Silas Beane's** interest in looking for constraints on the universe (to prove that a design is inherent), considering the Observer Effect and *anomalons*, and reading a preface to **Paul Davies'** <u>Goldilocks Enigma</u>, all of a sudden a very interesting idea popped into consciousness:

What if the Universe that we think we know is so unknown to us that we have yet to discover a higher order of Law or Process?

Earth as a Simulation or real 3D Earth is not really the issue – what if there is something that changes the laws of physics as we master them and which transcends static definition? What if the Universe is not only *anthropic*, but ***polymorphic***?

What if what we think is a discovery of a law that we assume had always been there and we just uncovered it, was in fact just now put into play?

Let me give an example from the world of computers.

Not too long ago, a PC at work (fully protected by the latest in anti-virus software) was hit by a virus. We attempted to 'clean' it but it defied all our attempts to do so -- much less identify it. Long story short, it was a **Polymorphic virus** that changed itself each time it ran. That reminded me of some **Trojan viruses** that are timed (with a date time stamp) to execute at some future date, or perhaps upon a unique set of keystrokes.

Now, what if an undiscovered function in the Universe is like any of those viruses – or even a combination of them? Remember we are probably dealing with a very intelligent and sophisticated Designer who runs the Universe, and for Whom setting up an Observer Effect or *anomalons* is no big deal.

Is the Universe *designed* to resist our discovering what it is? Are some constants programmed to change depending on our use of some other things, or after a certain passage of time? If there was no Big Bang and the Universe was created in a large and sophisticated Simulation to look the way it does, perhaps we are still an Experiment – to see if we can figure it out?

And as we move thru the wondering, hypothesizing, and testing, the Universe is *programmed* to morph slightly, ever so slightly, on cue, or in response to our specific actions, to keep us guessing and moving forward… so we find the Observer Effect and *anomalons*, the weird behavior of photons in the Double-Slit Experiment, **anisotropy,** and the non-local behavior of particles (Entanglement)… and last but not least, we never seem to get to the very bottom, absolute basic building blocks of the physical world.

So what are they doing at CERN with the Large Hadron Collider?

All matter is nothing but particles and energy in vibration that our common consciousness experiences. Yet the fundamental nature of Reality is not observable… because it doesn't exist. It is simulated. And what is *real* is just electrical signals interpreted by the brain. What if our world, even the Simulation, is just a hologram and we are programmed to interface with it, and think it is real?

Suppose there is no time, no locality (out in space, which way is "up"?), and objects just seem to have relationships, but little absolute fixed beingness. That is to say, even the atoms in a chair have huge spaces between them, and matter is not really solid…. How can it be? There appears to be no absolute, final, fundamental particle below which there is nothing else…. And lately Dr. Gates discovers "self-correcting code" in a lower level of the Universe!

It is almost unreal like a dream, yet it is pretty coherent and logical – just that aspects of our Reality (especially in the world of Quantum Mechanics) say that **we are not living in a real, solid, totally predictable world**. Yogis walk on water, avatars walk thru walls, gurus heal people, and advanced sages can just disappear at will… Savants can multiply three 6-digit numbers in their head and give the answer in 5 seconds – and be correct!

Contemplating that Earth could be a Simulation versus a 3D real planet in Quarantine pales by comparison.

And despite the best attempts in <u>The Science in Metaphysics</u> to specify how **Vision** really works, we also fail, but the implied solution comes closer to the truth when we invoke holographic principles, Infinite Mind and the personal subconscious.

Rewind: Oceans & Gravity

> In fact, there is a wild possibility that if Earth is a sphere, the oceans are held to it either (1) because the containing Barrier energetically 'pushes back' working like a reverse Gravity, or (2) the oceans being holographic are <u>programmed</u> to look and operate the way they do, because they are 'coded' that way.

What do we really know and why do some things defy understanding?

Why does hypnotizing someone let them demonstrate advanced abilities? Hypnotize Tom, blindfold him, touch a pencil eraser to his bare arm, and tell him that that is really a hot metal rod, and a blister/burn will immediately appear. And how could Tom see the watch thru Laura's body!? (Chapter 3)

What the ... is really going on!!?

Think you know? Watch out, the Man Behind the Curtain will move the furniture around and you will trip over what you thought you knew!

> A wise man once said "Not knowing is a very high state!" ... because then you are open to finding out.

Who showed up 3 times in my life to heal me so I could continue on this Earthly path? Why do houses and some people I once used to know, no longer exist?   Why does gold dust sometimes fall in very beautiful religious ceremonies?

And a real doozy from the world of NDErs:

> When you die, you return to the InterLife where a Being of Light shows you **a 'video' of your recent life**! Sight, sound, emotion, thoughts, feeling – all there! And we never saw Them recording us! OMG – even that secret tryst, that stolen donut, and the $200 I slipped under the door so a woman would not be evicted for non-payment of rent – They saw that!
> How can that be... unless we are in some sort of a Stage, or Simulation, where... everything... is... recorded...

> Next time you think about whether you should or shouldn't do something, remember:

Movie Time!

After years of study, it occurs to me that one of two things is true:

1. Earth is a **School** for the training of souls
   (and it doesn't matter whether it is a Flat Earth
   Simulation or a 3D real Planet in Quarantine, the
   effect for us is the same),

OR

2. This is a great **Amusement Park** designed for souls where almost anything goes (i.e., Freewill) and the Park Operator has built it to morph amuse, bewilder and challenge the best of us.

I used to think Earth was **an Asylum** where dysfunctional and damaged souls came to do rehab. Fortunately, when you pursue that line of thinking and research, it doesn't completely hold water. Calling Earth a **zoo** would come closer to what I've seen over the years, but there are too many good, and enlightened people here…

Most likely, it is just **a School with polymorphic processes** so that it can be whatever to whomever whenever something is needed… a very sophisticated all things to all people … which is an incredible Design. Too bad some of us can't do more with it.

Those who refuse to learn and be appropriate will find that Earth is more than a School – it is a **Prison** <u>for them</u>.

Nonetheless, it has been determined that 6% is the tipping point, or the trimtab, that is: enough enlightened souls on planet Earth can send society in a new, proactive direction… Perhaps the Universe is awaiting our reaching that point? Could that also be what another author called the **Omega Point**?

Appendix A: Bibliography

Books

Religion/Metaphysics/Spirituality

Castaneda, Carlos. *A Separate Reality*. New York: Washington Square Press, 1971.
_____. *The Active Side of Infinity*. New York: HarperCollins, 2000.

Capra, Fritjof.　*The Tao of Physics*. Boston, MA: Shambhala Publications, 1999.

Charles, R.H.　*The Book of Enoch the Prophet*. SanFrancisco: Weiser Books, 2003.

Elkins, Don and Carla Rueckert. *The RA Material, Book I*. Atglen, PA: Schiffer
　　　　Publishing/Whitford Press, 1984.

Golas, Thaddeus. *The Lazy Man's Guide to Enlightenment*. Salt Lake City: Gibbs-Smith, 1995.

Guiley, Rosemary Ellen and Imbrogno, Philip J. *The Vengeful Djinn*. Woodbury,
　　　　Llewellyn Worldwide, 2012.

Moody, Raymond A., Jr., MD. *Life After Life*. New York: HarperCollins, 2001.

Monroe, Robert. *Journeys Out of the Body*. New York: Doubleday, 1971.
_____. *Far Journeys*. New York: Random House/Broadway, 2001.
_____. *Ultimate Journey*. New York: Random House/Broadway, 2000.

Newton, Michael.　*Destiny of Souls*. Woodbury, Mn: Llewellyn Worldwide, 2002.
_____. *Journey of Souls*. Woodbury, Mn: Llewellyn Worldwide, 1994.

Rasha. *Oneness*. Santa Fe, NM: Earthstar Press, 2003.

Ring, Kenneth.　*Lessons from the Light*. Portsmouth, NH: Moment Point Press, 2000.

Roman, Sanaya. *Spiritual Growth*. Tiburon, CA: HJ Kramer, Inc., 1989.
_____. *Personal Power Through Awareness*. Tiburon, CA: HJ Kramer, Inc.,
　　　　　　1986.

Snellgrove, Brian. *The Unseen Self*. Essex, England: The C.W. Daniel Co., 1996.

Spong, John Shelby. *A New Christianity for a New World*. HarperSanFrancisco, 2001.
_____ *Why Christianity Must Change or Die*. HarperSanFrancisco, 1999.

The King James Study Bible. Nashville, TN: Thomas Nelson, 1988.

Wilde, Stuart. *The Prayers and Contemplations of God's Gladiators.* Chicago, IL: Brookemarke, LLC., 2001.
_____ *The Force.* Carlsbad, CA: Hay House, 2006.

Zukav, Gary. *The Dancing Wu Li Masters.* New York: Quill, 1979.
_____ *The Seat of the Soul.* New York: Simon & Schuster, 1990.

Scientific/Medical

Baugh, Carl E. *Why Do Men Believe Evolution Against All Odds?* Oklahoma City, OK: Hearthstone Publishing, 1999.

Behe, Michael J. *Darwin's Black Box.* New York: Simon & Schuster/Touchstone, 1996.

Braden, Gregg. *The Divine Matrix.* Carlsbad, CA: Hay House, 2007.

Brown, Walt. *In The Beginning: Compelling Evidence for Creation and the Flood.* Phoenix, AZ: Center for Scientific Creation, 1995.

Davies, Paul. *The Goldilocks Enigma.* NYC: Houghton Mifflin/Mariner Books, 2006.

Dong, Paul & Thomas Raffill. *China's Super Psychics.* NY: Marlowe & Co,. 1997.

Dubay, Eric. *The Flat-Earth Conspiracy.* POD book, publ. 2014.

Elvidge, Jim. *The Universe Solved.* Alternative Theories Press, 2007.

Greene, Brian. *The Elegant Universe.* New York: W.W.Norton & C0. 2003.
_____ *The Fabric of the Cosmos.* New York: Vintage Books. 2004.
_____ *The Hidden Reality.* New York: Alfred A. Knopf. 2011.

Hawking, Stephen with Leonard Mlodinow. *A Briefer History of Time.* New York: Bantam Dell, 2008.
_____. *A Brief History of Time and The Universe in a Nutshell.* (2-book volume). New York: Bantam Dell Books, 1996, 2001.

Hendrie, Edward. *The Greatest Lie on Earth.* Great Mountain Publishing, 2016.

Johnson, Phillip E. *Defeating Darwinism.* Downers Grove, IL: InterVarsity Press, 1997.

Kaku, Michio. *Hyperspace.* New York: Anchor Books, 1995.
_____ *Parallel Worlds.* New York: Anchor Books, 2005.

Krauss, Lawrence M. *The Physics of Star Trek.* New York: Basic Books, 2007.

LaViolette, Paul A, PhD. *Secrets of Antigravity Propulsion.* Rochester, VT: Bear & Co., 2008.
_____ *Genesis of the Cosmos.* Rochester, VT: Bear & Co., 2004.

Lloyd, Seth. *Programming the Universe.* New York: Random House, 2007.

Maltz, Dr. Maxwell. *Psycho-Cybernetics.* Publ. Psycho-Cybernetics Foundation, 1960.

McTaggart, Lynn. *The Field.* New York: HarperCollins/Quill, 2002.

Morris, John D., Ph.D. *The Young Earth.* Green Forest, AR: Master Books, 2006.

Pearce, Joseph Chilton. *The Biology of Transcendence.* Rochester, VT: Park Street Press, 2002.

Peterson, Dennis R. *Unlocking the Mysteries of Creation.* 6th edition. El Dorado, CA: Creation Resource Foundation, 1990.

Stout, Dr. Martha, *The Sociopath Next Door.* NY: Broadway Books, 2005.

Talbot, Michael. *The Holographic Universe.* New York: HarperCollins, 1991.

Watson, James D. *DNA.* New York: Alfred A. Knopf, 2003.

Wolf, Fred Alan, PhD. *The Yoga of Time Travel.* Wheaton, IL: Quest Books, 2004.

Yang, Jwing-Ming, Dr. *The Root of Chinese Qigong.* Roslindale, MA: YMAA

UFOs and ETs

Beckley, Timothy Green. *The Secret Space Program.* NJ: Global Communications, 2012.

Bramley, William. *The Gods of Eden.* New York: HarperCollins/Avon, 1993.

Cramp, Leonard G. *Space, Gravity and the Flying Saucer.* New York: British Book Centre, 1955.

Farrell, Joseph P. *The Cosmic War.* Kempton, IL: Adventures Unlimited Press, 2007.
_____ *Nazi International.* Kempton, IL; Adventures Unlimited Press, 2008.
_____ *Covert Wars and Breakaway Civilizations.* Kempton, IL; Adventures Unlimited Press, 2012.
_____ *Covert Wars and the Clash of Civilizations.* Kempton, IL;Adventures Unlimited Press, 2013
_____ *Saucers, Swastikas and Psyops.* Kempton, IL: Adventures Unlimited Press, 2011.
_____ *Roswell and the Reich.* Kempton, IL: Adventures Unlimited Press, 2010.
_____ *Genes, Giants, Monsters and Men.* Washington: Feral House, 2011.

Fowler, Raymond. *The Watchers*. New York: Bantam Books, 1990.

Good, Timothy. *Earth: An Alien Enterprise*. New York: Pegasus Publishing, 2013.

Greer, Steven M., MD. *Hidden Truth – Forbidden Knowledge*. Crozet, VA: Crossing Point, Inc., 2006.

Harbinson, W.A. *Projekt UFO: The Case for Man- Made Flying Saucers*. London: Boxtree Ltd., 1995.

Keith, Jim. *Saucers of the Illuminati*. Kempton, IL: Adventures Unlimited Press, 2004.

Lewels, Joe, Ph.D. *The God Hypothesis*. Columbus, NC: Wild Flower Press, 2005.

Mack, John E., M.D. *Abduction*. New York: Charles Scribner's Sons, 1994.
_____ *Passport to the Cosmos*. New York: Three Rivers Press, 1999.

Marrs, Jim.   *Alien Agenda*. New York: HarperCollins, 1997.
_____.   *Rule by Secrecy*. New York: HarperCollins, 2000.
_____.   *Our Occulted History*. New York: HarperCollins, 2013.

Missler, Chuck and Mark Eastman. *Alien Encounters*. Coeur D'Alene, ID: Koinonia House, 2003.

Pruett, Dr. Jack. *The Grandest Deception*. Xlibris Corp: Lexington, KY, 2011.

Story, Ronald D., Ed. *The Encyclopedia of Extraterrestrial Encounters*. New York: New American Library, 2001.

Tellinger, Michael. *Slave Species of God*. Johannesburg, SA: Music Masters Close Corporation, 2005.

Valleé, Jacques.   *Passport to Magonia*.   Chicago, Il: Contemporary Books, 1993.
_____.   *Dimensions*. Chicago, Il: Contemporary Books, 1988.
_____.   *Messengers of Deception*. Brisbane, Australia: Daily Grail, 1979.
_____.   *Revelations*. San Antonio, TX: Anomalist Books, 2008.

Von Daniken, Erich. *Arrival of the Gods*. London: Vega, 2002.
_____ *History is Wrong*. New Jersey: New Page, 2009.

History and Other Related Books

Childress, David Hatcher. *Technology of the Gods*. Kempton, IL: Adventures Unlimited Press, 2000.

Fomenko, Anatoly T. *History: Fiction or Science?*, Vol. 1. Isle of Man, UK: Delamere Resources, Ltd., 2003.

Fort, Charles   See Steinmeyer below.

Hawkins, David R. *Reality and Subjectivity*. West Sedona, AZ: Veritas Press, 2003.
_____ *The Eye of the I*. West Sedona, AZ: Veritas Press, 2001

Heinlein, Robert. JOB: *A Comedy of Justice*. NY: Ballantine Books, 1984.

Icke, David. *The Biggest Secret*. Wildwood, MO: Bridge of Love, 2001.
_____ *...And the Truth Shall Set You Free*. Isle of Wight, UK: David Icke Books Ltd., 1995.
_____ *The David Icke Guide to the Global Conspiracy*. Isle of Wight, UK: David Icke Books Ltd., 2007.
_____ *Children of the Matrix*. Isle of Wight, UK: David Icke Books Ltd., 2001.
_____ *Tales from the Time Loop*. Wildwood, MO: Bridge of Love, 2003.

Irwin, William. *The Matrix and Philosophy*. Peru, IL: Carus Publishing, 2002.

Keel, John A. *The Complete Guide to Mysterious Beings*. New York: Tor Books, 2002.
_____ *Why UFOs – Operation Trojan Horse*. New York: Manor Books, 1970.

Knight-Jadczyk, Laura. *High Strangeness*. Alberta Can: Red Pill Press, 2008

Kramer, Samuel Noah. *The Sumerians*. Chicago, IL: University of Chicago Press, 1971.

Northcutt, Wendy. *The Darwin Awards: Survival of the Fittest*. New York: Penguin Group/Plume Books, 2004.

Pye, Lloyd. *Everything You Know Is Wrong, Book I: Human Origins*. Lincoln NE: iUniverse/Authors Choice Press, 2000.

Sitchin, Zecharia. *Journeys to the Mythical Past*. Rochester, VT: Bear & Co., 2007.
_____ *The Twelfth Planet*. New York: HarperCollins, 2007.
_____ *The Cosmic Code*. New York: HarperCollins, 2007.
_____ *The End of Days*. New York: HarperCollins, 2007.
_____ *The Earth Chronicles Expeditions*. Rochester VT: Bear & Co., 2004.
_____ *Divine Encounters*. New York: HarperCollins/Avon, 1996.
_____ *Genesis Revisited*. New York: HarperCollins/Avon, 1990.
_____ *The Wars of Gods and Men*. New York: HarperCollins, 2007.
_____ *The Lost Book of ENKI*. Rochester, VT: Bear & Co., 2004.
_____ *The Stairway to Heaven*. New York: HarperCollins, 2007.
_____ *The Earth Chronicles Handbook*. Rochester, VT: Bear & Co., 2009.
_____ *There Were Giants Upon the Earth*. Rochester, VT: Bear & Co., 2010.

Steinmeyer, Jim. *The Book of the Damned; The Collected Works of Charles Fort.* New York: Tarcher/Penguin Group, 2008.

Stevens, Henry. *Dark Star.* Kempton, IL; Adventures Unlimited Press, 2011.
_____ *Hitler's Flying Saucers*, rev. ed. Kempton, IL; Adventures Unlimited Press, 2012.

Witkowski, Igor. *Axis of the World.* Kempton, IL: Adventures Unlimited Press, 2008.

Internet Sources

DNA and Genetics

Baerbel, "The Living Internet Inside of Us" is another translation of some of the Fosar-Bludorf work which can be found at Website: www.crawford2000.com

Baerbel, "Russian DNA Discoveries Explain Human 'Paranormal' Events" is another article edited and translated on various DNA aspects. Website: http://www.fosar-bludorf.com/index_eng.htm

"DNA study deals blow to Neanderthal breeding theory" is an article dealing with the possible interbreeding and non-interbreeding of Cro-Magnon and Neanderthals. Website: http://www.cbc.ca/health/story/2003/05/13/cro_magnnon030513.html

Dr. Barry Starr of Stanford Univ., "Whatever happened to those Neanderthals?" is a great article on possible interbreeding between Cro-Magnons and Neanderthals based on mtDNA, and speculates on the Neanderthal's mysterious disappearance. Also documents recent attempts to see just what was in Neanderthal nuclear DNA. Website: http://www.thetech.org/genetics/news.php?id=37

Grazyna Fosar and Franz Bludorf, "The Biological Chip in Our Cells: Revolutionary results of modern genetics" is an article written in the 95% perfect English of the German genetics researchers who also wrote the book Vernetzte Intelligenz (which is not available in English). The authors have an index with articles in English on
Website: http://www.fosar-bludorf.com/archiv/biochip_eng.htm

Related Websites worth visiting for further depth of the Fosar-Bludorf discoveries:
http://noosphere.princeton.edu/fristwall2.html
http://www.ryze.com/view.php?who=vitaeb
http://www.fossar-bludorf.com/index_eng.htm

Grazyna Fosar and Franz Bludorf, "The Cosmic Internet", article on group consciousness, how DNA acts as an antenna and communication device. Website: http://www.fosar-bludorf.com/vernetz_eng.htm

Grazyna Fosar and Franz Bludorf, "UFO Experiences and Hypercommunication" is another article seeking to explain the UFO abductee experience as one of hypercommunication (via DNA) between alternate realities, or parallel dimensions, as Jacques Vallee suggested years ago. Website:
http://www.bibliotecapleyades.net/ciencia/ciencia_hypercommunication01.htm

Rick Groleau, "Tracing Ancestry with MtDNA" article on NOVA Online website that explains how the father's and mother's DNA propagates, and how ancestry can be reliably determined, and what they discovered about the Neanderthals. Website:
http://www.pbs.org/wgbh/nova/neanderthals/mtdna.html

Tory Hagen, "Mitochondria and Aging" article explains how oxidants affect the mitochondria's ability to accurately reproduce and resist aging. Website:
http://lpi.oregonstate.edu/sp-su98/aging.html

Kean, Sam, "Who's the Fittest Now?" article in Mental Floss magazine for March-April 2009, p. 55-57, presenting the subject of Epigenetics.

"UltraViolet Light" article describes how UV is used to purify/sterilize, and how it can also negatively affect DNA. Website: http://www.Frequencyrising.com

Carl Zimmer, "The Search for Intelligence." Article in *Scientific American* magazine for October 2008, vol. 299, no. 4, pp 68-75. Effect of genetics vs environment on IQ.

Resistance to AIDS/HIV – Sample Report "About Resistance to HIV/AIDS" on the genetic testing website **23and Me**: https://www.23andme.com/health/Resistance-to-HIV-AIDS/ See also: Randy Dotinga article: "Genetic HIV Resistance Deciphered" , website: http://www.wired.com/medtech/health/news/2005/01/66198?currentPage=2

Dr. Fomenko, Anatoly

Wikipedia, "New Chronology (Fomenko)" is an excellent summary article of the major work (7 volumes) and discoveries of Dr. Anatoly Fomenko. It recaps his methodology and discoveries and statistically establishes that out traditional historical chronology has been seriously altered. Website:
http://en.wikipedia.org/wiki/New_Chronology_(Fomenko)
See also:
http://en.wikipedia.org/wiki/New_Chronology_%28Fomenko%29
and for the Parallelism image (Ch. 10):
http://en.wikipedia.org/wiki/Image:Fomenko_-_Roman_Empire_parallelism.jpg

Extraterrestial Genes in Human DNA

"Scientists Find Extraterrestrial Genes in Human DNA" is another article seeking to explain "junk DNA" and its probable origin and significance. Website:
http://www.bibliotecapleyades.net/vida_alien/esp_vida_alien_18n.htm

The Insider

"The Revelations of the Insider" is an article containing the blog during a 5-day visit by someone calling themselves an "insider" who had knowledge on most aspects of Earth history, science and religion. This was done anonymously via a proxy link to the GLP (Godlike Productions) forum in the Fall of 2005. The material is not copyrighted and can be reproduced as long as none of the original text is changed. Website: http://www.scribd.com/doc/403303/The-Revelations-of-an-Elite-Family-Insider-2005

Organic Portals/Mouravieff

"Matrix Agents: Profiles and Analysis (Parts I & II)" is an article that summarizes and clarifies information on Organic Portals, also called pre-Adamic beings. Much of this data is footnoted back to its original sources, including Mouravieff, Gurdjieff, Ouspensky, and the Cassiopaean Transcripts. Website: http://montalk.net/matrix/62/matrix-agents-profiles-and-analysis-part-i

"Organic Portals Theory: Sources" is a compendium of different writers' insights On the Organic Portal phenomenon. Particularly relevant are the significant Mouravieff quotes from Books II & III of Gnosis. Website: http://www.montalk.net/opsources.pdf

Bibliotecapleyades website product of Jose Ingenieros, has link to 3 volumes of original text of Gnosis work by Mouravieff. Book text is in English. Website: http://www.bibliotecapleyades.net/esp_autor_mouravieff.htm

Laura Knight-Jadczyk, "Commentary on Boris Mouravieff's Gnosis." Extensive article from her website that interweaves her analysis of Mouravieff's Gnosis book and its meaning for Man's spiritual development. Also included are relevant quotes from her Cassiopaean material. Website: http://www.cassiopaea.org/cass/mouravieff1.htm

Physics & Science
Black Holes, Light and Space

Crothers, Stephen J., "The Black Hole Catastrophe and the Collapse of Spacetime" is a recent article (Oct 3, 2008) by a physicist, upsetting current astrophysical conclusions and shows how Black Holes have never been found (because they don't exist), and how Light **anisotropy** has been verified. Website: http://www.thunderbolts.info/thunderblogs/guest.htm

African Uranium Mine

Kean, Sam, "Nice try, Einstein" in *Mental Floss* magazine, vol. 7 issue 5, Sept/Oct 2008. Article points out that a uranium mine in Africa contains uranium and isotopes not found elsewhere; old deposits were above ground as if they had been dumped there. Also the properties of same deposits show deviation in traditional constants (i.e., rate of decay). Website: www.mentalfloss.com

Sea Monster/Plesiosaur

40ANA blog website article, "Sea Monster or Shark?" Posted by The Moviebuff at 7:16am on 9/1/2006. Shows pix of the carcass caught in the trawler's net. Website: http://40ana.blogsppot.com/2006/09/sea-monster-or-shark.html

Serpents, Reptiles & DNA

Paul Von Ward, "Aliens, Lies and Religions" article on great Belgian website that discusses the author's book Gods, Genes and Consciousness. The issue of serpents and DNA is clarified as well as other AB (Advanced Being) issues. Website: http://www.karmapolis.be/pipeline/von_ward_uk.htm

Simulation / VR

Simulation

Interview with **Nick Bostrom** at the Future of Humanity Institute at Oxford University, England wherein he explains the idea that if Mankind does not go extinct, and our science continues to develop insight and control over our world, and computers gain incredible processing power, that we will be able to run simulations on our history – much as he theorizes advanced future humans may be doing to us. How do we know we are souls or really sentient? It may all be the detailed aspects of some very large computer program.
http://www.simulation-argument.com/si...
You-Tube:
http://www.youtube.com/watch?feature=player_detailpage&v=nnl6nY8YKHs

In a related article published by Cornell University, **Prof. S.J. Gates** reports that while examining very high-level equations dealing with the structure and organization of the universe, it was found that the equations demonstrate self-correcting code called Block Linear **Self Dual Error-correcting Codes** – suggesting that the universe is not only designed, but that it may be a logical mathematical construct... a simulation. http://arxiv.org/abs/0806.0051 is the original Cornell article.
You-Tube:
http://www.youtube.com/watch?feature=player_embedded&v=ZPju_NFwVXs

SimCity

An open-ended city-building game in 2D, game players are able to model potential cities with all the ramifications of a city: pollution, waste collection, sewers, and SimCity Societies has 3D graphics. http://en.wikipedia.org/wiki/Simcity

Videos of Interest

Forbidden Planet. MGM classic from 1956; debuts Robby the Robot.
The X-Files (TV series, 1993-2002): Twentieth Century Fox.
K-Pax. Universal Pictures, Lawrence Gordon et al. 2001.
Millenium. Gladden Entertainment. 1989.
Hangar 18. Republic Entertainment. 1980.
Capricorn One. Associated General Films. Lazarus/Hyams prod. 1978.

Groundhog Day. Dir. Harold Ramis, Columbia Tristar. 1993.
Men In Black. I & II Dir. Barry Sonnenfeld, Columbia Pictures. 2000.
The Matrix. Dir./Written by The Wachowski Bros., Warner Bros. 1999.
The Mothman Prophecies. Dir. Mark Pellington, Screen Gems/LakeShore
        Entertainment. 2001.

Taken. (TV miniseries) Stephen Spielberg, Dreamworks. 2002.
V, the TV series (1983-85, and 2009-11). WarnerVideo, Kenneth Johnson
        Production.
The Truman Show. Peter Weir, Paramount Pictures. 1998.
The Young Age of the Earth. Aufderhar, Glenn. Earth Science Associates /
        Alpha Productions. 1996.

They Live. Dir./Written by John Carpenter, Universal Studios. 2003.
Prometheus I. Ridley Scott, 2oth Century Fox. 2012.
The Thirteenth Floor. Columbia Pictures, Roland Emmerich. 1999.
The Day the Earth Stood Still. Twentieth Century Fox, Erwin Stoff et al, 2009.
The Forgotten. Revolution Studios. 2004

Iron Sky. Timo Vuorensola, Ger/Fin release via Paramount Pictures, 2012.
Paul. Universal Studios, Greg Motola. 2010.
Lucy. EuropaCorp & Universal Studios, Luc Besson, 2014.
2012. Sony Pictures, Roland Emmerich. 2010.
The Fourth Kind. Universal Pictures, Olatunde Osunsanmi. 2010.
The Adjustment Bureau. Universal Pictures, George Nolfi. 2010.

Knowing. Summit Entertainment, Alex Proyas. 2009.
Dark City. New Line Cinema, Alex Proyas. 1998.
Source Code. Summit Entertainment, Duncan Jones. 2011.
eXistenZ. Canadian Television Fund, David Cronenburg, 1999.
Johnny Mnemonic. Tristar Pictures, Robert Longo. 1995

***

Appendix B: Postscript

With the update for Flat Earth in this 4th book in the Set of 6, my job is done. This has not been a Trilogy nor a Quadrilogy… it is merely a Set of six books that relate what Earth might be and what we are doing here. The six books go together and cover most 'dots' out there, suggesting a plausible nexus.

Virtual Earth Graduate (VEG) was the first book and because it was so huge (at 868 pp), it was split into a companion Book 2: Transformation of Man (TOM).

I thought that was it, and then I was visited again, and given a super bunch of data for Chapter 4 of VEG. And if you want to know the rest of the story about UFOs, where they are, how they came to be, what they are and how they work, Chapter 4 of VEG was continued in Book 3: The Earth Warrior (TEW) – which was done as a docu-novel.

> Currently, TEW is being considered for a movie, and we are 'hassling' over what parts of the 336-page book should be in the 120-page Script.
> (The written script is registered with WGA.)

Again, I thought I was done. Surprise.

Baldy visited me again and suggested that I take Chapters 12 and 13 from VEG and do a Book 4, Quantum Earth Simulation (QES), and make it clear what Earth really is… although he did not say Sphere nor Flat Earth, he gave me 3 pieces of information (see Introduction page 11 Insert box) and I interpreted what he gave me. He implied that Earth is not what I have assumed and suggested I research it further. Book 4 is not a rehash of VEG's two chapters, it is a re-focused look in more depth such that those who were tongue-in-cheek about Earth being a Simulation from reading VEG, now have more to think about.

QES' title contains the word *Quantum* because Jim Elvidge in Chapter 7 explains that simulations (and our reality) must be quantized. It is a *Simulation* because of all the anomalies and glitches that occur: constants that change for example, and are characteristic of simulations with finite calculations.

Book 5 was <u>The Science in Metaphysics</u>, a look at how New Thought and its practitioners' beliefs often have a counterpart in metaphysical truths. This was dedicated to the New Thought people who soundly rejected it. Surprise.

And Book 6 was <u>The Anunnaki Legacy</u> – a global survey of the common myths and knowledge that the Anunnaki, or the Watchers, or the Shining Ones, or the Ancient Ones (probably the same beings) left with Man. Similar motifs of sacred Trees, The Flood, Serpent Wisdom and Kundalini (think: Göbekli Tepe), Goddess Wisdom and the Divine Feminine, and a look at Alchemy's roots and how that applies to the secret knowledge Man was given.

And that does it. The six books look alike because they are a Set. All deal with the reality of where we are and why, and who's doing what to us.

*　　*　　*

Encyclo-Glossary

1-Sec Drop -- this is a direct communication from a higher being into one's mind and memory/knowledge base. It is not a voice, not automatic writing. It takes a very brief split second and one knows that it is happening, and then it can take anywhere from 10 seconds to 20 minutes to examine what one was given. It is information that is usually complete and appears to the recipient to be something that s/he already knew and is now aware of. Similar to an insight or revelation, except that it has an energy signature about it that you know it is being "dropped" into you. (Reminiscent of **V2K** but there are no words 'spoken.')

100th Monkey Effect – When one animal in a group discovers some new behavior and finds it serves him, it is said that the behavior is not learned as much as passed on as soon as the energy reaches a critical level so that their group soul can recognize and 'appropriate' the behavior. This was the case with a few monkeys on an island who discovered that washing their fruit before eating it avoided the problem of sand in the mouth. More and more monkeys on island 1 began doing it, and while they had no way to communicate the new behavior to the monkeys on islands 2 and 3, after about 100 monkeys were doing it on island 1, the others on islands 2 and 3 also began doing it (as confirmed by zoologists who were present studying the islands).

Anisoptropy -- the tendency of light to have a different speed depending on which way it is projected (in the same medium)... as if it is flowing with lines of force one way, and against them if turned 90°. (see VEG Ch. 8 and 9.)

Anunnaki – one of the early, original ET visitors to Earth who interfered in the natural progression of the bipedal hominids here, and created some of the first 'humans' in Africa and Sumeria. Because of their technology and power, they were looked upon as gods. Supposedly from the planet Nibiru, but more likely Orion or Sirius systems. (See **Zechariah Sitchin.**)

Astral Realm – note that there are levels in the Astral realm, and in particular, the one that most concerns Man, is the Level I (**Chart 4** in Chapter 4) which is a kind of intra-dimensional space – more than 3D and yet not really 4D, and this is inhabited by Man's oppressors, the STS Gang. The normal 4D STS/STO entities occupy the higher 4D and lower 5D Astral realms and cannot see 3D Man.

Attractor – energy in the form of an idea, person, or thing that draws other things, ideas or people together based on similar and strong resonance.

Baldy – a human visitor, 6'4", bald head, big blue eyes, perfectly proportioned. As he could just pop in and visit me, and knew what I was thinking, I suspect he was an **Insert** of the gods who run the Earth Simulation. He was not an Angel or ET.

Beings of Light – often referred to as Angels, or today's Watchers, they guide and protect Man. They are also known to provide the life review that NDErs speak of, and they are the *'Inspecs'* that Robert Monroe spoke of.

Bionet – a term coined in Chapter 9 to describe the hyperdimensional network of communication in the body. Like the Internet, *chi* is carried in meridians of energy to all parts of the system, from the chakras, and tells the cells and organs what to do. The Bionet is manipulated during Accupuncture.

Brain Waves – a measurement of consciousness.
        Beta cycle: 12 – 19+ Hz (normal waking consciousness)
        Alpha cycle: 8 – 12   Hz (relaxed, aware state; pre-sleep)
        Theta cycle: 4 – 8 Hz   (sleep)
        Delta cycle: less than 4 Hz (deep sleep)

Catalyst – anything like an event, an idea or a word, that causes change in a person; the threat of being fired for bad performance at work is a catalyst to perform at one's best. Illness is a catalyst to see what is wrong, or what energy is blocked, in one's body.

Chakra – a vortex of energy formed in the body wherever two or more chi meridians come together; same as a vortex on the earth with its ley lines. (Sedona, AZ is known for several of these.) These are also referred to as 'energy centers' as they transduce energy from the air/water/Sun around a body and draw it into the body thru the chakras. There are 7 main charkas in the body and 1 above the head, and 1 below the feet. There are many more, minor charkas all over the body.

Chi – energy particles, also called *ruach*, orgone, mana, prana or **ki** – without chi in our food, air and water the human body could not exist. The chi is a force that travels along meridians (pathways) in the body that link the etheric aura (1st level of the aura) to the physical body; it can be directed by the mind to specific parts of the body for healing.

Cognitive Dissonance – the result of hearing/reading something new that does not fit into one's reality, or in what one thought was their reality; the effect is to create confusion followed by denial of new concepts. More specifically, when a new idea conflicts with an established idea that one already thinks they know, the result is 'dissonance', and rejection. When people were told 500 years ago that the Earth was round, they experienced great cognitive dissonance… which led to denial.

Coherence – resonating alike; attracted to each other by similar resonance. Two energy waves are coherent if they have the same shape, size, and strength.

Construct – 3D Construct – the energy 'Envelope' that holds the replicated Earth; also called the Sphere. Built in 4D, then phase-shifted so that no lifeforms in 4D are visible from Earth. It is subject to, and operates using 3D Laws… souls do not have their normal 4D abilities so that the School is effective.

Cspace – the area bounded by the inner 3D Construct (containing Earth), and the outer **Konstruct** – another high energy shell to protect the Earth Experiment. It is said to be somewhere between 10-50 million miles between the two energy shells.

Déjà Vu – the experience of having done, seen and/or heard something before; as though one is reliving a prior moment in their current lifetime. Relates to reliving a fractal simulation. See **Recycling**.

Elite -- those humans who are mostly descended from the Anunnaki hybrids; as a group they may be augmented by the Remnant Insiders who stayed behind when the main contingent of Anunnaki went home. They are generally not the enemy. See **PTB**.

Energy Vampire – a person, OP or ensouled, who subconsciously starts an argument, gets the other person angry, and the instigator takes the other person's energy through the Law of Energy Potentials. Energy always flows from the higher potential to the lower and this applies to car batteries, as well as humans. So the instigator creates a fight, not to win or lose (they don't care), they will walk off with some of your energy, and they quit the argument when they have it. They are up, and the victim is usually tired.

Entrain – to induce a state in B like in A; usually done by music, movies, and words, but can be done by powerful thoughts and beliefs. A hypnotist entrains a subject into a desired state; Hitler's harangues entrained the crowds into the Nazi mindset he wanted; and classical music entrains the listeners into a relaxed (Alpha) state.

Entropy – the tendency of all things in the universe to wind down and die; also called the Second Law of Thermodynamics. The enemy of Evolution.

Era – occasionally the Higher Beings have to clean up and reset the Simulation, usually what this book referees to as a Wipe and Reboot. When Man is restarted after a Wipe and Reboot, the Era will have some dominant theme in the Greater Script that the activities of Man are to experience and handle. Our current Era began about AD 800-900.

Flow – often referred to as The Flow. This is the rising energetic vibrational entrainment into the higher 4th and 5th dimensional realms. It has increased awareness, compassion, Light, and STO aspects for service and is available to all who seek to align themselves with a Higher Way. It was created by the Higher Beings and is supported by an archetype that masters on the Earth reinforced and made available to all spiritual growth aspirants.

Free Will – an illusion. The more one grows spiritually, the more one does the will of the Father of Light. Baby souls, or those who insist on their own way, think they have free will but the Father is merely letting them experience the results of what they do... their **Script** controls much of what young/baby souls can do. As Jesus said "Not my will, but Thine be done." Advanced souls have surrendered their will by eliminating their ego.

Galactic Law – the ethics and rules as set forth by the Galactic Council and adhered to by all subordinate councils for the maintenance of order. It includes a Non-interference directive, responsibilities of 'creator races', transportation/communication protocols, terra-forming procedures, and energy creation/disposal to name a few.

gods – this is a god with a small "g". The Anunnaki were called gods because of their power and control over Man. In addition, Sargon, Moses, Gilgamesh, Alexander the Great and a number of Anunnaki offspring who were half-human half-Anunnaki (Inanna and Marduk) were considered god-like.
Used as "the gods who run the Simulation…" These are the Higher Beings and they are benevolent. Part of the Hierarchy between The God and Man, they are Angels, Arch-angels, Masters, Teachers… also called Higher Beings.

Godhead – a collection of higher souls, and Soul Groups, in closest proximity to God, like spokes on a wheel where the hub is God Himself. The Godhead works directly with the Oversoul for each Soul Group and sometimes the two are hard to distinguish. The basic hierarchy is: God – Godhead/Oversouls – Soul Groups – Angels/Neggs – and souls.

Gods-in-Training – when Man graduates from the Earth School, he can be useful to the Father of Light in various places in the Multiverse. One of those places is to undergo an apprentice position in overseeing the Earth and its souls – under the tutelage of more advanced 'gods' who give direction and training. The gods-in-training still make mistakes, just as Man does, and while often minimized by karmic override, these are allowed in part as an aspect of the new gods' training.   This is therefore sometimes a source of things going wrong in an Earth person's life. If a god-in-training abuses his power, he is recycled to Earth.

Granularization – the extent to which a material or system is composed of distinguishable pieces or grains. A picture at 300 dpi (dots per inch) has a greater granularity than a picture at 1200 dpi – because the viewer can zoom down further in to the 1200 dpi picture before it gets grainy and one can see the pixels.

Ground of Being – who and what you really are; your PFV is the physical reflection of the sum of your STO/STS quotient. If you, your soul essence, were to be removed from your body, the energy being that you are would have a certain vibration level (also reflected in the color of your aura) – higher or lower depending on how much Light you hold, how compassionate you are, whether you seek to serve (STO) or be served (STS), what issues (stuck points, agendas and attachments) you still carry with you, and in general, it refers to the "quality" of Light & Love that you are. Ultimately, it reflects the highest actions/thoughts that you are capable of.

Higher Beings – Light Beings above the Astral and reincarnative levels (1-6) and who are responsible for the operation of these lower 6 levels, reside on the 7th level themselves; may intervene in 3rd – 4th – 5th – 6th dimensional affairs when the Greater Script of the Father of Light, or the One, requires it to keep the Multiverse working. Also colloquially called **"the gods"** who run the HVR Sphere or Simulation. The Higher Beings are not the Beings of Light (angels) nor ETs nor aliens.

Higher Self – also called the Oversoul, this is the coordinating entity of each Soul Group and acts to oversee Scripts, events, lessons – and coordinate with the souls of the same

Soul Group, and with other Oversouls who manage other Soul Groups in the same Godhead. Each Godhead has multiple Oversouls that interface with the multiple Soul Groups.

Hybrids – this is any human-looking but 'upgraded' version of Homo *sapiens* which may or may not have a soul. It can be the Anunnaki hybrids – part Anunnaki, part human, and their bloodline. Or it may be Homo *noeticus* that the Greys have been so busy developing to restart civilization after the big Change event in the near future. Most are very intuitive, psychic and look to be the next step in the development of Man.

Insert – an object or person inserted into the Earth Simulation; the **object** is to make humans wonder, like the Antikythera device, or to educate them. The inserted **person** is to reveal some special art, music, philosophy to inspire Man or a scientific discovery to benefit him. Moroni to the Mormons, Jesus to the Christians, Buddha to the Asians …e.g.

Insiders – Anunnaki hybrid Remnant still on the Earth (may include Enki). The pro-active ones who try to help mankind and block the Dissidents (qv).

Interdimensionals – those beings in 4D and above who normally have very little interest in Man, and may be STO or STS. The STOs are often curious and observing. The STS version has been known to use OPs for unknown agendas. Also a generic term for the 4D STS Controllers inasmuch as they operate between dimensions. Possibly **Djinn**.

InterLife – where souls go when they die, after passing through the **Tunnel** to the **Light**. Also called the Other Side, and sometimes appears to be Heaven. It is where the **Script** is designed, souls are counseled by the Masters and Teachers, souls are rehabilitated after a rough lifetime on Earth (or elsewhere), and it is where the Heavenly Biocomputer referred to in Chapter 13 resides. This is also where reunions with other members of one's **Soulgroup** happen.

Karma – *Aka* **The Law of Karma**. – originally the concept of "meeting oneself", or "what goes 'round, comes 'round." It does not mean being stabbed in this lifetime because one stabbed someone else in a former lifetime. The original, true concept was that of the Universal Law of Cause and Effect, and it forms the basis of one or more aspects of your Life Script. Karma can also be a manipulative issue in the Virtual Reality of Earth if the life review is done by a Negg posing as a Being of Light.
Note that Karma applies only to Earth; other souls who do not come to Earth do not have to deal with Karma.

Konstruct – an energy shell like the **3D Construct *Gegenschein*** (qv) which is the outside barrier and keeps the **Cspace** and 3D Construct in place. This is the outer shell that Dr. Hogan in the GEO600 experiment heard making noise (See Chapters 7 and 8.) Also called the Firmament.

Law of Confusion – when RA was asked a question that violated someone else's right to privacy, or asked something that would be giving advanced level information that the person had no context for, RA would comment that the question could not be answered because it "violated the Law of Confusion." We are to work thru confusion and seek the answer(s) on our own; we have the 'right' to be confused and are expected to work thru it, or ask, thereby absorbing the lesson and information on a level that makes the lesson/info part of us.

Law of One – the concept that we are all connected at a higher level, mostly thru our Higher Selves, and we are all part of the One, the Father of Light – if you have a soul. This does not apply to OPs. The Law of One also includes freewill and love. Telling someone else what to do, how to live, etc is a violation of the Law of One, a violation of freewill whose flipside is called the Law of Confusion.

Light – an intelligent aspect of God; sometimes referred to as the Force. It may be used interchangeably with Heaven. There are biophotons of light that support the operation of DNA and sustain bodily operations.
Note that Light (large L) is a conscious aspect of the God force, which force can have a brilliant light about it. The light (small "l") is everyday, regular light.

Lightworker – any entity, physical or Astral, that uses Light as an energy source to do its work. (Includes angels, demons, Neggs, energy healers and Higher Beings.) Caution: it does not always imply STO behavior.

Loosh – a term coined by Robert Monroe in his chapter 12 of Far Journeys. It is energy produced by 3D living beings that is allegedly harvested by 4D entities in the astral for sustenance. Loosh is bountifully produced by humans who go into states of deep **fear** or anger or lust – they radiate the energy after being manipulated to produce it – like a grain of sand in an oyster produces a pearl that is harvested. See **Energy Vampire**.

Matrix – synonym for the HVR Sphere, but not to be confused with the matrix as shown in the movie *Matrix*. The underlying Grid of the Universe; probably synonymous with Dark Matter, or The Field.

Memes – a concept, or idea, that generally has spread through a population – an idea that may spread like a biological virus – such as a belief in ghosts, or a belief that black cats bring bad luck…or, if you go out in the rain and get wet, you can catch a cold. There are positive and negative memes. (See also "**100th Monkey Effect**.")

Modular Programming – instead of writing a monolithic, single-code program, that might be 400 pages long, like an Operating System, the major sections of code are broken up into modules principally by function – all Input is performed in one module, all Output in another. Special calcs are kept in another module, and error codes and user-interface displays are kept in another section. This facilitates making changes to code because it only is done in one place in the overall Operating or Application System. And the code can be called by another program if a special input routine is wanted, and the code already exists— why rewrite it? See **Object-Oriented** programming.

Morphic Resonance – said of a plant or animal that takes its physical shape from the *morphogenetic* field that establishes a 'morphic' (shape) resonance with the object's energy. The plant's shape is entrained by the morphic resonance with the morphogenetic field (pattern) that governs how living things take shape, according to Rupert Sheldrake.

Morphogenesis – Rupert Sheldrake conceived of the presence of a 4D field around living things that influences the shape they take – kind of an Astral Template that governs height, width, color and other aspects of the oak tree for example, such as when and where it sends out its branches, how fast and how far.

Multiverse – the universe we live in is one of a number of universes comprising a Multiverse… multiple universes interconnected forming a coherent larger universe consisting of multiple levels (realities), and can involve parallel universes or dimensions in 'superposition' (or stacked).

NDE – a Near Death Experience where the person appears to die, and their body is pronounced clinically dead, but they come back to life and relay their experience of meeting a Being of Light with whom they have a Life Review, and they usually come back a changed (better) person. The NDE effect often produces a spiritual transformation in the person.

Neggs – the 4D 'dark' angelic beings operating in the Astral realm around the Earth, whose sole purpose is to apply the negative lessons specified in one's Life Script. Thus they are "**NEG**ative Guide**S**." They work with the Beings of Light (Angels).

They are programmed to afflict mankind – they are appointed to effect the negative parts of one's Script (aka **catalyst**). They provide catalyst and feedback inducing Man to change and grow. They work <u>with</u> the Beings of Light (Chapter 6) because they, too, are Beings of Light who <u>volunteered</u> to serve the negative agenda and they were 'reoriented' to Darkness to maximize their effectiveness. They still carry a small, suppressed connection to their original Light down inside and they will be restored to their original condition when their service is complete.

NPCs – these are the other characters in a Virtual Reality game; they are not programmable or operable by the player – the Game or operating system uses them to play a part in the Drama. They are called Non-Player Characters. Same as **OP**s.

Object-Oriented Programming – is used along with Modular Programming to minimize code-writing, and to control the way parts of programs can call each other, and how subcode can be used. Each module of code is part of a Class, which has data and functions attached to it, so when calling a top Class, everything below that Class (Object) comes into the program as well.

OPs – Organic Portals -- (pronounced "Oh Pee") human beings, flesh and blood (Organic part), and they can serve as a portal for 4D entities (Neggs and 4D STS Controllers) to operate thru them. They also are **not fully human as they lack a soul** and that is because they have incomplete DNA and only the first 3 chakras are wired to function; they cannot access higher energy centers. They do not have a connection with anything higher than they are (no Higher Self as an ensouled person has), and as a result, they have no conscience (think: Charles Manson). Due to their somewhat robotic nature, they can be used by the STS Gang to manipulate and/or influence ensouled humans in 3D. They are often playing the role of **NPC**'s (as in a video game) in our world. Dolores Cannon called them **Backdrop People** (Convoluted Universe, books 4 & 5.).

PFV – **Personal Frequency Vibration** -- the day-to-day, overall vibratory rate (resonance) of the soul energy sustaining the human body. When a person is angry their aura 'glows' red, and the PFV can drop to a lower (denser) vibration than when a person feels a lot of love and the aura 'glows' rose and the vibration reflects the energy of the heart charka (higher, lighter energy). The PFV also denotes which charka is dominant in the person; a person living from their higher charkas has a higher PFV than one engaged in sex, violence and pettiness (lower chakra activity). The aura typically reflects what one is feeling, yet the base PFV does not change; when the person is at rest, the base PFV is consistent from day to day as it reflects the overall level of soul growth. Also known as that person's "energy signature" as recorded in objects (Psychometry).

Phase-Shifted – refers to 3D and 4D entities or 3D and 4D timelines which cannot see the other even though they may occupy the same space. For example, there may be a 30° phase shift, or a 60° or 90° shift (the most common). Think of 2 Sine waves almost on top of each other (congruent and coherent; now move one wave to where it's trough is below the other wave's peak – they are 90° phase-shifted.

Pixellation – Take a picture that was set at 300 dpi resolution and start expanding it (zooming in) until you see the little squares that comprise the colors and shapes of the picture. (See Chapter 7, Elvidge). Also referred to as digitization of the picture.

Placeholder – an OP-like version of a real ensouled human living on another timeline (parallel universe) that already split into two, with duplications of people between the timelines. If the ensouled human did not replicate to the new TL, the other people who went to the new timeline still need/expect that 'body' for their everyday world activities to function, and his absence in their lives would be noticed. And so minus a soul, John Doe exists as a kind of 'synthetic' human in the <u>new</u> timeline. A Sim or NPC.

Points of Choice – there are pre-programmed points in a person's life where important choices must be made, and they are found in a person's Script. Examples are whether to move to Florida or stay in California, whether to accept what looks like a great new job, or whether to get married. Sometimes the choice results in a **timeline bifurcation** into a fractal subset so that another aspect of you can see how that turned out. See **Timeline**.

Pre-Soul – also called First Time Soul, allegedly the initial stage of an animal that leaves 2D and enters the 3D human soul realm (metempsychosis); this is not a complete soul, but a potential one if the entity applies itself as a 1st time human. Typically, only the first 3 chakras are functional, and thus there is not enough 'soul energy' to create an aura. (See also **OP**s.)

Prime Directive – a requirement in our Galaxy for those races who can create life and modify existing life genetically – often referred to as a 'creator race.' They are responsible for overseeing the welfare: safety and education of their creation. This is why a **Remnant** of the Anunnaki stayed behind (now known as the **Naga**.)

Prophylactic Fantasy – describes the world of denial that some people live in. 'Prophylactic' because they feel safe in <u>their version</u> of the world, and they reason that nothing really destructive has ever happened to them, nor can it.

'Fantasy' because they do not accept the real world and its negativity; they see their world as they want it to be and think that they can exert a 'force' that makes it that way.

PTB – the earthly human Powers That Be; the 3rd dimensional STS people running the world for their Anunnaki Dissident masters (control group still here). They are also influenced by corrupt DNA. Puppets. Many of them are OPs (soulless). See **Elite** – not the same thing, just a higher level of control.

Qualia – individual instances of subjective, conscious experience, that a machine cannot have. It is the way a person interacts with the way a red rose, for example, impresses them, and may involve recalling past events with red roses, their smell, and whether the person likes the rose or not—it is a very qualitative reaction.

Quantization – refers to the organization, structure and arrangement of quanta in a video game. These are the discreet elements of whatever the video display is showing – below the level of pixels. The particular arrangement and grouping can create **granularization** or not.

Recycle – short-circuited version of reincarnation: to come back into the same body, same lifetime, hence experiences **Déjà Vu**. Implies the inability to move forward into new realms and experiences in the greater **Multiverse,** as with **Reincarnation**.

Reincarnation – the spiritual growth aspect of a soul moving thru the different realms in the Multiverse (not just back to Earth) for the purpose of experiencing and gaining knowledge and wisdom. On the other hand, a repeated lifetime limited to Earth is more of a recycling.

Resonance – vibrating alike: such that two tuning forks A and B side by side, with A struck hard to set it vibrating, when put next to B which was not vibrating, will set tuning fork B vibrating at the same frequency as tuning fork A. This also happens with people in close proximity: a very negative person can 'detune' (bring down) a room of people and some people may actually feel ill and not know why (as they pick up the negative person's vibes). See Entrainment.

Reset – synonymous with **Wipe & Reboot**—resets a part or all of a VR game, may also reset the scenery in a simulation from a master file.

Satan – allegedly the leader of the demonic spirits which were the deceased Nephilim and/or their offspring. This titular role may have been filled by the Nephilim (fallen Igigi) leaders known as Samayaza or Azazyel, or the Gnostic favorite: Ialdabaoth, in a former Era. The **Egyptian Set, or Sata**, was synonymous with one of the three just named. A convenient mythological character to personify Man's need for duality in the universe.

Schumann Resonance – natural frequency of earth's vibration/resonance: 7.8Hz.

Script – instead of pure Karma, when one is born, one is given a Script covering what basic events are to happen in one's life, which one is expected to overcome; they may be positive or negative, and how one meets them and handles them determines how one is progressing towards the goal of getting out of the earth experience. It often has Options programmed into it (**Points of Choice**) where the soul must make a significant choice. It is a test of soul growth. A personal Life Script is usually subject to the Greater Script of the Father of Light and works within it.

The Script says generally what your life is about, but **it does NOT tell you what to do or say.** Caroline Myss called it the Sacred Contract... see Book 5 (TSiM).

ShapeShifting – the ability to control what people see... the being doing the shape-shifting does not actually change any of his atomic structure – just the way his appearance is perceived, and perception is holographic. So to effect a different appearance, the being just produces new interference waves that the observer 'sees' differently. Commonly done by 4D and above entities while in 3D.

Sheep – people who are barely conscious, and refuse to think for themselves. They want someone to tell them what to do and when to do it, and they go along with whatever they are told. They are easily manipulated by the Media.

Also called **'Sheeple'** and may be OPs or 'dense' ensouled humans.

Sim – an NPC or OP – Placeholder. A 'simulated' human. All caps (SIM) referred to a Simulation that had been replicated (as SIM II).

Soul Aspect – all souls can 'split' themselves to experience different realms; as when a timeline splits, one part of the soul stays with the original TL and another part replicates to the new TL. Each soul has aspects in different TLs, dimensions, worlds, and realms, etc. and at a point in the future, they reunite to the Soul Group. Not a **Fragment** (see next).

Soul Fragment – some souls may fragment **due to trauma** and then special therapy is often needed to coax the missing fragment to rejoin its source. Some fragments are held by family members, past lovers, and even by the Neggs themselves.

Soul Group – each soul was part of a group of like souls (same core vibration PFV which usually synchs up with a specific archetype) and these split up to better experience the Creation – souls will eventually reunite in their original group when their explorings are done. The Soul Groups reunite with the **Godhead** from which they came.

Soul Merge – as in the case of the author, to undertake a special project where a 3rd level soul has volunteered to serve in a capacity that it alone can't do, and so a Merge is performed to give that 3rd level soul the extra knowledge and strength of the merging soul (who is of the <u>same soul group</u> -- usually from a higher level) and together they perform some task that the Higher Beings must have approved – before the Merge can happen. Not synonymous with Walk-in.

Soul Migration – the concept that animals can progress to first-time human beings with 'baby' souls and the full-fledged human soul must be earned thru successive incarnations. As they would also have only the lower 3 chakras operative, they may be mistaken for OPs. (See Metempsychosis on Wikipedia.)

STO – Service To Others; altruistic behavior, self-sacrificing.
STS – Service To Self; selfish behavior; 'Me-My-Mine' syndrome.

Subquantum Kinetics – is an approach to microphysics with roots in general system theory, nonequilibrium thermodynamics, and nonlinear dynamics. It represents quantum phenomena differently than Quantum Physics (QP) and works with the concept of the **Ether** (VEG Chapter 9) which is composed of subquantum units called etherons (as opposed to QP's quarks). It is simpler than Quantum Physics and explains the issues that QP is still wrestling with: wave-particle dualism, strings, singularities, and the cosmological constant.
It also embraces and explains Tesla's work better than QP. (Refer to Chapter 4 of **Dr. Paul LaViolette**'s book <u>Secrets of Antigravity Propulsion</u> for a more complete description in layman's terms.)

Terraforming – an advanced technical process whereby a whole planet is set to its original, or a near-new, pristine condition following some catastrophe or pollution, or both. The ecology is balanced, the air, land and water are unpolluted, and in the case of planet Earth, it can once again support lifeforms. See also "**Wipe and Reboot**."

Thoughtform (**TF**) – any thought that many people subscribe to and which reflects a widely held belief, esp. one imbued with a lot of fear, or hate, generates a TF which after a while (depending on the amt of energy put into it) takes on a 'life' of its own; **man is a creator and thoughts are things**. If enough people fear and believe in werewolves, there will be thoughtform 'werewolves' … which are not real entities but are attracted to those who fear and believe in them (like attracts like). (QV: **Tulpa**.)

Any unwanted TF can be cancelled and should be before it attaches itself to a person's aura and then 'feeds' off the person's energy – like a parasite. TF have no conscious volition of their own, they are reactive and go to wherever (1) they are attracted by sympathetic vibration, and (2) where the person's aura is weak.
Carl Jung called these TF's Archtypes.

Timeline (**TL**) – the linear coherent vector on which all souls and Placeholders (OPs) of a certain frequency range have their being; a reality timeline that linearly moves forward creating causal events. It is not permanent and is subject to entropy if a bifurcation results from a rise in consciousness and attendant agreement coherently shared among the souls seeking to live in a higher consciousness in TL2 is preferable to the negatively polarized TL1. If there is not enough agreement (energy) to sustain the new TL2, it dissolves.
If a dimension has only one TL, the TL is the dimension, but dimensions can have multiple TLs. There is a TL where Hitler won, for example.
And timelines may create, 'run' and dissolve **fractal** subsets (within the larger TL framework) for special purposes (qv).
(…cont'd…)

Note that Timeline and Simulation are not synonymous but a Simulation may exist on a Timeline and be replicated to another Timeline.

Unconscious – unaware, not a very high level of perception. A person who is 'asleep' spiritually and is not aware that there are more than the 5 senses. Can also mean 'spacing out' with eyes wide open. Standard condition of the **Sheep**.

V2K – "Voice to Skull" -- a microwave enhanced transmission of words directly into a person's head, as if they actually hear the words, without any external devices or hearing apparatus. Developed by the US Army to communicate with a soldier on the battlefield, to the exclusion of other soldiers, it was perfected during the mind experiments with Helen Schucman while she transcribed the *Course in Miracles* book. Who sent her the information is not known. (Ch. 11, VEG)

Vibration/Vibe – the energy state of a person, place or thing. Everything puts out an energy 'signature', which is how pyschometry works… objects record the energy of the person that held/owned the object, and places often hold the residual energy of events that happened there: some sensitive people cannot visit Gettysburg as they feel the negative energy from all the hate and fear created in that place – even thought it was long ago, it still holds some energy that has not completely dissipated. (See **PFV**.)

Visual Spatial Acuity – the ability to see fine detail; visual term reflecting the number of rods/cones in the retina. Similar to **pixels** in computer printing, display screens and digital cameras.

Wanderers – higher souls from other realms who have volunteered to incarnate on Earth in troubled times to serve as the Light leads: they may anchor the Light, write books, lead New Thought churches, heal or perform other services to benefit Mankind. Usually 6th level beings (souls). The Indigos and different forms of "Starseed" are part of this group.

Wipe and Reboot – an end to a current **Era** of Man on Earth, followed usually by a terra-forming (resetting the environment back to clean and balanced), followed by the Re-seeding of Man on the planet. **A Reset**.

The term is borrowed from the computer world where when a PC is non-functional (i.e., locked up and displays the dreaded BSOD [Blue Screen of Death]), it is necessary to "Wipe" the hard disk – reformat it – and reload the operating system and application software… i.e., "Reboot" the system and start all over again.
Whereas the PC gets a clean start as if nothing happened, each new Era for Man still includes whatever objects were created in the prior Era – i.e., pyramids, huge walls, and Stonehenge.

Zechariah Sitchin – the late Middle Eastern scholar, speaking several languages, who translated the Anunnaki/Sumerian tablets. Chapter 3 is mostly dedicated to a summary of his findings about Man's origins. His claim to fame was *The Earth Chronicles* series of 8+ books that revealed the Sumerian – Anunnaki connection.

ZPE – Zero Point Energy – the natural state of the Dark Energy/Dark Matter in the Universe. Also called the Ether, it is the zero-point energy of all the fields in space, which in the Standard Model includes the EMF, Higgs Field, and Fermionic Fields. It is the energy of the vacuum of space, which in quantum field theory is defined not as empty space but as the ground state of the fields. A related term is *zero-point field*, which is the lowest energy state of a particular field.

Quantum Earth Simulation

Chapter Endnotes

Introduction
Hendrie, *The Greatest Lie on Earth*, p. 241.

Chapter 2   Endnotes

[2] http://en.wikipedia.org/wiki/Simcity
[3] Ibid.
[4] Ibid.
[5] Ibid.

[6] Robert Heinlein, *JOB: A Comedy of Justice.* P. 419.

Chapter 3 Endnotes

[7] Michael Talbot, *The Holographic Universe*. Pp 19-20.
[8] Ibid., 20.
[9] Ibid., 54-55.
[10] Ibid., 31.

[11] Lynne McTaggart. *The Field*. P. 80

[12] Paul Dong & Thomas Raffill. *China's Super Psychics*. P. 7.
[13] Ibid., 53.
[14] Ibid., 108-09.

[15] James G. Friesen, MD. *Uncovering the Mystery of MPD*. 59, 115, 143.

[16] Op Cit, Talbot, 141.
[17] Ibid., 142.
[18] Ibid., 143.
[19] Ibid., 144-45.
[20] Ibid., 140.
[21] Ibid., 159.
[22] Ibid., 159.
[23] Ibid., 158.
[24] Ibid., 50.
[25] Ibid., 285.

[26] Dr Paul La Violette, *Genesis of the Cosmos*, 273-283.

[27] Op Cit, Talbot, 160.

[28] Op Cit, McTaggart , 85.
[29] Ibid., 85.

30 http://www.colinandrews.net/Cloud-Radar-Circle-Australia-2010-0116.html
See also: **credit: http://www.news.com.au/lifestyle/real-life/bueau-of-mereology-cant-explain-mysterious-patters-on-radar-system/story-e6frflri-1225848774377**

Chapter 4 Endnotes

31 Steinmeyer, Jim. *The Book of the Damned; The Collected Works of Charles Fort.* (New York: Tarcher/Penguin Group, 2008), 381.
32 Ibid., 381-82.
33 Ibid., 382.
34 Ibid., 838-839.

35 Monroe, Robert. *Journeys Out of the Body.* (New York: Doubleday, 1971), 73-74.

36 Monroe, Robert. *Far Journeys.* (NewYork: Random House, 2001), 93.
37 Ibid., 93-106.

38 Monroe, Robert. *Journeys Out of the Body.* 117-119.

39 Monroe, Robert. *Ultimate Journey.* (NewYork: Random House, 2000), 183-184.
40 Ibid.
41 Ibid., 274-275.

42 Bostrom, Nick. See: Http://www.simulation-argument.com/

43 Elvidge, Jim. *The Universe Solved.* Alternative Theories Press, 2007. See Ch's 1-7.

44 Monroe, Robert. *Far Journeys.* 102-106.

45 Greene, Brian. *The Hidden Reality.* (New York: Alfred A. Knopf, 2011). 288.

46 Op. Cit, Elvidge, 238.

47 Wilde, Stuart. *The Prayers and Contemplations of God's Gladiators.* (Chicago: Brookemarke, LLC, 2001), pp 7-8.
48 Ibid.,
49 Ibid., 11-15.

50 Icke, David. *And The Truth Shall Set You Free.* (Isle of Wight, David Icke Books, Ltd.,1995), 9.

51 Vallée, Jacques. *Dimensions*. (Chicago: Contemporary Books, 1988), 222.
52 Ibid., 272.
53 Ibid.
54 Ibid., 277.

Chapter 6 Endnotes

[55] Brent Silby, "The Simulated Universe". See
http://www.scribd.com/doc/3015396/Simulated-Universe-by-Brent-Silby
[56] Ibid.

[57] This very scenario was done on a TV show called *The Carbonaro Effect*, and another magician called Criss Angel performed the same trick back in 2006-7 on TV as well. Are they really magicians or a couple of advanced humans who can do psychic things? (This was addressed further and Michio Kaku's input was given at the end of Ch 9 in VEG from his book *Hyperspace*).

Chapter 7 Endnotes

[58] Nick Bostrom, "The Simulation Argument" See
on http://www.simulation-argument.com/si...
and http://www.youtube.com/watch?feature=player_detailpage&v=nnl6nY8YKHs

[59] Ibid., and Wikipedia: , http://en.wikipedia.org/wiki/Simulated_reality, p. 3.

[60] Nick Bostrom, "The Simulation Argument." See:
http://www.simulation-argument.com/matrix.html
(Also: Times Higher Educational Supplement, May 16, (2003).)
[61] Ibid.
[62] "The Simulation Hypothesis." See:
http://en.wikipedia.org/wiki/Simulation_hypothesis
[63] Ibid.

[64] Wikipedia: http://en.wikipedia.org/wiki/Cognitive_computing

[65] Brian Weatherson, "Are You a Sim?" see:
http://www.simulation-argument.com/weatherson.pdf
Also: *Philosophical Quarterly* (2003), vol 53., 425-31.
[66] Ibid.

[67] Elvidge, Jim. *The Universe Solved*, p 194.
[68] Jim Elvidge, article from his website:
The Singularity, Infomania, and Programmed Reality , December, 2008.
http://www.theuniversesolved.com/evidence.htm

[69] Op Cit, Elvidge ., p. 195.
[70] Ibid., p.32.
[71] Ibid., p. 197.
[72] Ibid., pp 297-08.

[73] Jim Elvidge, article from his website:
Nanotech and the Physical Manifestation of Reality, March 2008.
http://www.theuniversesolved.com/evidence.htm
[74] Ibid.

[75] Posted by: by http://theghostdiaries.com/life-in-the-matrix-new-evidence-supports-the-simulation-theory
also see Gate's original paper at: http://arxiv.org/abs/0806.0051 via Cornell University site.

[76] Greene, Brian. *The Hidden Reality*. Pp. 281-82.
[77] Ibid., pp 284-85.
[78] Ibid., p. 288.
[79] Ibid., pp 288-89.
[80] Ibid., pp 291-92.
[81] Ibid., p. 306.

[82] Lloyd, Seth. *Programming the Universe*. Pp. 6-7, 31.

[83] Wikipedia: http://en.wikipedia.org/wiki/D-Wave_Systems
[84] Ibid., p. 54.
[85] Ibid., p 54.
[86] Ibid., pp. 149-51.
[87] Ibid., p. 154.

[88] Op. Cit., Elvidge, p. 117.

[89] OP Cit., Lloyd, p. 166.

[90]

https://en.wikipedia.org/wiki/Titan_(supercomputer)#/media/File:Titan_supercomputer_at_the_Oak_Ridge_National_Laboratory.jpg

And   https://en.wikipedia.org/wiki/Tianhe-2

[91] https://en.wikienpedia.org/wiki/Supercomputer.
[92] Ibid.

[93] See the following article:
http://beforeitsnews.com/story/1658/888/NL/Scientific_Evidence_The_Universe_Is_A_Holographic_Projection_Around_The_Earth.html

[94] Op Cit, Lloyd, 166..
[95] Ibid.

[96] Wikipedia, http://en.wikipedia.org/wiki/Simulated_reality
[97] Dvorsky, George, "**Physicists say there may be a way to prove that we live in a computer simulation**"
on http://io9.com/5950543/physicists-say-there-may-be-a-way-to-prove-that-we-live-in-a-computer-simulation

[98] http://en.wikipedia.org/wiki/Greisen%E2%80%93Zatsepin%E2%80%93Kuzmin_limit

[99] Science Channel, Through the Wormhole with Morgan Freeman, 5/20/15 episode, "Do We Live in the Matrix?"

[100] Greene, Brian. *The Fabric of the Cosmos*. P. 425.

[101] Grabianowski, Ed, **"You're living in a computer simulation, and math proves it"**
on http://io9.com/5799396/youre-living-in-a-computer-simulation-and-math-proves-it
[102] Ibid.
[103] Ibid.

[104] Silby, Brent, "The Simulated Universe" on http://www.scribd.com/doc/3015396/Simulated-Universe-by-Brent-Silby
[105] Ibid., p. 3.

[106] http://en.wikipedia.org/wiki/Simcity

Chapter 8 Endnotes

[107] http://en.wikipedia.org/wiki/Omphalos_hypothesis
[108] Ibid.
[109] Ibid.

[110] Alan Turing.
http://en.wikipedia.org/wiki/Alan_Turing#Early_computers_and_the_Turing_test

[111] http://en.wikipedia.org/wiki/Consciousness_Explained
[112] Ibid.
[113] Ibid.

[114] http://en.wikipedia.org/wiki/Roger_Penrose
[115] Ibid.

[116] Lynne McTaggart. *The Field*. Pp 91-94.
[117] Ibid. 93.
[118] Ibid., p93. The original quote was from 1994:
Jibu, Hagan, Hameroff et al. "Quantum Optical Coherence in Cytoskeletal Microtubules: Implications for Brain Function", *Biosystems*, 1994: 32: 95-209.
[119] Ibid., 94.
[120] Ibid., p 95.
[121] Ibid., p. 95
And original quote from: Laughlin, "Archtypes, neurognosis, and the Quantum Sea."

[122] Dr. Nick Bostrom in www-simulation-argument.com/matrix.html

[123] http://www.newscientist.com/article/mg18524911.600-13-things-that-do-not-make-sense.html?page=1#.VWU-NVh0xHg
Also see: http://www.newscientist.com/special/13-more-things

Chapter 9   Endnotes

[124] Wolf, Fred Alan, PhD. *The Yoga of Time Travel*. (Wheaton, IL: Quest Books, 2004), 103.

[125] Ibid., 105.
[126] Ibid., 103 .
[127] Ibid.

[128] Morton, Chris and Ceri Louise Thomas. *The Mystery of the Crystal Skulls*. (Rochester, VT: Bear & Co., 2002), 268.

[129] http://en.wikipedia.org/wiki/Virtual_particle

[130] Op Cit, Wolf, 122.

[131] http://www.bing.com/videos/search?q=geminoid+hi-1%2c&FORM=VIRE1#view=detail&mid=F7E14E30215E225772C5F7E14E30215E225772C5

[132] http://pinktentacle.com/2010/04/geminoid-f-remote-control-female-android/
Also see Decodedstuff.com

[133] http://en.wikipedia.org/wiki/Wright_brothers
[134] http://en.wikipedia.org/wiki/Antikythera_mechanism
[135] http://en.wikipedia.org/wiki/Baghdad_Battery

Chapter 10   Endnotes

[136] Robin Hanson," *How To Live in a Simulation*. "
http://www.jetpress.org/volume7/simulation.htm

See also
Nick Bostrom, "Are You Living in a Simulation?"
http://www.simulation-argument.com/matrix.html

See also
Hans Moravec. "Simulation, Consciousness, Existence"
http://www.frc.ri.cmu.edu/~hpm/project.archive/general.articles/1998/SimConEx.98.html

[137] Ring, Kenneth. Lessons from the Light. (Portsmouth NH: Moment Point Press, 2000), 173-77.
[138] Ibid., 184.

[139] http://en.wikipedia.org/wiki/Simulated_reality

[140] See the following:
http://en.wikipedia.org/wiki/Simulated_reality
http://www.simulation-argument.com/matrix.htmlhttp://www.jetpress.org/volume7/simulation.htm

also see:
David Davenport. "Computationalism: The Very Idea."
An overview of Computationalism.
Computer Eng. & Info. Science Dept.,
Bilkent University, 06533 Ankara –Turkey.

[141] Michael Talbot. *The Holographic Universe*. (NY: Harper, 1991), 65-66.

[142] Rene Descartes, http://en.wikipedia.org/wiki/Ren%C3%A9_Descartes#Philosophical_work

[143] Morpheus in *The Matrix*. Warner Bros., 1999.

[144] Christopher Grau. Philosophers Explore The Matrix
Oxford University Press. pp. 157–158

[145] Op Cit, Descartes.

[146] Op Cit Michael Talbot, 65-66.
[147] Ibid.

[148] Timothy Freke, Lucid Living. (United Kingdom: Sunwheel Books/Books for Burning, 2005). (No pp given; booklet is 40 pp long, unnumbered pages.)

[149] Newton PhD, Michael. Destiny of Souls. (Woodbury, Mn: Llewellyn Worldwide, 2002), 103-104.

Chapter 11 Endnotes

[150] Wikipedia: https://en.wikipedia.org/wiki/Flat_Earth

[151] https://en.wikipedia.org/wiki/History_of_optics

[152] Nancy Davis, The Zuni Enigma, pp. 90+, 138-145.

[153] Edw. Hendrie, The Greatest Lie on Earth, p86.

[154] Eric Dubay, The Flat-Earth Conspiracy, pp 124-129.

[155] Bennett/Percy, Dark Moon, pp. 392-93.

[156] Eric Dubay, YouTube video : (1:28:50)
Eric Dubay's The Flat Earth Conspiracy Documentary
https://www.bing.com/videos/search?q=flat+earth+eric+dubay&&view=detail&mid=D7248F1ED9CFFF6C3A2ED7248F1ED9CFFF6C3A2E&FORM=VRDGAR

there are some very good demos of why the Earth is probably flat – lighthouses, Polaris, tides vs flat water, and sunsets... very shocking that we have never noticed that the Sun is not 93 mil miles away! If the Earth is spinning at 1000 mph [as the scientists say], how can airplanes fly and land without difficulty?

[157] https://en.wikipedia.org/wiki/Aurora

[158] Koran: N.J. Dawood, The Koran, Penguin Classic. P. 592.

[159] Op Cit., Dubay, p. 242.
[160] Ibid. P. 244.
[161] Ibid., p. 243.
[162] Op Cit, Hendrie, pp. 234-238.
[163] OP Cit Dubay., p. 248.

[164] http://www.pbs.org/newshour/rundown/study-invisible-shield-space-protects-earth-killer-electrons/
And
https://www.sciencedaily.com/releases/2014/11/141126133829.htm

[165] Wikipedia: https://en.wikipedia.org/wiki/Auguste_Piccard

[166] Charles, R.A. *The Book of Enoch the Prophet*. (San Francisco, CA: Weiser Books, 2003), viii-ix.

Chapter 12   Endnotes

169 Wohlberg, Steve. *End Time Delusions*. (Shippensberg, PA: Treasure House, 2004), 39-40.
170 Ibid., 39-41.
171 Ibid., 127.
172 Ibid., 127-128.
173 Ibid., 50.
174 Ibid., 46.
175 Ibid., 44-45.

176 Spong, John Shelby. *A New Christianity for a New World.*, 8.
177 Ibid., 122-123.

178 Pagels, Elaine. *The Gnostic Gospels*, 127.
179 Meyer, Marvin. *The Gospel of Thomas*. (New York: HarperCollins, 1992), p.53.

Chapter 13   Endnotes

[178] Paul Davies, The Goldilocks Enigma. p. 105.
[179] Ibid., 27, 131.
[180] Ibid., x-xi.

[181] Jim Marrs, <u>Alien Agenda</u>. Pp 1-12.
[182] Ibid., 6-7.

[183] Professor James Gates, YouTube: Physicist Finds Computer Code in String Theory:
https://www.youtube.com/watch?v=cvMlUepVgbA

[184] https://www.youtube.com/watch?feature=player_detailpage&v=YGxVGtkTa4s

[185] Author P.K. Dick, YouTube: Your Life is a Computer Simulation:
https://www.youtube.com/watch?feature=player_detailpage&v=YGxVGtkTa4s
This YouTube ref. is a treasure trove of PK Dick, James Gates and Thomas Campbell (physicist) speaking on Dr. Nick Bostrom.

[186] Gregg Prescott article in *In5D Guest* and Waking Times:
Earth's Quarantine Force Field Discovered by NASA? , Dec. 3, 2014.
[187] Ibid., with ref to
http://www.betawired.com/science-fiction-like-electron-shield-found-around-earth/1418358/
may also see:
http://lasp.colorado.edu/home/?post_type=mag-seminars&p=16204

Mini Index

Items not identified above can be found in the Table of Contents.

The Anunnaki Legacy

Transformation, Kundalini, Hyperborea, Göbekli Tepe, Hopi, Maya,
Gods, Goddesses & the Divine Feminine, Skygods & Creation,
The Flood, Sacred Trees, Alchemy and Serpent Wisdom.

by TJ Hegland

Virtual Earth Graduate

TJ Hegland

Reflecting: (inserts left to right, all covered in the book):

Physics (atom), Genetics (DNA), Dragon, Ubaid Statue of Anunnaki, UFO (TR3B), and *Castillo* at Mayan Chichen Itzá.

Center: Earth with Soul/Earth Graduate

Background: Electromagnetic Dark Energy Matrix

NOTES